STRATHCLYDE UNIVERSITY LIBRARY
30125 00079126 8

KU-214-083

AN INTRODUCTION TO MINING

THE ROCK MECHANIC

(With apologies to William Henry Davies)

What is this life if full of care
We have no time to stand and stare.
No time to stand beneath the backs*
And with a compass trace the cracks.
No miner or geologist we
If the cleat we cannot see.
How can we expect to support the roof
If with closed eyes we stand aloof?
Toil with bolt or prop or crib,
And if it still falls then tell a fib,
'It was the other shift'.

L.J.T.

*Roof, for a coal miner, but it doesn't rhyme.

AN INTRODUCTION TO MINING

Exploration, Feasibility, Extraction, Rock Mechanics

L. J. THOMAS

B.SC., PH.D., C.ENG. (U.K.), M.AUS.I.M.M., M.I.MIN.E.,
FIRST-CLASS CERTIFICATE OF COMPETENCY
(COAL MINES) U.K. AND N.S.W.

Senior Lecturer in Mining,
University of Sydney

HICKS SMITH & SONS

SYDNEY 1973

First published 1973 by
Hicks Smith & Sons Pty Ltd,
301 Kent Street, Sydney, 2000
225 Swan Street, Richmond, 3121
91 Elizabeth Street, Brisbane, 4000
238 Wakefield Street, Wellington, N.Z.

Registered at the G.P.O., Sydney
for transmission by post as a book

ISBN 0 454 01730 8 (cased edition)
ISBN 0 454 01890 8 (limp edition)

Printed at The Griffin Press, Adelaide

CONTENTS

SECTION C. ROCK MECHANICS

PREFACE

Australia has in the last few years become a notable producer of minerals. This island contains what are probably the world's two largest iron ore mines, both situated in the Pilbara region of Western Australia. Hamersley Iron should be producing about 37.5 million tonnes of ore by 1975 and the Newman Consortium about 35 million tonnes. Present mining is of high grade haematite with about 64 per cent iron.

The aluminium industry has reserves of more than 2000 million tonnes of bauxite in the Weipa area of Northern Queensland alone, and certainly more than 3700 million tonnes available in Australia to provide for its expansion. The Gladstone alumina plant (Queensland) will have a rated capacity of 2 000 000 tonnes annually by 1972, making it the biggest in the world, and it is anticipated that the plant will achieve some 2 400 000 tonnes annually in practice.

There are also enormous reserves of coal for that rapidly growing industry. A few years ago what was then the world's largest dragline operated at Moura opencut and this machine with a $100m^3$ bucket is still one of three of the largest successfully operating machines. In some underground mines coal production outputs of 900 to 1000 tonnes per shift from a continuous miner are obtained regularly and several mines are producing 1 million to $1\frac{1}{4}$ million tonnes per year with manpowers of 100 to 300 men. Only about four per cent of coal from New South Wales is at present won by longwall methods, but by using equipment specially strengthened to support the hard sandstone roofs, high productivity is now being obtained.

For heavy mineral sands (rutile, ilmenite and zircon) Australia is the world's largest producer and has pioneered some dry-mining techniques. Australia is also the third largest producer of lead, exceeded only narrowly by the United States and the U.S.S.R. In 1969 the lead content of Australian ore production was 445 000 tonnes. Mount Isa Holdings produced about one third of this total from its mine at Mount Isa in Queensland. This mine also produces vast quantities of copper and zinc and it is probably the largest underground metal mine in the world. Nickel production in Australia is rising rapidly; excellent deposits of uranium have recently been found in the Northern Territories. Almost every commercial metallic ore is mined in some part of Australia.

In 1970 more than $170 million was spent on exploration for minerals and petroleum. In the five years to the end of June 1970 private capital investment in mining ventures totalled $1750 million and another $860 million was spent on extraction, refining and foundry plant. Currently another $1600 million is committed to expansion of existing projects, or to new ventures, such as the $14 million asbestos project in New South Wales.

The Australian population is concentrated mainly along short stretches of the coastal fringe, and the interior of the continent is virtually waterless. Many mining developments have to establish their own communities, frequently building air-conditioned homes at a cost of $40 000 per unit. They may have to provide their own roads, water supplies and railways to the coast, and even provide their own ports.

Engineering equipment may have to be freighted 3000 km across a continent and all food may have to be air-freighted in to a remote mine. Consequently, the mines have to be planned to produce a large output from a small manpower, especially if the ores are low grade. This has caused a rapid improvement in mining technology.

To cope with the expansion of mining many overseas engineers have been brought in. Overseas technology has been adapted to local conditions for new minerals. The fruitful minds of many engineers have, from necessity, produced innovations and increased productivity.

It is against the background of a vigorous and expanding industry that this book is written. Its prime purpose is to provide basic material for mining students in their first year of acquaintance with the mining profession. Since the book assumes only a general grounding of scientific knowledge, and no specialist engineering training at this level, it is also suitable for anyone who wishes to understand aspects of the mining industry.

To make the book more comprehensive for users who will not study geology an introductory section on exploration and mineral reserve estimation is included. This is followed by an explanation of how a feasibility study is made to assess whether or not a deposit should be exploited. The various methods by which a mineral can then be mined are given in more detail. The final section is about applied rock mechanics in the sense that the theory is omitted and I have only given the general rules of how to avoid being squashed by an errant lump of rock, and how to lower a chicken farm in a potential subsidence area without breaking the eggs. References are given to more advanced theoretical texts.

For the more serious reader the book can act as a refresher on basic ideas. It is assumed that the serious student will be reading technical journals and manufacturers' data contemporaneously with this book so that my simple sketches and descriptions of equipment can be translated into reality. For my purposes it does not matter who has made the electric shovel for an opencut mine, nor to a great extent whether the bucket is of 10 m^3 or 100 m^3, or how many amperes go into its kilowatts.

The reader is given its main principles and applications, and the mining engineer who passes this year of instruction will learn more of electrical power and machinery and mine production at a later date.

The dabbler in stock markets will, I hope, be able to understand enough of the problem to dabble with less risk, but not necessarily more profit. Who can predict the actions of those who rush in to buy where wise men fear to mine?

I should like to acknowledge here the help I have received in writing this book, both direct and indirect. Where I have taken material from recent publications I have quoted the source. However, in the course of time one reads perhaps 200 or so journals a year, some only skimmed through. I may have thus subconsciously absorbed and reproduced sketches and material familiar to other authors. There is also of course the material that has been hashed and rehashed by many people; its origin is lost and best covered by the name 'anon'.

Much of the material has been provided by friends within the mining industry, and gathered during visits to the mines. It would be a long list to quote and I prefer to say a general thank you to the mining industry both in the United Kingdom and Australia, and to overseas visitors, for the generous provision of information. I would however like to acknowledge the particular help given with material and drawings by Associated Minerals Consolidated Ltd, The Coal Cliff Collieries Pty Ltd, Cobar Mines Pty Ltd, and Mount Isa Mines.

L. J. Thomas
January 1973

SECTION A. EXPLORATION AND FEASIBILITY STUDIES

Chapter 1. Introduction

Geological Terminology

There are some general geological terms that should be understood for exploration and mining purposes. The simplest form of rock strata is that in which each stratum (layer) of rock is superimposed on another. These are *conformable beds.* Movements of the earth's crust and subsequent weathering and rock displacement provide for mineral formation and relocation into the site where it will be found.

The original layers of rock were formed by sedimentary or chemical processes with material from higher land masses being redeposited in lower areas. The beds could then be distorted into folds to produce *synclines* or *anticlines* as shown in Figure 1.1. The scale of these movements is such that a single fold producing a *synclinal basin* may stretch for hundreds of kilometres or may be measured with a wavelength of a few metres.

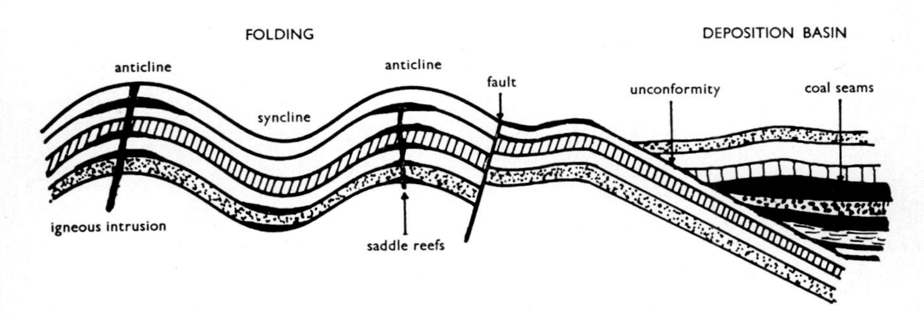

Fig. 1.1 Geological terms for strata

Where the folding is severe the rock is frequently heavily fractured, particularly where it is under tension on the crest of an anticlinal stratum or at the base of a synclinal stratum. These weakened zones provide a path for *igneous intrusive rocks* or mineral rich solutions to permeate the host rock and produce ore bodies. It is for this reason that many ore bodies have a typical form, such as the *saddle reefs* shown in Figure 1.1.

A simple *fault* is shown in the figure. In practice a fault of this magnitude would have many associated smaller faults and fractures that can again give rise to an area favourable for subsequent mineralisation.

An *unconformity* is also shown; in this case produced by an old shore line. These shore lines were often the location of the deposition of coal seams in the younger geological eras. It can be seen that it is important to distinguish whether a coal seam has terminated at the edge of a basin, or whether it has been interrupted by a fault or lost by erosion at an *outcrop*. Outcrops occur when a series of beds are cut away by general erosion of the surface, or dissected by rivers, so that the edges of beds are exposed.

After strata have been deposited and mildly disturbed they may be broken up by *orogenic movements*. These are the large movements of the earth's crust that are associated with mountain building and sometimes cause eruption or intrusion of igneous rock with subsequent mineralisation.

Syngenetic deposits are those mineral deposits formed contemporaneously with the parent rock and enclosed by it. They can be igneous or sedimentary. *Epigenetic deposits* are formed subsequently to the enclosing rock, often by a solution (the mother liquor) penetrating into rocks and depositing minerals. If the solution is a result of magmatic (igneous) action it will probably be hot and under pressure and will deposit non-ferrous minerals. If the solution is a result of sedimentary action it will percolate downwards causing either chemical changes or direct deposition. *Lateritic deposits* are formed by the leaching away of soluble minerals leaving behind in the *laterite* a valuable ore such as nickel or bauxite (aluminium ore). Thus these are normally surface deposits although some such as the iron ores have been changed and buried so that they are no longer lateritic.

Ore and Gangue It should be noted here that just as a weed is a flower in the wrong place, so an ore is a valuable mineral that is required, and gangue is waste that is not required at that mine. However gangue from one mine, such as limestone for instance, may be mined at another location to produce cement. Alternative names exist for the gangue, such as *mullock*, or simply *waste*.

Disseminated and Massive Ores Where the ore occurs in relatively pure form, for example as almost 100 per cent lead sulphide (galena), in even small amounts, it is said to be massive. If however the ore is spread thinly through a body of rock so that individual specks of the ore are surrounded by a high percentage of rock then this is a disseminated ore (Figure 1.2). Thus if an ore body is sampled over a narrow width of massive sulphides it could show a value of lead in the ore sample of 50 to 60 per cent (lead in pure galena is in fact 86.6

per cent). Over a greater width it might only show about 20 per cent lead because the sulphides had become disseminated, or dispersed. The disseminated sulphides, *assayed* separately (i.e. analysed for metal content), may only show as little as 0.2 to 0.5 per cent lead.

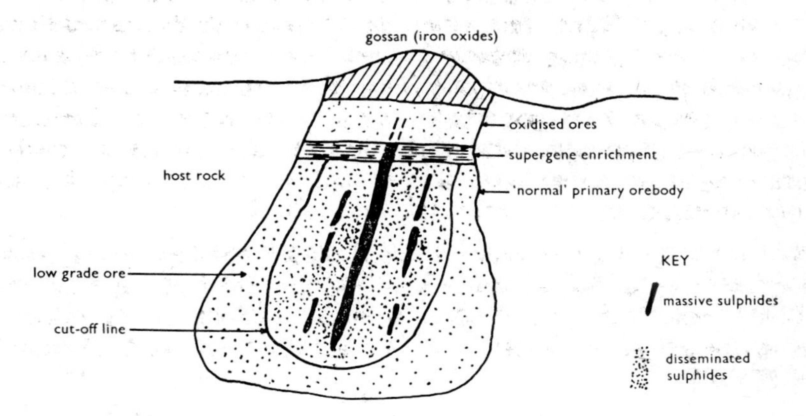

Fig. 1.2 Geological terms for an ore body

Cut-off Grade A gradual transition of high grade to low grade ore is common and it is not always easy to determine the physical limits of an ore body. Consequently it is necessary to determine the lowest value of metallic or non-metallic mineral content which can be mined economically and use this as the physical limit of the orebody. However this limit will obviously vary with the market value of the ore and the mining methods: Chapter 4 gives more details.

Gossans Where orebodies outcrop to the surface, particularly in geologically mature areas free of glacial deposition, the portion near to the surface may be leached by impure percolating waters and subsequently oxidised. Many metallic ores contain a proportion of iron sulphides and these do not leach away but remain as iron oxides (haematite, goethite and jarosite). The colour of the oxides varies slightly with the original mineral content, but the general end result is a yellowish, brownish, or reddish ferruginous capping known as a gossan. This is generally situated directly over the ore body in arid areas, such as in Australia, but in some cases it is transported from the outcrop and has to be traced back.

Supergene Enrichment The minerals leached away from the outcrop often cause a zone of secondary enrichment (generally in the form of sulphides) at a depth of 60 to 120 m from the surface (Figure 1.2). This enrichment zone then lies on top of the primary syngenetic orebody. Consequently if one is mining or prospecting from the surface for non-ferrous metallic orebodies there is frequently a transition from *oxidised ores,* which may need special extractive metallurgical processes, to the supergene enrichment zone showing high metal contents, and then to the orebody proper which may have a lower metal content again. Thus prospecting with shallow drill holes can give very misleading results.

Placer and Alluvial Deposits When the syngenetic deposits are brought to the surface by geological methods, and then mechanically eroded, they can suffer one of two fates. They may go into solution for subsequent chemical redeposition, or they may break up into fairly stable particles which are transported varying distances. Heavy mineral particles, such as gold, settle close to their source. The original falling area, known sometimes as a rock scree when situated against a mountain slope, is the *eluvial* region. Further transportation produces the *placer* deposits of such as gold and tin, although there appears to be no precise separation of this term from that of *alluvial* deposits. The alluvial deposits may be found in existing river beds, or in old river beds covered with new deposits such as basalt flows. These buried deposits may be called *deep leads* as in the state of Victoria, Australia, where much of the gold of the 1850s rush was found.

Transportation to the sea followed by wave action produces beach deposits. *Mineral sands* is the term applied to the heavy sand deposits of rutile, ilmenite and zircon, although they are sometimes colloquially called beach sands. These can be on either existing or old shore lines. Other minerals such as tin or diamonds also are found in beach deposits.

Bedded Deposits Chemical and physical deposition of consolidated minerals can produce bedded deposits which may extend over large areas. These are minerals such as coal, iron ore, salt or potash deposits, and phosphates. Limestone required for cement or blast furnaces is also a bedded mineral deposit. Although the dip (slopes) of these bedded deposits can be steep, and the beds are broken by faults or other geological disturbances, their lateral extent is usually larger and more continuous than is that of the metallic non-ferrous deposits.

Mining Terminology

Mining terms vary from one country to another, and even from one district to another.

Perhaps fortunately for the English-speaking world the spread of ‘Cousin Jacks’ from Cornwall gave most metal mining areas in the 19th century a large proportion of their mine captains, and the main terms for metal mining are thus common. In a similar way the spread of English coal miners did the same for coal mining. Unfortunately the fact that the coal miners and metal miners came from different parts of England, and different types of deposit, meant that the same object could have two names.

In the following pages only the most frequently used terms are given so that descriptions of mining methods can be understood. Other terms can be learned from journals or continued practice. The British Standards Institute[1] has published a glossary of standard terms with short definitions for coal mining. *World Mining* has published a multi-language glossary of terms[2] without definitions.

It is convenient to group the terms for illustrative purposes but no significance should be attached to the grouping or sketches. The good mining engineer has an inventive mind and must adapt methods to circumstances.

Large Deep Orebodies Large orebodies are often known as massive, but the term massive should not be confused with massive and disseminated ores in a geological sense. In fact a massive orebody is more likely to contain disseminated ore minerals. A vein-type orebody will probably contain bands of massive ore minerals. Figure 1.3 shows a typical arrangement for a large orebody of irregular shape.

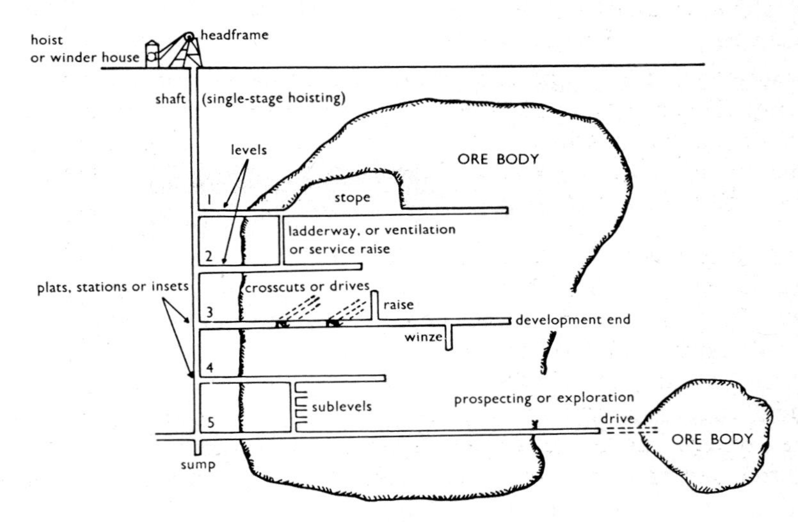

Fig. 1.3 Mining terms for large deep orebodies

A *shaft* is sunk alongside or through the orebody for access. In practice there must be two openings to allow air to be forced through the mine workings. Provided that there are at least two means of access to the mine further shafts may be partitioned for two-way airflow within them, but this is not permitted in coal mines. It must be noted that if the shaft is sunk through the orebody then some of the ore is sterilised. A cone of ore surrounding the shaft and widening with depth must be left until the rest of the ore is extracted. Otherwise the movement of rock caused by ore extraction will damage the shaft and necessitate expensive repairs, and lost production.

Waste rock, ore, men, and materials, have to be raised and lowered in the shafts. There are two types of conveyance, *cages* and *skips.* Cages are used for man-riding and to wind mine cars containing ore, waste rock or materials. To add to confusion the *mine cars,* which are virtually metal boxes on wheels, varying in sophistication from a tin can to a Rolls-Royce, are also frequently called skips. The shaft skips are loaded from special measuring pockets in the shaft wall (Figure 1.4); at the surface they will discharge into an ore bin or onto a conveyor.

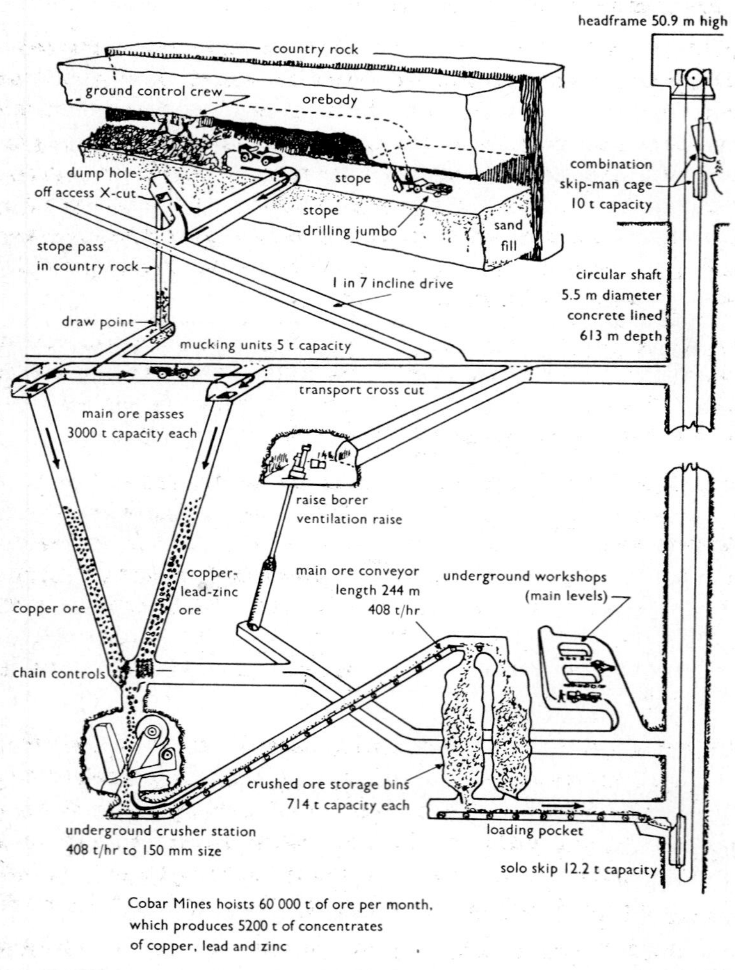

Fig. 1.4 Ore flow at Cobar mines (*original by courtesy of Cobar Mines Pty Ltd*)

A large shaft may be equipped with a pair of skips for ore and rock hoisting and a single cage for man riding. The area around the shaft at ground level is known as the *brace* in metal mining, and *pit top,* or *bank,* in coal mining.

At regular intervals down the shaft, openings are made and *levels* are driven into the orebody to enable it to be worked. The openings are called *plats* in Australia, *stations* in South Africa and the U.S.A., and *insets* in Britain. The interval between the levels is a function of the mining method and may vary from about 25 m to 60 m. At Mount Isa for instance the level interval is approximately 55 m with every second level being a haulage level. Improved drilling and raising techniques have enabled the interval between main levels to be increased in newer developments in mines.

Within the orebody it may be necessary to have more frequent access for working purposes, and *sublevels* will be driven at intervals between the main levels. It is also necessary to make vertical connections between levels for ventilation access for men and materials (known as *service raises*), and for dropping ore or rock through (*ore passes*). The connections may be made by driving upwards, known as *rising* or *raising*, or working downwards, known as *winzing*. These *developments* are made in a vertical plane and must be connected to each other in a horizontal plane by *crosscuts*, or *drives*.

After the orebody has had some or all of its development openings made, partly in barren rock and partly in the orebody, then systematic extraction of the profitable mineral starts. The working area in which mining is taking place is known as a *stope*.

Vein-type Orebodies A large orebody may in fact consist of clusters of smaller orebodies of a vein-type, and development may be on a regular grid. Alternatively the orebody may be one predominant vein which is long and narrow on its *strike* (horizontal) direction and which extends to the *dip* for a considerable depth. Technically the strike is at right angles to the full dip, but within an orebody the strike may follow a sinuous line and the dip frequently varies, so that general trends have to be taken.

In Figure 1.5 it can be seen that a vertical shaft rapidly loses contact with an inclined orebody and increasingly long levels have to be driven. It can also be appreciated that mining operations would weaken the *hanging wall* i.e. the rock lying above the ore body, and cause it to collapse. As a result the major developments are normally made in the *footwall*, below the orebody, where the rock is more stable.

In the early stages of working the shallowest ore an *inclined shaft* may be driven down from the surface and progressively deepened to follow ore extraction. However haulage speeds are generally lower and maintenance costs are higher in inclined shafts, so that with increased depth a change will be made to vertical shaft winding. An ore pass system will probably be developed to drop ore and rock to widely spaced main haulage levels.

In areas such as the Coeur d'Alene silver/lead district in the U.S.A. and gold mining areas in the Witwatersrand, South Africa, where workings in veins or reefs have extended to 2500 or 4000 metres, it has been found that single shaft hoisting is uneconomic. Consequently the shafts are broken into lengths of about 1000 metres to 1500 metres and the surface winding gear is duplicated underground. Even with modern technology this is still necessary, or may be more convenient for inclined orebodies where shafts are progressively deepened. For near-horizontal bedded deposits on a single horizon, where the full depth has to be sunk before production of mineral can start, then single lifts of over 2000 metres are made.

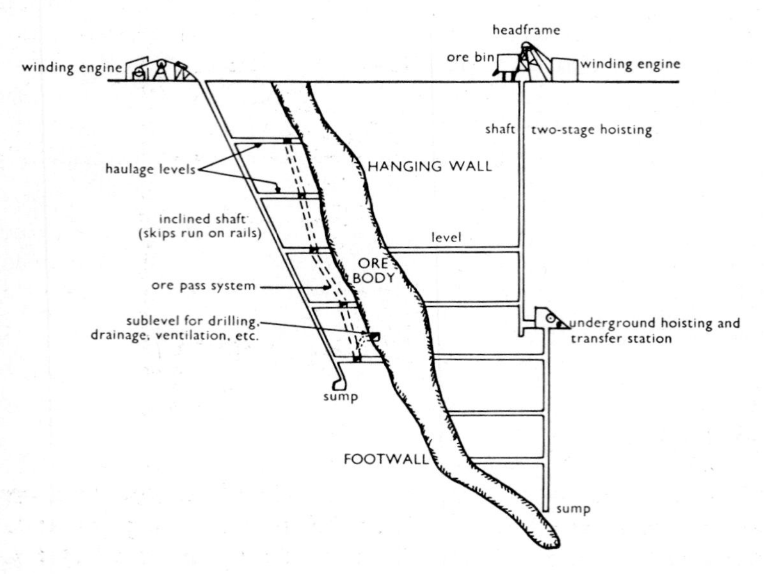

Fig. 1.5 Mining terms for vein-type orebodies

Large Shallow Orebodies Shallow orebodies are worked as open pit mining operations wherever possible. The scale of equipment is generally smaller in metalliferous orebodies than in shallow bedded deposits because of the need for *benching* downwards (Figure 1.6). To start the removal of each successive layer a *centre cut,* or *drop cut,* is made in the pit bottom and a bench taken outwards.

Because of the need to keep the pit slopes stable the overall inclination of the pit walls is generally about 45° and bench heights are generally about 15 m. It can be seen that as the pit becomes deeper then more waste rock (overburden, or spoil) has to be taken in ratio to the ore won. Eventually economics force a change to underground mining methods.[3]

Haulage out of the pit is at first by truck in a spiral around the benches. It may then become more economic and convenient to drop the ore through a crusher to an underground haulage for transport out of the mine. The access tunnel driven down through the strata to the bottom of the open pit is called a *drift*. A drift is usually driven as a straight line in plan to reach from the surface to the desired underground horizon and is used for rail or conveyor haulage at grades of up to 1 in 2½. With locomotive haulage the grades are restricted to less than 1 in 15.

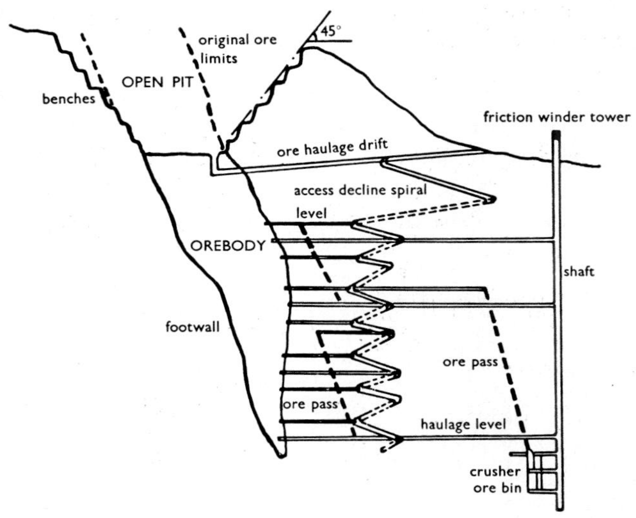

Fig. 1.6 Shallow orebodies

With a change to underground methods of mining two types of haulage are possible. Down to about 200 metres it is possible to haul ore economically by low-profile diesel trucks up an *access spiral*. This access is driven at a grade of about 1 in 7 to 1 in 9 and in addition to ore haulage allows diesel driven rubber-tyred loaders and ancillary equipment to be taken into the mine and brought out for servicing. For large scale mechanised mining this is probably cheaper than the alternative of hauling ore and waste rock to a shaft and then hoisting it to the surface. However below 200 m to 250 m a detailed study will probably show that shaft hoisting is cheaper than trucking up an incline, although the spiral incline will be continued for vehicle access and through ventilation.

Coal Mining The terms used in coal mining are more pragmatic, and where they are not the same as in metal mining then they tend to be self-descriptive. Most of the terms will be dealt with in the chapters on coal mining methods. Noticeable differences are that one is usually dealing with a regular layer of rock stratum, which when it is coal is called a *seam*, rather than with an irregular orebody.

Consequently the seam has a roof and a floor rather than a hanging wall and footwall. In roadways and working areas (known as *stalls, bords,* or *faces,* not stopes) the roof is called a roof. In metal mining the roof is called the *backs*. Short vertical shafts are sometimes called staple shafts, and, occasionally, raises.

Surface mining of coal is called opencast mining (or strip mining in the United States) whereas the metal miner refers to open pit mining. The difference lies in the fact that surface mining of coal extends for several kilometres at a relatively shallow depth, filling its hole behind itself, whereas surface mining of an orebody is rarely above a few hundred metres radius, but proceeds downwards leaving a large hole.

The initial opening into a coal seam for opencast mining is known as a *box cut*. From the box cut coal removal proceeds down the dip, or across the mining lease. The layer of waste rock over a coal seam is called *overburden* and is removed from in front of the advancing highwall and thrown into the mined area, as in Figure 1.7.

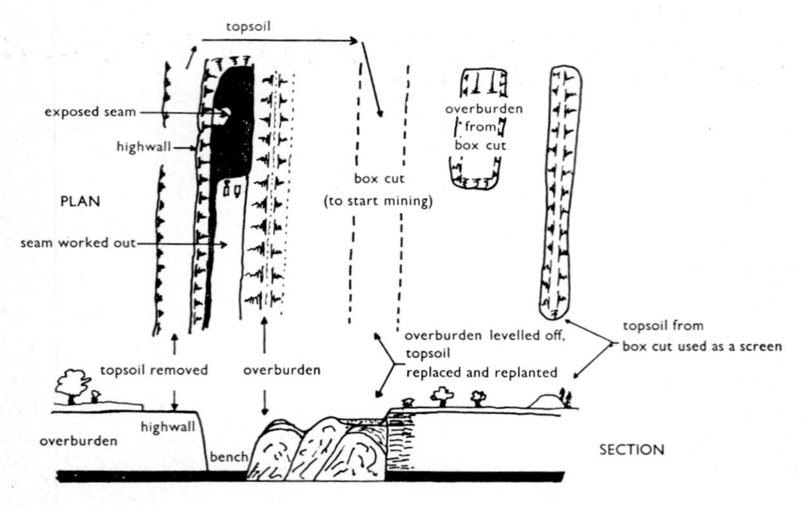

Fig. 1.7 Opencast mining of coal

Other Bedded Deposits The terms used when working deposits of phosphates, limestone, bauxite, etc. depend on whether they are mined underground or on the surface. If underground, they tend to follow coal mine terminology because of the similar working methods. On the surface the terms depend on whether the mining is a straightforward earth moving equipment enterprise, on civil engineering lines, as for bauxite (aluminium ore), or whether it is quarrying hard rock, or mining a seam under varying depths of overburden. Generally the language is similar for all operations, especially if things go wrong!

References

1. ANON. *B.S. 3618 Glossary of Mining Terms.* Sections 1 to 11. British Standards Institution, London, Various Dates.
2. ANON. *International Glossary of Technical Terms.* Vol. 1. World Mining, San Francisco, 1968.
3. CALLAGHAN, R. P. *Development of the Prince Lyell orebody.* Australian Mining, p. 60, July 1971.

Bibliography

1. An interesting description of the detailed activities of a group of lead-silver-zinc mines will be found in
ANON. *Broken Hill Mines—1968.* Monograph Series No. 3, The Australasian Institute of Mining and Metallurgy, Melbourne, 1968.
2. Practical examples of surface and underground workings are described freely in
BOLDT, J. R. and QUENEAU, P. *The Winning of Nickel.* Methuen, London, 1967.
3. The historical background to early mining in Australia can be found in
BLAINEY, G. *The Rush That Never Ended.* (2nd Ed.) Melbourne University Press, Melbourne, 1969
and in the same author's books on individual mines.
4. A background to the social history of mining, still valid in many aspects, is given by Emile Zola in his book *Germinal.* I have the Penguin Classics translation, but other editions are also available.
5. A comprehensive list of terms and definitions associated with the mining industry can be found in *A Dictionary of Mining, Mineral and Related Terms.* U.S. Department of the Interior, Bureau of Mines, 1968.

Chapter 2. Exploration

Likely Mineralised Areas

It is probably true to say that major exploration is usually associated with non-ferrous metals and oil. The consumption rate of non-ferrous metals in relation to the size of deposits is such that a normal mine design life is a 'rule-of-thumb' ten years. Large mines such as the group at Broken Hill have of course been in existence for over 75 years, but the early leases are exhausted and the mines are spreading outwards and downwards. The need for replacement of exhausted orebodies forces a continual exploration for new sources. The same problem exists for oil.

In more recent years the increasing consumption of steel, and rising costs, have prompted a search for large high-grade deposits of iron ore that can be mined cheaply by open-pit methods. These large iron deposits are easily detected by surface or airborne magnetic methods.

Deep boring for oil in sedimentary basins, and water bores, produce a useful knowledge of the general rock structures and enable the less exotic minerals such as potash deposits, salt and coal to be found. In any case, coal is frequently found as outcrops, and is usually worked at first for local fuel. Increasing mechanisation, and lowered transport costs because of bulk carriers, have brought the large coal deposits in Australia and Canada, known of for many years, into world prominence. The easier working conditions in thick seams have enabled costs of production to be reduced below those of European and United States mines. New coal sources in Australia have been found by step-out exploration from existing coal mines.

Choice of Areas

In countries such as Britain the geology has been well charted, rock outcrops are easily seen, and the geologist already knows the most promising areas for minerals exploitation, if indeed they have not already been exhausted. Some mining, such as the opening of Wheal Jane tin mine in Cornwall, involves the reworking of old areas using modern techniques to recover previously unprofitable ore.

In other areas, such as Australia and Canada, there are vast relatively unexplored areas. In Australia they are covered by dry red featureless semi-desert or arid scrub. In Canada they are covered by glacial moraine and frequently by snow or thick forests and lakes. Under these circumstances modern exploration cannot rely on the lone prospector with a pick, although he comes into his own in the end stages of exploration, and can still accidentally stumble on to the surface outcrop of a mineral.

The first thing is to decide one's objective and then select the areas most likely to be productive. For oil or gas, one is seeking sedimentary areas with structures likely to trap and retain the fluids. For coal one is also looking for the younger sedimentary rocks, often as basins that have preserved the coal from subsequent erosion. Sedimentary basins are also the locations of phosphate, salt, and potash deposits.

Gold, silver, nickel, copper, lead and similar minerals are likely to be found in association with major orogenic movements such as those along the line of the North American Rocky Mountains, and the South American Andes. In Australia, the prospectors found alluvial gold in the streams of the Great Dividing Range and traced it back to the quartz reefs from which it came. Tin and copper were found in the mountainous west coast of Tasmania.

However, the earliest mountains are now worn flat and are covered by barren surface material, and even when the broad outlines of the potentially mineralised areas are known it is by no means simple to locate the prospective rocks within them. The early copper mines in Australia were found more by accident than design, because of the staining of their outcrops, and the lead mines of Broken Hill were found by pastoralists herding sheep, who examined with no great interest the few low rocky hills on a vast plain.

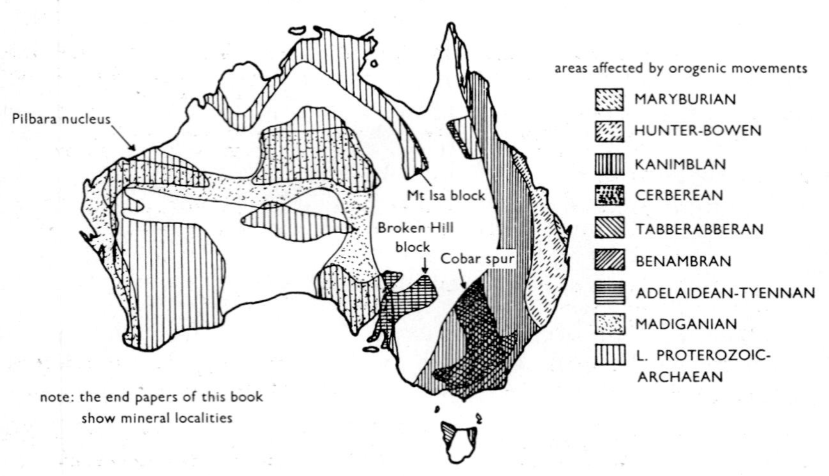

Fig. 2.1 Areas likely to be mineralised in Australia

Location of a metalliferous province is helped by knowledge of the geological periods in which mineralisation is likely to occur, and the rocks which are associated with mineralisation. Liddy[1] has summarised some of this information for Australia. Table I gives the major periods

TABLE I.

PERIODS OF OROGENIC MOVEMENT IN AUSTRALIA

Geological Age		Name	Remarks
CAINOZOIC	Pliocene	Kosciusko	Uplift and heavy faulting; no mineralisation
MESOZOIC	Upper Cretaceous	Maryburian	Minor mineralisation in S.E. Queensland
	Permo-Triassic	Hunter-Bowen	Widespread mineralisation in N.E. Australia: in particular in New England
	Epi-Lower Carboniferous	Kanimblan (Herbertonian)	Widespread mineralisation in eastern Australia; major mineralisation in coastal Queensland
	Epi-Upper Devonian	Cerberean	Mainly in Victoria; gold, etc.
PALAEOZOIC	Epi-Middle Devonian	Tabberabberan	Major mineralisation: Cobar, Mt. Lyell, Zeehan, Walhalla-Woods Point, Renison Bell, etc.
	Siluro Devonian	Bowning	Some gold, copper and tungsten in Victoria and New South Wales
	Epi Ordovician	Benambran	Major gold and tin in Victoria and S.E. New South Wales
	Middle-Upper Cambrian	Adelaidean-Tyennan-Jukesian	Variable but no significant mineralisation
	Upper Proterozoic (1000-1100 million years)	Madiganian	Minor mineralisation in Northern Territories, South Australia and Western Australia
	Middle Proterozoic (1400-1800 million years)	Houghtonian-Mt Isa-Broken Hill	Major Precambrian mineralisation of continent
	Early Proterozoic (1800-2000 million years)	Hall's Creek-Pine Creek	Some mineralisation in N.W. Australia
PRECAMBRIAN	Late Younger Archaean (2200-2300 million years)	Kalgoorlie-Rum Jungle	Widespread gold in Older Greenstones
	Younger Archaean (2600-2900 million years)		Some pegmatite mineralisation in Western Australia
	Older Archaean (2700-2900 million years)	Pilbaran	Some gold in Older Greenstones; in particular sulphide and arsenical types

of mountain building movement in Australia, and the areas of Australia likely to be mineralised are shown in Figure 2.1.

Table II shows the likely association of metals with magmas. The magmas are molten igneous rocks which penetrate the earth's crust in characteristic forms. It is noticeable in the table that several of the metals are repeatedly grouped together and in practice many of them occur together in one orebody. Consequently, a mine rarely produces only one pure metal. Most lead mines produce varying quantities of zinc and silver; silver is frequently contaminated with lead; copper ores frequently contain gold, and vice versa. A mine such as Mount Isa produces within its leases copper, lead and zinc from adjacent orebodies, plus contained gold and silver values. A nickel mine can make extra profits from platinum and palladium. Table III shows a grouping of frequently associated minerals.

TABLE II.

ASSOCIATION OF METALS WITH MAGMAS

Magma Type	Rock Type	Associated Metals
ULTRABASIC	peridotite	platinum, chromium, nickel, diamond, asbestos, corundum, iron
	pyroxenite	nickel, copper, iron, titanium
	anorthosite	titanium, iron
BASIC	gabbro	iron, titanium, copper
	norite	platinum, palladium, nickel, copper
INTERMEDIATE	quartz-diorite	gold, silver, lead, zinc, iron, copper, arsenic
	monzonite-diorite	titanium, iron, copper, gold, arsenic
	granodiorite	mercury, antimony, gold, iron, silver, lead, zinc, copper, arsenic, molybdenum, tungsten, bismuth, tin
	syenite	iron, molybdenum, zinc
	nepheline-syenite	iron, titanium, zinc, gold, copper, zirconium
ACIDIC	granite	silver, cobalt, nickel, antimony, uranium, tellurium, tungsten, lead, zinc, gold, tin, molybdenum, iron, arsenic, bismuth, copper
	pegmatite	molybdenum, copper, tin, tungsten, bismuth, beryllium, rare earths, uranium, lithium, iron, tantalum, columbium, titanium

Airborne Exploration

The easiest way to investigate large areas is to use airborne exploration equipment; fixed wing aircraft are cheaper to use than helicopters, which are kept for close ground flying in rugged country and for ferrying ground based equipment and personnel.

TABLE III.

TYPICAL PARAGENETIC RELATIONSHIPS

Two Members	Three Members	Four or More Members
galena-sphalerite (lead-zinc)	galena-sphalerite-pyrite	galena-sphalerite-pyrite-quartz
pyrite-chalcopyrite (iron-copper)	pyrite-chalcopyrite-quartz	pyrite-chalcopyrite-galena-sphalerite-quartz
gold-quartz	gold-quartz-pyrite	gold-quartz-pyrite-sphalerite-galena-arsenopyrite
gold-tellurium	gold-tellurium quartz	gold-tellurium-quartz-pyrite
cobalt-nickel	cobalt-nickel-pyrrhotite	cobalt-nickel-pyrrhotite-quartz (iron)
cassiterite-wolframite (tin-tungsten)	cassiterite-wolframite-quartz	cassiterite-wolframite-quartz-tourmaline
Molybdenite-bismuth	molybdenite-bismuth-quartz	molybdenite-bismuth-quartz-wolframite-pyrite
cinnabar-pyrite (mercury-iron)	cinnabar-pyrite-quartz	cinnabar-pyrite-quartz-calcite
magnetite-chlorite	magnetite-chlorite-garnet	magnetite-chlorite-garnet-pyroxene-pyrite
chromite-serpentine	chromite-serpentine-olivine	chromite-serpentine-olivine-pyroxene

Really large-scale exploration in thousands of square kilometres is only likely to be done for a government, or with the help of government subsidies. For example, the Australian Commonwealth Bureau of Mineral Resources, Geology and Geophysics has flown reconnaissance airborne magnetometer survey lines across the continent, and continually conducts regional aeromagnetic and ground gravity surveys over large segments of Australia. These are usually areas of the 1:250 000 series index map sheets which are one degree of latitude wide and one and a half degrees of longitude long (i.e. approximately 17 000 km^2). Geological mapping of Australia is also proceeding, sheet by sheet. By studying these and other government geological and geophysical reports, mining companies can select areas to lease for both regional and local mineral search.

For airborne exploration, and in fact for any area where outcrops are concealed or have been deeply weathered, the geologist or mining engineer has to rely on geophysics and geochemistry rather than surface geological mapping. *Geophysics* is the application of physics to the study of the earth. Ore bodies frequently differ in physical properties (magnetic susceptibility, electrical conductivity, density, etc.) from the rocks in which they are embedded. By systematic surficial measurement of physical properties of rocks broad geological structures can be evaluated and large portions of the survey area can be eliminated as unlikely to contain specific types of ore. Detailed searches can then be directed to limited areas where geophysical methods may directly locate mineral concentrations. It must be remembered that elimination of areas only means that there is no indication of an orebody within the limits of depth of penetration and sensitivity of the instruments used. Improvements in instrumentation may make it worthwhile to re-survey abandoned areas.

Particular improvements have been made in the newer science of *geochemistry*, which detects changes caused by metallic orebodies in the chemical composition of air, ground water, soil or botanical specimens. The order of accuracy and scale of changes is in parts per million, that is less than 0.0001 per cent, or even 1 in 10^{-11} for some elements such as mercury. In effect, one grain of dandruff dropped into a test tube of sample solution would blot out any trace of the ore.

Geophysical measurements which vary markedly from the general background values, and thus may indicate the presence of orebodies, are described as *anomalies*. Each anomaly represents an area where some property of the earth's crust is different from its general surroundings. However the anomaly could be due to conducting overburden, faults, cavities in the rock, and so on, rather than to the desired mineral. A geophysical survey aircraft carries many different instruments to continuously record variations in physical properties. Correlation of these variations indicates which are significant and those anomalies are then prospected in more detail on the ground.

Barringer[2, 3] has suggested suitable groups of instruments for different orebodies. These are given in Table IV. The individual instrument

TABLE IV.

ESTABLISHED AND POTENTIAL APPLICATIONS OF AIRBORNE SURVEY METHODS

Orebody or Geological Requirement	Instruments
Porphyry copper exploration	RP+EP+GR+Mag+Merc
Lead-zinc exploration	RP+EP+GR+Mag+Merc
Precious metal exploration	RP+EP+GR+Mag+Merc
Massive sulphide exploration	RP+EP+Mag+Merc+EM
Manganese and haematite exploration	RP+EP+GR+Merc+AP
Uranium exploration	RP+EP+GR+Mag
Structural geological mapping	RP+EP+Mag+AP
Mapping of Pleistocene and surface geology	RP+EP+AP

KEY

RP = radiophase EP = E-phase (RP and EP are both VLF-EM)

Mag = airborne magnetometer GR = gamma-ray spectrometer

Merc = airborne mercury spectrometer

EM = airborne electro-magnetometer VLF-EM = very low frequency EM

AP = air photos

types are briefly described below. A fuller description of the principles of the instruments is given in such books as those by Parasnis[4], or Keller and Frischknecht[5]. Their information can be brought up to date with the articles by Barringer[2] and Paterson[6], amongst others.

MAGNETIC METHODS

Certain types of ore, especially magnetite, ilmenite and pyrrhotite-bearing sulphide deposits, produce distortions in the earth's magnetic field (pyrrhotite is an iron sulphide). Some iron-rich manganese and chromite ore may yield magnetic anomalies. Most of the magnetic effects of ores are due to their iron content, and ores are very frequently contaminated with iron sulphide, even if they are nominally non-ferrous deposits.

The ferrimagnetic materials have two magnetic properties. One is that the earth's magnetic field effectively turns the orebody into a large magnet, which in turn warps the normal field, thus producing an anomaly. The other is that the ferrimagnetic materials often have a residual magnetism, due to their original formation, in their own right. This residual magnetism may act at any angle to the earth's field and may weaken or strengthen the original anomaly.

Unfortunately many geological structures other than orebodies can produce magnetic anomalies. Faults, contact zones, igneous intrusions and other features also give changed values, so that skill is needed in interpretation, and correlation with other geophysical data is necessary. This is even more important in airborne than ground work, because the accuracy and depth of penetration of airborne *magnetometers* used to measure the variation in the earth's magnetic field are less than those of surface magnetometers.

Ground and airborne magnetometers may be of the fluxgate type, which measure relative variations in the total magnetic field, or in one component of the total field, or of the proton precession type which measure accurately the total magnetic field. Their construction is beyond the scope of this book. Hood[7] has made a detailed review of modern instruments.

The unit of magnetic prospecting is the *gamma*: one gamma is equal to 10^{-5} gauss. In the S.I. system of units the gauss has been replaced as a unit of magnetic flux density by the Tesla, T, where 1 Tesla $= 10^{-4}$ gauss: therefore 1 gamma $=$ 10 Tesla. When a survey is made over an established grid on the ground, or continuously along aircraft flight paths, the value of the magnetic field in gammas is measured. The differences in magnetic intensity between the theoretical (or arbitrary) background level and the measured values are plotted on a grid plan or flight path map. The data is then contoured (as for a physical map) to link interpolated points of equal value. The contours are called *isoanomaly lines*, or *isogamma lines*, and are drawn at convenient round values such as 1, 2, 3, etc. or at 100, 200, 300, the unit being gammas (Figure 2.2). In addition to the isoanomaly lines a cross-sectional profile of the values is drawn to aid interpretation (Figure 2.2).

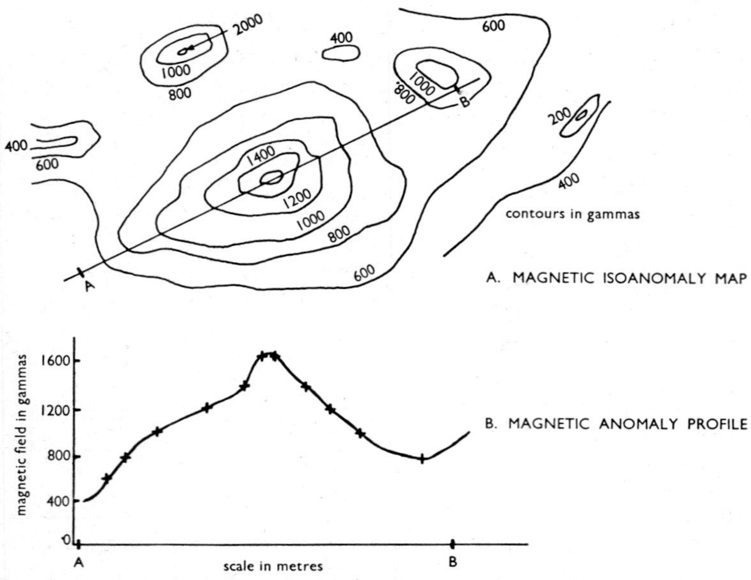

Fig. 2.2 Isoanomaly curves and profiles

Qualitative interpretation of the magnetic data may indicate lithologic variations, structural controls, and the shape and extent of magnetic orebodies. Quantitative interpretation of contour maps and profiles involves the use of catalogues of charts, manuals of standard curves and computer modelling to yield the depth and geometry of the magnetic body as well as the contrast in magnetic susceptibility.

ELECTRO-MAGNETIC METHODS

These methods have been developed over many years and advances in computer data processing have enabled a high degree of sophistication to be achieved. There exist a wide variety of electromagnetic methods all of which are used to delineate rock masses of high electrical conductivity.

When an electromagnetic field, produced by an alternating current flowing in a coil or straight cable, is propagated through the ground it induces electric currents in any conductor in its path. These secondary currents in turn produce their own alternating secondary electromagnetic field which opposes the primary field. The lower the resistance of the conductor, then the stronger will be the opposing current.

Thus, if the induced field is passed through a good conductor such as an orebody containing graphite, pyrrhotite, pyrite, chalcopyrite, galena or magnetite, a strong secondary field is set up. If this field is measured with a receiver, and compared with the theoretical value for the given distance from the transmitter, then a difference will be found. The anomaly can be quoted as a difference in parts per million of measured field to expected field at the receiver, or given as a ratio of receiver signal strength to transmitter signal strength in percentage. In either case there is no unit, only a numerical value.

In addition to the highly conductive ores, the low conductivity ores such as haematite, zinc blende, chromite, etc. may also be detected if they occur in association with significant amounts of the highly conductive ores. However, electromagnetic disturbances can also be caused by faults, zones of crushed rock, fissures with conductive saline water, and other geological features, so that skilled interpretation is needed.

Several systems of airborne electromagnetic (AEM) prospecting are in use. Paterson[6] lists them and gives a comparison of their respective performances and suitability. This material is beyond the scope of an introductory book, but some brief details are given below.

The properties of an induced secondary magnetic field that can be measured are its dip angle (tilt angle) and its horizontal and vertical components. The primary field can be produced by a relatively close stationary source (known as a fixed source), or the transmitter can be mounted on the aircraft as a moving source, or it can be a very low frequency (VLF) radio transmitter at theoretically infinite distance. Combinations of these factors provide the various commercial systems. Another system, AFMAG, is similar to the VLF systems but uses the lightning discharges from electrical storms as a signal source.

The *dip angle* method relies on the fact that the electrical field from the transmitter, which is polarised vertically or horizontally by design, is tilted in the presence of a good conductor. The receiver coils measure the amount of tilt, that is the dip angle, to define the anomaly. The amplitude of the field must also be measured at the same time.

The other systems measure the *in-phase* and *quadrature* components of the resultant field. The *in-phase* component is the proportion of the resultant or secondary field that can be considered as acting in the same direction or directly opposite to the primary field. This is sometimes called the *real component.*

The *quadrature* component, sometimes called the *out-of-phase* or the *imaginary* component, is the proportion of the field that is at 90° to the direction of the primary field. The quadrature component is usually arranged to be the physically vertical component for measurement and some EM systems measure this component only. In all of the systems it is the secondary field which gives rise to the measured anomalies. Figure 2.3 shows how anomalies may be recorded as profiles over an orebody. Corresponding isoanomaly lines are also prepared.

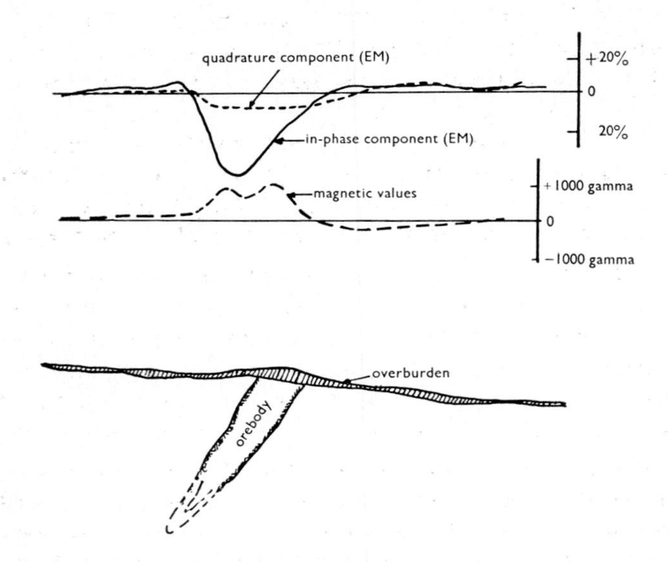

Fig. 2.3 AEM and magnetic profiles over an orebody

There is another system of recording electromagnetic anomalies, and that is to measure the decay time of an electric pulse sent through the earth: this is the *time domain* method. The Barringer INPUT system is one such method (INduced PUlse Transient). The presence of conducting bodies affects the decay time of the curve. The principal use of time-domain systems is to explore for horizontally bedded deposits and for aquifers.

There is a major problem in all EM methods in that the presence of a major orebody may be masked by eddy currents produced by conductive overburden. Whereas EM methods have worked successfully in Canada under glacial overburden, their present usefulness is limited in Australia and further development work is needed. The reason is that in semi-arid regions with rainfall less than about 0.35 m, and a relatively flat topography, development of salt pans and saline lakes is common. Again, the soil profile often develops a hard pan, or calcrete horizon, which is rich in mineral salts and is highly conductive with even the small amounts of moisture found in desert conditions. This conductive layer masks the effect of any orebodies underneath it. As an example of a difficult region one could take the nickel areas of Kalgoorlie-Kambalda in Western Australia, which are overlain by a series of vast salt lakes. Part of W.M.C.'s Kambalda mine (the Silver Lake ore shoot) is actually under Lake Lefroy.

RADIOACTIVITY

The radioactivity of rocks is dependent on the amounts of radio-isotopes of uranium (238) and thorium (232), and of their daughter products such as radium that they contain. These ores decay with the emission of alpha, beta and gamma rays; the factor that is easiest to measure is the rate of emission of gamma rays. The rate is obviously highest above large radioactive orebodies, but not necessarily in proportion to ore types. For instance a uranium orebody can only be located by the gamma rays emitted by the radium daughters (such as radon 222) of the continual radioactive decay process (see Reference 4 for example).

The fact that most rocks have a background radioactivity and emit gamma rays at characteristic rates enables airborne gamma ray spectrometers to be used for structural geological mapping as well as for exploration for radioactive ores. Just as the Geiger counter was increasingly replaced by the scintillometer, so the scintillometer has been increasingly replaced by the gamma ray spectrometer for airborne work.

When gamma rays impinge on certain substances, such as crystals of sodium or caesium iodide, their energy is translated into a light pulse of proportional energy. This property is the basis of the scintillometer which amplifies the light pulses electrically and counts them. Gamma ray energies are measured in million electron volts (mev) and are characteristic of their nuclide sources. The characteristic emissions appear as photopeaks above general background levels, with potassium 40 at 1.46 mev, thorium at 2.62 mev and a uranium series peak at 1.76 mev. The spectrometer is designed to discriminate between these photopeaks by changing them to electrical pulses and then either measuring pulse heights or amplitudes. Initial airborne exploration from low flying aircraft can detect radioactive anomalies qualitatively if the surface weathering is not too thick. A quantitative search is then made on the ground.

CHEMICAL METHODS

Barringer [2, 3] has developed a mercury spectrometer which can detect about 10^{-11} parts by weight of mercury vapour in air (actually nanograms/m^3). Earlier studies by other workers and by the U.S. Geological Survey had shown that not only were mercury values in the atmosphere higher over mercury deposits but, under certain meteorological conditions in sparsely vegetated areas, other ore deposits such as porphyry coppers and precious metals could also show anomalous amounts of mercury vapour.

Increased instrument development has shown that most base metal deposits also have slight increases over the normal background values. Research is still being carried out in these fields.

IMAGERY

Aerial photography has been used for many years and often enables structural forms to be determined so that associated mineralisation may be located. The techniques have lately been extended by side-looking radar and satellite photography, but both can be too expensive for normal prospecting use. Colour infra-red photography may enable the greens of vegetation to be distinguished so that the tonal differences caused by metallic salts in the soil may be located.

Thermal infra-red imagery also makes it possible to locate abnormally warm areas of the earth, which might be due either to oxidising ore-bodies or to hydrothermal areas with possible attendant mineralisation.

Surface Exploration

After the airborne exploration has located promising anomalies, or mapped the structural occurrence of beds such as the greenstones that are frequently mineralised, then the hard physical work of geological exploration begins. There is no substitute yet for footwork in rough physical terrain, although four wheel drive vehicles are used in open country to transport men and equipment, and much of the geophysical equipment is used while mounted on the truck.

In mountainous areas, equipment and personnel are frequently ferried in and provisioned by helicopter. Conzinc Riotinto of Australia (C.R.A.), whilst prospecting around New Guinea and proving the Bougainville copper prospect, used a converted ship as a mobile base. The ship was fitted with a helicopter platform and equipped with laboratories and living quarters. It then moved around the islands flying personnel in and out for prospecting work as necessary.

Where rock outcrops are frequent, as with steeply dipping beds, location of bedded deposits can be done by direct physical observation, although some geophysical work will still be done to determine the extent of the concealed formations. In the absence of outcrops, then a routine of geophysical and geochemical exploration will be followed to determine areas suitable for subsequent drilling.

GEOPHYSICAL EXPLORATION

The magnetic, electromagnetic and radioactivity methods used on the ground surface include all the airborne methods, but with some limitations and some improvements. The scale of exploration is more limited, but because instruments can be carefully set up on a firm surface, and their map location more precisely defined, then their accuracy can be greatly increased. Also, because they are a hundred metres (more or less) closer to the earth's surface, then their penetration into the ground can be correspondingly increased in many cases. Consequently, anomalies located from the air can be carefully checked to determine boundaries, and promising geological structures can be examined for deeper mineralisation missed from the air.

In addition to the methods described earlier five more are available. These are the self-potential, electrical, induced polarisation, gravity and seismic methods, although they are not all equally useful.

The *self-potential* (SP) methods are useful as indications of near surface anomalies because they are cheap, simple and easy to use. If two non-polarizable electrodes (porous pots with copper-copper sulphate electrodes) are driven anywhere into the ground and connected to the terminals of a sensitive voltmeter a small electric voltage is found to exist between the terminals. The voltage is chemical in origin and will be less than about a hundred millivolts as a background value. However, above some sulphide orebodies, particularly those containing pyrite, chalcopyrite and pyrrhotite, the chemical activity may produce potentials of several hundred millivolts to a volt, measurable as a systematic pattern that produces a plotted anomaly.

The self-potential method cannot be employed on surfaces such as dry barren rock or frozen ground, which are bad electrical conductors. It also does not definitely identify an ore body, but if there is a large negative self potential present in the ground then it is one more factor to indicate that there may be an orebody below the surface.

For *electrical* prospecting methods, an electric current is sent into the ground and the resulting distribution of potentials (voltages) is measured by a pair of electrodes driven into the ground and connected across a sensitive voltmeter. This method, as for SP, requires satisfactory electrical contact between the electrodes and the ground.

When an electrical current is passed through the earth its theoretical paths through homogeneous ground are known. These can be graphically depicted as smooth curves between point electrodes, or as parallel straight lines orthogonal to parallel linear electrodes (i.e. a rectangular grid effect) as in Figure 2.4. The current paths crowd together in the

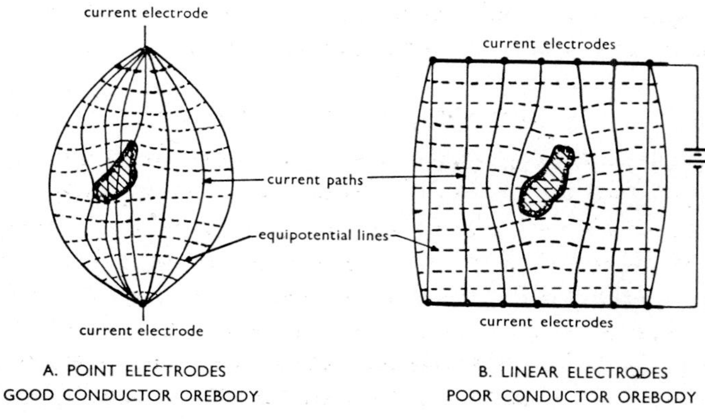

Fig. 2.4 Electrical prospecting

presence of an orebody that is a better conductor than the average ground, leaving a relatively empty area just around the orebody. Where an orebody is a worse conductor than average then the current paths twist away from the body and crowd together on either side. Cutting the solid-drawn current paths at right angles are the dotted *equipotential* lines. If the measuring electrodes are placed along these lines then no potential difference will be recorded. The behaviour of the equipotential lines is opposite to that of the current paths. They spread apart over a good conductor, because there is a low voltage drop across it, and crowd together over a poor conductor because it has a high voltage drop.

The presence of an orebody can be detected by plotting the behaviour of the equipotential lines or by measuring the potential drop gradients along the theoretical current paths. The latter method using an *earth resistivity* (ER) technique is the most common. The earth resistivity is measured as a gradient along the current path, usually taken along a straight line between the current electrodes. It is the ratio of drop in voltage/current, $\Delta V/I$ per unit length, and therefore has a value in ohm-metres, and is frequently denoted by R. However, it is not the average resistance of the volume of earth between the measuring probes, because its value is as much dependent on the configuration of the current and measuring electrodes, as on the resistivity of the ground. Commercial instruments generally give a direct read-out of the value of $\Delta V/I$.

An attractive feature of resistivity methods is that they are sensitive to small changes and can map the details of ore formations, and of shear zones, faults, changes in strike, etc. where these are associated with changes in resistivity.

A natural corollary to resistivity surveys are the *induced polarization* methods. It is found that when the current source of the electrodes is cut off the voltage does not drop immediately to zero but decays at a continuously increasing, but measurable, rate. Similarly, when the current is switched on the maximum voltage is not reached immediately but the value increases steadily towards its peak for several seconds or minutes. Figure 2.5 shows a diagrammatic curve of residual voltage decay. The effect can be likened to that of an electrical condenser. The plates of the condenser are provided by the interfaces of metallic minerals and surrounding rock; the more mineral particles there are, then the greater the induced polarization effect (IP).

Because the mineralisation produces a condenser effect it can be sought in two ways. There is the transient decay of induced polarisation produced by a direct current voltage, and there is the effect of frequency variation with alternating current. If a constant voltage a.c. current is applied and the frequency of the current is increased then

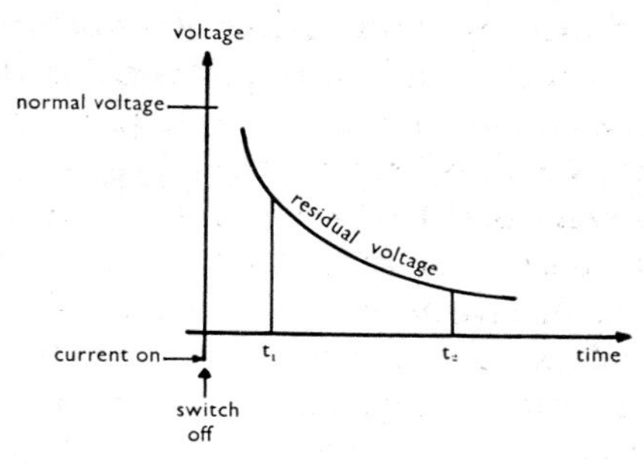

Fig. 2.5 Induced polarisation voltage decay

the resistivity of the rocks shows an apparent decrease. The usual frequency range in IP is from 0.1 to 10 Hz.

Exploration for minerals can therefore be by time-domain or by frequency-domain IP methods. In the *time domain IP* method a pulse of d.c. is sent through the ground for from 3 to 15 seconds and the residual voltage decay is monitored. The measurement can be of the residual voltage V_t at a time t after current cut off. This is expressed as a ratio of V_t in millivolts to the normal voltage V_o in volts. The IP effect is then (V_t/V_o) millivolts per volt. Alternatively the value can be expressed as $100V_t/V_o$ percent with both voltages in the same units. Another time-domain measurement uses a time integral. Figure 2.5 shows the time integral as the area under the voltage decay curve between times t_1 and t_2. The value is usually given as millivolt-seconds per volt; the transient voltage integral is divided by the normal voltage.

Frequency-domain IP systems compare the earth resistivity $\Delta V/I$ at two different a.c. frequencies, usually 0.1 and 10 Hz. Thus, if the resistivity is ρ, where $\rho = \Delta V/I$, then the IP frequency effect, f.e., is given by

$$\text{f.e.} = \frac{P_{0.1} - P_{10}}{P_{10}} \times 100\ \%$$

An alternative parameter is metal factor, which is arbitrarily set at

$$\text{Metal Factor} = \frac{P_{0.1} - P_{10}}{P_{0.1} \times P_{10}} \times 2\pi \times 10^5$$

If the resistivities are in ohm metres then the metal factor is in units of reciprocal ohm metre, Ω^{-1} m^{-1}, or, colloquially, mho/metre. Because of a general overlap of values for each type of ore in conjunction with each type of electrolyte it is not possible to identify ores in the field from metal factor values, but only to determine anomalies.

Gravity anomalies within a restricted area are relatively small and are expressed in milligals (1 mgal is one thousandth of a Galileo). One Galileo, or 1 gal, is an acceleration of 1 cm/sec^2, i.e. 1/980 of the earth's average gravitational field, so that a milligal is roughly one millionth of normal gravity. An effective gravimeter measures to 0.1 mgal and needs corrections for altitude, topographical features (e.g. nearby mountains), latitude, instrumental drift, and even the tides, in precision surveys. Because only very large orebodies could be detected the main use of gravimeters is for major structural surveys.

As *seismic* methods are not frequently used in ore prospecting it is superfluous to give a description of them here. Parasnis[4] gives details of the methods involved. Seismic prospecting is more a tool of the oil prospector who is looking for major geological structures. Small scale seismic methods, such as the hammer seismograph, may be used to determine depths of overburden in coal basins. However, pattern drilling can do this and provide mineralogical specimens at the same time.

There are some specialist applications. The National Coal Board in the United Kingdom used sparker seismographs to survey the depth of alluvium to bedrock under the waters of the Firth of Forth. This enabled safety margins to be worked out for undersea coal mining.

GEOCHEMICAL EXPLORATION

Airborne geochemical work is insignificant when compared with the effort put into surface geochemical sampling. The raw materials for sampling are not only surface soils and stream and pond water, but vegetation itself, both living and accumulations of plant debris. Of course the normal sampling of gossans and rock is also geochemical work.

Geochemical methods are used to determine values of elements that are higher than a normal background value. Likely areas have to be located before sampling begins. Although chemical elements are being analysed, the orebody being sought is not necessarily identified at once, because it may be one of a paragenetic group (see Table III).

The samples taken can be roughly classified. Rock samples may show halos of low grade mineralisation that indicate an orebody is close by. Sampling of gossans may show high trace values of the original mineral. Sampling of stream sediments may indicate that the original source rock had higher-than-average metal values. The stream water itself may show high trace-element values, indicating trace elements in the watershed area.

Trees in particular are useful. Their roots penetrate the subsoil and the sap system pumps up trace elements in to the leaves along with normal plant nutrients. The leaves act as mineral concentrators. Samples of leaf mould from under trees, or samples of living leaves, can show that the subsoil contains anomalous mineral values, whereas the surface soil has had these minerals leached away by normal run-off. The concentration effect is doubly welcome because the analyses have to be accurate to better than one part per million (ppm).

After the analyses have been made then chemical isoanomaly lines are drawn to indicate promising areas. It may be necessary to tackle this in two stages. The original isoanomaly maps may show high values corresponding with (say) a series of granite outcrops with no major mineralisation, although existing orebodies are known in the area. If, then, the observed isoanomalies are smoothed, so that the profiles are rounded instead of stepped, and then the smoothed values are subtracted from the original values, secondary anomalies may be found that were not obvious from a subjective examination. The secondary anomalies may be orebodies, or yet another granite dome that is just below the surface. This form of anomaly smoothing has become much easier with the development of suitable computer programs.[8]

MATHEMATICAL PROSPECTING (GEOMATHEMATICS)

Expenditure on exploration is a major item in the financial budget of large mining companies. Sums of $1.5m to $3m, representing 3 per cent to 12 per cent of turnover, may be spent annually. For small companies the figures may be even larger as a percentage of earnings.

Because of the large sums involved the companies try to improve the probability of finding orebodies. They tend to concentrate on the more easily found very large shallow orebodies, which by nature are usually low grade. They then compensate for the low grade by working with highly mechanised methods. Today, $2 000 000 000 worth of ½ per cent copper porphyry on the surface is better than $20 000 000 worth of 10 per cent ore buried several hundred metres underground, even if the profit per ton on the high grade ore is greater. It is the total profit from bulk output per year that is important.

Sophisticated mathematical techniques are used to increase the probability of ore discoveries. However, it is unlikely that any mathematician will ever find an orebody unless a geologist leads him to it. The role of the mathematician is to apply statistics and probability techniques to the processes of decision making. The modern mining engineer and geologist should have a course on statistics included in their educational curriculum. An article by Agterberg and Kelly[9] gives some of the modern applications of geomathematics and lists some current textbooks on the application of statistics to geology.

Mathematical techniques have been used in an attempt to predict the likelihood of finding orebodies in unit cells of the earth's crust. A geologist might say that gold values will occur in greenstone areas of an archaean shield and will start to map all the greenstone outcrops. The mathematician would take the block of land, divide it into (for example) 25 km x 25 km squares, and compute the probability of finding the correct combination of parameters to indicate an orebody of given size within the block. The assumptions made by him for this to be done are gathered from a survey to determine which are the statistically significant factors of known orebodies.

After predicting a convenient block size for exploration, the statistician then computes the grid dimensions for diamond drilling.[10] There are three ways of looking at this: either the whole block can be systematically drilled, which at $30-60 per metre depth of drill hole could be very expensive, or each anomaly discovered can be systematically drilled, or each anomaly or promising outcrop can be randomly drilled.

It is in deciding between systematic and random drilling, and the spacing of random drilling, that the mathematician can provide most help. It must be recognised that his help is not infallible, but he will quote the odds on finding ore for a given number of holes on a given pattern in relation to the size and shape of orebody. If a company is faced with a choice between drilling several anomalies in their mining lease area, with only a limited budget available, then the statistician can cooperate with the geologist to decide which anomalies should be drilled, and how many holes should be drilled and where.

There are other facets of sampling in which the application of statistics will help. For instance, when sampling is carried out underground in an exploration drift cross-cuts may also be driven to give samples from the full stope width. If the samples from the cross-cuts are compared with the samples from the exploratory drift as two separate populations, and are found to belong to the same population at a chosen level of significance, then the cross-cuts could be discontinued with considerable financial savings.

The United States Bureau of Mines has applied statistical techniques extensively to sampling programs, mainly on the grouping of sample data from adjacent drill holes or spot samples to give the most accurate results. This is dealt with in Chapter 3.

Application of Techniques

As a final note of caution it should be realised that the quickest way to bankruptcy is to rush in with a diamond-drilling program. At a minimum of about $35 per metre depth, six holes to 300 m would cost $63 000, and this with about $30 000 for analyses and some incidental expenses would write off $100 000 from a budget with a minimal chance of return.

The geologist or mining engineer should decide what he is looking for: nickel in Western Australia would be associated with magnetics, copper in Queensland would not. A normal program might run as follows.

1. Search the Bureau of Mineral Resources files and read old geological papers.
2. Carry out extra regional airborne work, such as high-flying magnetic surveys and colour air photography to identify rocks and gossans.

3. Identify promising anomalies and make a preliminary visit for a broad geochemical survey, say from chip samples and sampling at stream junctions, and to check on aerial identification of surface features. A helicopter is useful at this stage.
4. Carry out any further low level airborne geophysical surveys; for instance magnetics on $\frac{1}{2}$ km spaced lines at about 100 m height, over the earlier anomalies.
5. Make detailed geochemical surveys and a geological map.
6. Sample outcrops by back-hoe across the gossans or other exposures.
7. By this stage the geologist should now be able to pin point the most promising areas to drill as in Chapter 3.

In the absence of a massive effort to catch a rising market one could assume that steps 1 to 6 would take about a year for a cost of (say)—

geologist and vehicle for year	$20 000
aerial survey	5 000
geochemical survey	5 000
helicopter hire	10 000
	$40 000

For about half the cost of those 'wildcat' drill holes the company now has a geological map of the anomalies and a stratigraphical interpretation which will enable drill holes to be put down where they will yield maximum information.

Where an area has already been mined near the surface, then a geologist should make a survey of the workings and prepare a structural map. This will enable holes to be drilled into the likely continuation of favourably mineralised areas.

The Search for Coal and Bedded Deposits

Most of this chapter has been concerned with a search for ore deposits, but the search for coal is conducted on similar, but perhaps easier, lines. A sedimentary basin is found and outcrops are checked for signs of a coal-bearing sequence or favourable oil seepages. Usually, however, weathering processes will have removed the traces, except in sea cliffs or deep river valleys. Drilling for bedded deposits is easier because they exist over a wider area and two or three holes will provide evidence as to whether further work is justified.

Limitations of Exploration

At this stage perhaps the previous methods of exploration should be put into perspective. A case study[11] from Western Mining Company's Kambalda Mine in Western Australia illustrates the point that even while the most sophisticated techniques are being applied, orebodies are still found as much by hard slogging with an occasional element of luck, as by elaborate techniques.

Gold was first discovered in Western Australia in 1885 in the northern part of the State and by 1895 all the major gold deposits in an area of over two million square kilometres had been discovered. In the 1930s and more recently some extensions to previously known areas were found.

The extensive prospecting for gold only produced a few base metal finds. Those were small deposits of copper ore. In over 80 years of prospecting along hundreds of kilometres of potentially mineralised outcrops only a few deposits of any ore were found, plus several unprofitable shows.

In 1962, Western Mining Corporation, which has gold interests at Kalgoorlie and Norseman, extended detailed geological mapping south of the Kalgoorlie gold deposits along the mafic-ultramafic belt. (This is a portmanteau term for the highly ferro-magnesian ultrabasic rocks —see Table II.) The belt had previously been defined on regional aeromagnetic maps produced by the Australian Commonwealth Bureau of Mineral Resources. By 1964 the mapping had almost reached the Kambalda Dome. In 1954 a prospector looking for uranium had some green-stained minerals from this area analysed and anomalous nickel values were found, but no action was taken. In 1964 the same prospector with a partner brought samples of limonitic (gossan) material with 1 per cent nickel content to the Western Mining office. Inspection of the area showed small limonitic outcrops of about 0.3 m width along a strike length of 450 m. This is not a very large object in a country where red dust stretches for thousands of kilometres and water for the Kalgoorlie area is piped in from 600 km away!

Subsequent mapping showed a domal structure with contact outcrops over a length of 21 km. Induced polarization, ground magnetometer and geochemical soil surveys were made around the contact zone. A number of IP anomalies were found, some due to nickel sulphide mineralization and others due to pyrrhotite in sediments.

A decision was made to drill several holes to investigate the sulphides giving rise to the limonitic outcrops. The first hole intersected 2.7 m of massive sulphides containing 8.3 per cent nickel. The next two holes on a 120 m grid away from the discovery hole were barren, and at the same time another hole into a strong IP anomaly on the other side of the dome intersected barren sulphides in sediments. As the authors of this paper[10] said, 'It is interesting to speculate what might have happened if these subsequent barren holes had been drilled first', bearing in mind that no-one had found nickel in this area before. Subsequent exploration has shown nickel deposits at Kambalda to a value exceeding all of the gold mined in 75 years from the Kalgoorlie area.

After this find nickel exploration started in earnest along the ultramafic belt, and a geologist found gossans about 300 km north of Kambalda that were later proved as the Poseidon deposits at Mt Windarra (Laverton). Again the gossan was found before geophysical work other than regional aeromagnetics was carried out.

To stress the point that it is the detailed surveys that count it is germane to note that the Peko-Wallsend Company's gold-copper deposits at Tennant Creek were found when re-drilling leases another company had abandoned. Much of the current large scale prospecting is re-surveying areas already known to be mineralised. The earlier prospecting in countries such as Canada, Australia and America was done in areas of difficult access. Prospectors travelled by foot along the stream beds or mountain tops and found orebodies often more by luck than by judgement. The modern prospector now uses his judgement in these same areas to find new orebodies.

References

1. LIDDY, J. C. *Ore guides in exploration.* Australian Mining, p. 66, March 1971.
2. BARRINGER, A. R. *Airborne exploration.* Mining Magazine, **124**, p. 182, March 1971.
3. BARRINGER, A. R. *Remote-sensing techniques for mineral discovery.* Proc. 9th Commonwealth Mining and Metallurgical Congress, Vol. 2, p. 649, Inst. Mining and Metallurgy, London, 1970.
4. PARASNIS, D. S. *Mining Geophysics.* Elsevier, London, 1966.
5. KELLER, G. V. and FRISCHKNECHT, F. C. *Electrical methods in Geophysical Prospecting.* Pergamon Press, London, 1966.
6. PATERSON, N. R. *Airborne electromagnetic methods as applied to the search for sulphide deposits.* CIM Transactions, Vol. LXXIV, pp. 1-10, 1971. (Canadian Institute of Mining and Metallurgy Bulletin, p. 29, January 1971.)
7. HOOD, P. *Magnetic surveying instrumentation—a review of recent advances.* Economic Geology Report No. 26, p. 3, Geological Survey of Canada, Ottawa, 1970.
8. HARRISON, J. P. *New tools to cut exploration costs.* Australian Mining, p. 79, Sept. 1971.
9. AGTERBERG, F. P. and KELLY, A. M. *Geomathematical methods for use in prospecting.* Canadian Mining Journal, p. 61, May, 1971.
10. DOUGHERTY, E. L. and SCHEEVEL, R. W. *Computer exploration techniques.* Mining Magazine, **123**, p. 102, August 1970.
11. WOODALL, R. and TRAVIS, G. A. *The Kambalda nickel deposits in Western Australia.* Proc. 9th Commonwealth Mining and Metallurgical Congress, Vol. 2, p. 517. Inst. Mining and Metallurgy, London, 1970.

Bibliography

A comprehensive review of current exploration techniques and practical results can be found in

Proceedings of the 9th Commonwealth Mining and Metallurgical Congress, Volume 2 *Mining and Petroleum Geology.* Institution of Mining and Metallurgy, London, 1970.

Mining Geophysics. Vols. 1 and 2. Society of Exploration Geophysics, Tulsa, Oklahoma, 1965.

Mining and Groundwater Geophysics, 1967. Geological survey of Canada. Economic Geology Report No. 26. Geol. Soc. Canada, Ottawa, 1970.

Chapter 3. Determination of Ore Reserves

After an anomaly, or a presumed orebody, has been found it is necessary to define its limits, to see whether it is in fact an orebody, and to determine the mineral content of any ore present. The definition of an *ore* is rather vague, but it is generally taken to be a body which comprises one or more minerals that can be profitably extracted. It is unfortunate that the word profit occurs because it means that an orebody today might by definition not be an orebody tomorrow if the price of the mineral commodity falls. However, it can generally be understood that an orebody is a collection of mineral in one location and of such a concentration that a mining company will seriously investigate the possibility of mining it.

The investigation usually consists of detailed geophysical and geochemical surveys as described in Chapter 2, followed by a program of surface sampling or drilling as appropriate to evaluate the orebody both geologically and mineralogically. It is not only necessary to know the size and shape of the orebody, and its mineral content, but also to know whether the physical or chemical separation of the mineral from its enclosing waste material is both possible and profitable. Some potential orebodies have been known for years, but will not be mined until a profitable extraction process has been developed.

Beneficiation, or *extractive metallurgy,* is the process of upgrading low quality ore to rich *concentrate,* pure metal, or a clean product. Some of the worst problems are in recovering metal values from ores, particularly sulphides. These sulphides may be present in such proportions as to give 0.5 to 10 per cent or so of pure metal in the rock as mined. This is upgraded at the mine surface to give a concentrate with as high a metal value as possible before it is sent to a smelter to recover the metal. The concentrate may have a metal value of around 11 per cent if nickel, to 50 per cent if zinc, or over 70 per cent if lead, all as the metallic sulphide, before it leaves the mine. An ore that is difficult to upgrade is said to be *refractory.* It should be noted that a refractory ore will not make refractory bricks. The former is difficult to work with, whereas refractory bricks or linings are those specially made to resist high temperatures in retorts or furnaces. For those who wish to acquire a general acquaintance with the methods adopted in extractive metallurgy two introductory books are listed at the end of this chapter.

Classification of Ore Reserves

There are three generally accepted groups of ore reserves, although their old fashioned definitions should be changed to comply with modern exploration techniques.

PROVED RESERVES

Sometimes known as measured ore. These are reserves which have been carefully delineated and evaluated. The tonnage and grade should be known to an accuracy certainly better than 80 per cent, but more usually defined as better than 95 per cent. In older textbooks these reserves were given as ore which had been blocked out on all four sides in underground workings, but modern drilling and sampling techniques used with statistical reasoning means that underground proof is not as necessary. Proving on four sides was no guarantee that barren rock was not present in the middle of the block.

PROBABLE RESERVES

Also known as indicated ore. This is ore for which tonnage and grade are partly computed from measurements and partly from geological projection and evidence from widely spaced boreholes. These reserves are generally added into the working reserves of a mine with the certainty that they exist and the probability that they can be worked at a profit. Older definitions implied that they should have been penetrated by exploratory drifts and be sampled on two or three edges. The limit of accuracy should be an error of less than 20 per cent.

POSSIBLE ORE

Can also be called inferred, or potential ore. These are ores for which no specific samples are available. Their presence is inferred from geological structures and geophysical anomalies. The shape of the orebody may indicate that it ought to continue further before bottoming or pinching out. The stratigraphical evidence may indicate that bedded deposits still exist beyond a major fault, perhaps at a lower level.

However, until a sampling or drilling program is started, possible ore must not be added in to the value of a mine, and should not be considered as reserves. A company report should refer to this as mineralisation.

Sampling Shallow Deposits

Where deposits outcrop at a shallow angle, or are spread over a wide area with only a thin cover of overburden or subsoil, then the sampling process is easy and cheap.

COSTEANING AND TEST PITS

The early methods consisted of hiring a gang of labourers to dig trenches known as *costeans* across the line of the outcrop at frequent intervals, and to sink small pits into the orebody at chosen points. The sequence of strata would be mapped and samples would be taken for analysis. The same methods are still used, but are generally mechanised. A bulldozer will now strip the topsoil and a front-end loader or a back-hoe will be used for trenching. A large mechanised earth-auger for shallow holes, or a clamshell or orange-peel type bucket sinker will be used for test pits. These machines can be bought, or hired from civil engineering contracting firms.

For alluvial gravel deposits, and lateritic bodies, the test pits are taken down to bed rock or to the water table and as far below as easy pumping can take them. To go beyond the water table a change is made to drilling; however, drilling of loose rock is very difficult, even with casing in a hole, so that hole digging equipment is preferred if it can be made to work.

Even with hard rock deposits the need to obtain bulk samples for metallurgical purposes may necessitate costeaning and pitting. In harder rock the test pits are more properly called exploration winzes. An example of the need for bulk samples can be taken from the Woodsreef asbestos deposits at Barraba, New South Wales. After diamond drilling exploration had been carried out, the topsoil and overburden was stripped from the open-pit area. Transverse lines were marked across the orebody and sampling trenches were dug along the lines to recover several tons of material for processing in the pilot-scale mill.

SHALLOW DRILLING

For hard rock orebodies, and for those orebodies which extend downwards for more than a few metres, it will be necessary to obtain samples by drilling. A diamond drill, as described in the next section, could be used to obtain cores, but it would normally be more economical to use *percussive* drilling. Percussive drilling costs in 1972 for down-the-hole drills are around $10-15 per metre, both in Australia and North America. This is one quarter to one half the cost of diamond drilling. Percussive drilling is also two to four times as fast as diamond drilling.

Although the term percussive drilling includes the older cable tool rigs these are rarely used in modern prospecting, which favours the use of lightweight portable rigs.

Earlier percussive drilling machines used compressed air for motive power, and a water flush for cooling and removal of debris. With improvements in portable compressors to give pressures from 0.8-1.3 MPa an air flush is now more usual for surface drilling. The percussive drill consists essentially of an air-driven piston to give continual impacts on the penetration bit at the end of a string of drill rods, and a rotation motor which turns the rods to avoid the bit becoming stuck in the hole. The two operations are usually independently controlled. Compressed air is blown down through the drill rods and lifts the fine rock chips and dust up through the annulus between the rods and the hole.

For the holes of 10 to 15 cm diameter that are usual for percussive exploratory drilling the modern equipment is the down-the-hole (DTH) drill (Figure 3.1). The piston is fitted directly above the bit and it is thus at the bottom of the drill hole. The string of drill rods in the hole only has to provide the rotary motion and thus does not waste any of the impact energy. The bit used can be a simple chisel or cross shape tipped with a hard tungsten carbide insert. For very hard and abrasive rocks, such as the iron ores, the bit is pestle-shaped with tungsten carbide studs inserted into it.

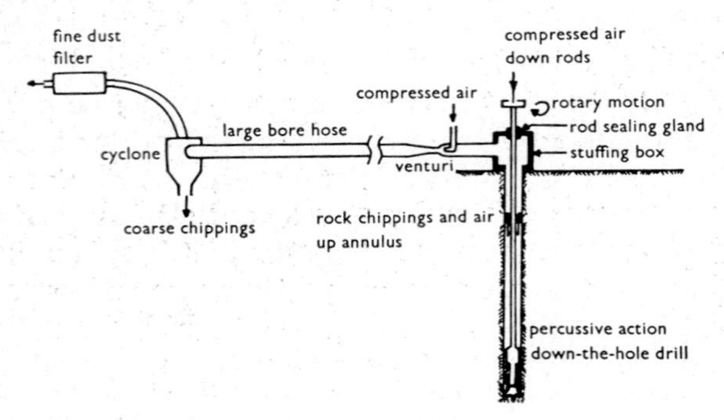

Fig. 3.1 Down-the-hole drill

It is essential to keep the bit cool and to flush the broken rock out of the hole. Generally if the particles are blown out of the hole then sufficient air is passing to keep the bit cool. In practice, an upward air velocity of about 25 m/sec is necessary to lift the bit chippings, the lowest limit is about 15 m/sec. For example, a 108 mm hole with 83 mm o.d. drill rods in it needs about 10.3 m^3/min of air at 550 kPa (5.6 kg/cm^2) gauge pressure with 1000 kPa available to clear blockages.

Air flush drilling is particularly convenient in the dry regions of Australia where water can be a precious commodity. The air flush will also work in wet holes although it is messy. Damp holes can be a nuisance because sticky clay balls can form and cause a blockage, this is overcome by adding a fine mist of chemical spray to the air blast. A thick foam similar to a shaving aerosol is formed when the chemical and the water combine and the foam lifts the drilling debris to the surface. Foam drilling can also be used in weak grounds where the full air blast would scour the walls and cause them to collapse.

If large diameter holes are drilled with air or water for bulk sampling, then reverse circulation would be used with the debris being sucked or forced up the hollow drill rods. This is the only practicable way of providing a sufficiently high velocity to lift the rock particles.

Down-the-hole drill

RECOVERY OF SAMPLES

The essential part of exploratory drilling is that the material broken out of the borehole must be recovered for analysis. The debris can be flushed out with water, run over fine mesh screens to separate the larger particles, and then settled in a tank to recover fine particles.

With air blast a collar is placed around the top of the drillhole and a suction fan or a venturi is used to force the rock debris through a cyclone. Coarse particles drop through the cyclone and the fine dust from the top outlet is passed through a filter before the air is finally discharged.

Several of the larger chips of rock can be collected on site and mounted in a plastic button. The mounted chips are polished smooth and a *mineragraphic* examination will be made under a microscope. The disposition of ore minerals within the rock can be measured and the minerals identified.

The main mass of rock cuttings are sampled at predetermined depth intervals of about 0.3 m to 1.5 m. In barren sections where identification only is needed a grab sample will be taken. Where the borehole is in ore then the whole of the cuttings (fine and coarse) for the sample length (say about 20 kg) will be thoroughly mixed and put through a sample splitter. About 1 kg will be sent to the assayer to determine metal values for ores, P_2O_5 content for phosphates, or other desired properties. Some of the sample may be kept for later check assays, and the rest will go for metallurgical tests for beneficiation. A case study of shallow drilling with DTH drills, checking by diamond drill coring, and final statistical treatment can be found in U.S.B.M. Report of Investigations No. 7486[1].

Sampling Deep Deposits

The surface extent of a deep orebody that outcrops to the surface can be determined by costeaning and shallow percussive drilling. In many cases the top 60 to 120 m of the orebody may be oxidised or leached so that it is not representative of the rest of the orebody. It is also necessary to know the vertical extent of the ore so that its volume and tonnage can be calculated.

DEEP DRILLING

Diamond drilling is the usual way of recovering rock samples from depths below a few tens of metres, and even from shallower holes. Although diamond tipped bits can be used on many types of drill rigs, and cores need not be taken, if a geologist or mining engineer refers to a diamond drill hole he means a hole from which a core of rock has been extracted.

The basic diamond drill consists of a driving motor, gear train, rope hoist, fluid circulating pump, and combined feed and drive head (Figure 3.2). The motor can be diesel or petrol driven on surface rigs, and scrubbed diesel, electric, or compressed air underground. The size obviously varies with the duty required which can call for coring to over 2500 m. For hole drilling underground a rock chamber is excavated to correspond to a surface mast or tower and the rope sheave is bolted into the top of the chamber. For surface drilling the holes may have to be inclined or vertical and masts have to cater for this. The mast may be mounted on a swivel for a light drill, or be back-stayed into an A-frame for larger holes. Vertical holes can be drilled with an hydraulically pumped-up telescopic twin-leg mast which is held erect by wire guy-lines.

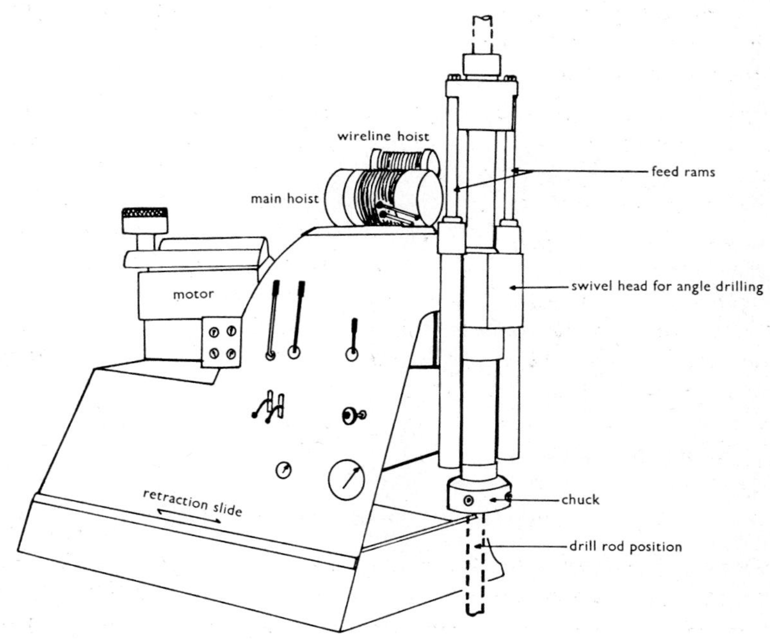

Fig. 3.2 Diamond drill

The item which marks out a diamond drill as different from others is the method of handling the rods which are tubular, and in screwed lengths. They are driven by inserting them through a tubular guide equipped with a mechanical or hydraulic chuck similar to that used on a lathe. The guide and chuck is equipped with hydraulic thrust rams, generally with a stroke of about 0.6 m. The rams are double acting and provide the thrust needed to force the drill into inclined or upwards drilled holes, and to control the drilling pressure in down holes. The hole is then bored in 0.6 m increments, releasing and recycling the chuck and feed rams each time.

Portable diamond drill
A tower can be erected from tubular scaffolding, or may be prefabricated.

Because the chuck and feed are in line with the hole they obstruct rod pulling and casing operations. Consequently the machine is designed so that when the water swivel and top rod are clear of the hole the rig can be retracted about 0.5 m on its base, or else the drilling head can be swung aside on a hinge to clear the hole. The former is preferable because it does not expose the drive gears.

A water swivel is mounted on the top rod, and water, or water thickened with drilling mud, or water with 0.5 to about 1 per cent soluble oil is pumped down through the rods to cool the annular diamond-tipped bit and to flush the rock chippings to the surface. It is also possible to use air or foam flush as for DTH drills.

HOLE SIZES AND CODING

Although a geologist likes large rock cores for structural geology inspection the diameter of the core is generally kept as small as possible to minimise drilling costs. For a given rating of drill (kilowatts in the S.I. unit system, previously horsepower) and the smaller the diameter of hole drilled, then the greater the depth that can be drilled, or else the smaller and more portable the rig.

It is obviously convenient to have a standard range of rock core diameters because then handling techniques can be standardised and the drilling equipment and core treatment equipment can be standardised as well. There were formerly two ranges of drilling equipment, one for metric and one for imperial units. With impending world-wide metrication the British Standards Institute BS4019 has been accepted as a draft international standard for an integrated range of drilling equipment.

TABLE V.

DIAMOND DRILLHOLE SIZES

Code	Core Diameter		Bit Diameter	
	mm	in	mm	in
XR	18.3	$\frac{23}{32}$	29.8	$1\frac{11}{64}$
XRT	19.1	$\frac{3}{4}$		
EX	21.4	$\frac{27}{32}$	36.9	$1\frac{29}{64}$
EXT	22.2	$\frac{7}{8}$		
AX	29.4	$1\frac{5}{32}$	47.2	$1\frac{55}{64}$
AXT	32.5	$1\frac{9}{32}$		
AXWL	23.8	$\frac{15}{16}$		
BX	42.1	$1\frac{21}{32}$	59.1	$2\frac{21}{64}$
BXWL	33.3	$1\frac{5}{16}$		
NX	54.8	$2\frac{5}{32}$	75.0	$2\frac{61}{64}$
NXWL	43.7	$1\frac{23}{32}$		

Note: In the absence of other information the metric sizes are a direct conversion from the Imperial units.

The consequent likely range of popular core diameters is given in Table V. These are in fact the sizes generally in use now outside the continent of Europe. The EX and AX are possibly most used, but XRT may be used for fill-in holes after an initial program of drilling larger holes. The letter coding system is such that the first letter, E, A, B, etc., gives the hole size and the second letter gives the core barrel series. A third letter gives the core barrel type. The older range of X series single and double wall barrels were standard for some years but the W range of wireline drilling equipment has assumed major significance recently. Thus for example an AWT core shows A hole size, wireline equipment and a thin wall core barrel.

CORE BARRELS

The early core barrels consisted effectively of an extension of the drill rods with only a single wall. This disintegrated all but the strongest cores by snapping them off and twisting them round. They were also difficult to empty and return. Double core barrels were developed with the inner portion either rotating with the rods, or mounted on a bearing to remain stationary, essentially like that in Figure 3.3. Non-rotating core barrels could give better than 95 per cent core recovery even in weak friable strata such as coal. Triple core barrels have also been developed in which a split inner lining enables the core to be slid out intact from the double barrel, and x-rayed in situ, or rolled carefully onto a core tray instead of being tipped out.

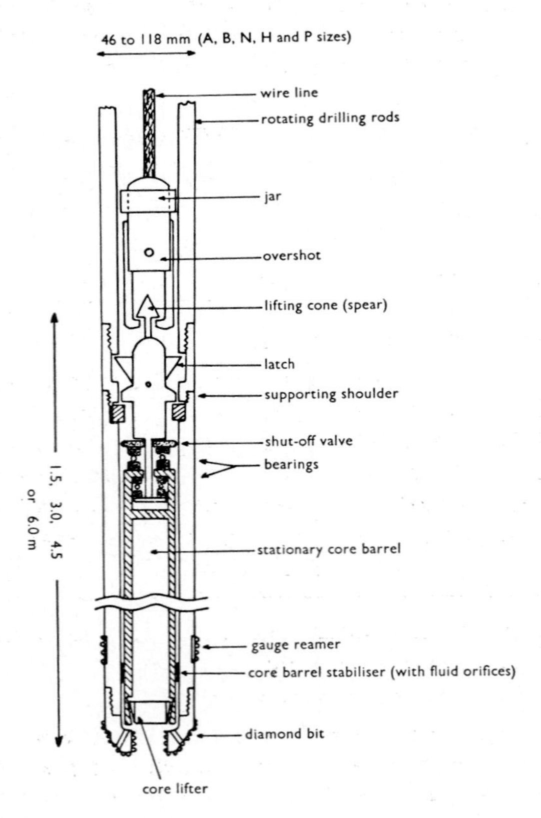

Fig. 3.3 Wireline core barrel

Special corebarrels are used for air flush drilling and wireline core drilling. Occasionally in very weak rock a narrow throat diamond core bit is used with thick drilling mud to allow a mud covered core to slide into the barrel and stay intact.

WIRELINE DRILLING

The Longyear Drilling Company introduced the wireline core barrel about 1950. Since then it has been highly developed and is made by most drilling companies. Its need results from the excessive time taken to draw drilling rods from deep holes to remove each full core barrel. There are occasions too when the core would jam in a barrel, before the barrel was full, and be destroyed.

With a wireline core barrel the drilling rods stay in the hole and the core barrel is retrieved by an overshot lowered down through the rods on the end of a wire line (hence the name) after the water swivel has been removed. Figure 3.3. shows a simplified arrangement of the core barrel. Because of its complicated nature and restricted fluid passages the smallest diameter obtainable is the A range. Further details of drilling techniques can be obtained from the articles and book listed at the end of the chapter.

TREATMENT OF BOREHOLE CORES

When a borehole core has been recovered from a hole it is extremely valuable. It represents the ore grade and structural geology of several hundred thousand tonnes of ore or coal and should be carefully treated. There is an Australian code of practice[2] A.S. CK14, 1967, which deals with the methods of handling borehole cores from coal seams. It provides an ideal to aim for.

Briefly when the cores have been retrieved from the core barrel they should be laid out carefully in a special box. This is a flat metal or wooden tray, about 1.2 m x 0.3 m and of appropriate depth (about 10 cm). It is fitted with grooves to match the core size. The beginning and end of each barrel length is marked with the hole depth at start and finish: there may occasionally be gaps. It is preferable too to paint arrows on the core to show top and bottom. The core boxes should then be carefully stored. It should be noted that some states or countries require all geological records to be reported to their geological survey department and there may be a government core storage warehouse to which cores must eventually be sent.

When the cores have been recovered, preferably under skilled supervision, they should be cleaned and geologically logged. Where facilities are available it may be possible to carry out non-destructive strength tests (elastic moduli by sonic methods) for rock mechanics purposes.

Coal cores may also be x-rayed for ash and mineral content. If the cores are mineralised then they are split longitudinally: one half is kept for geological records and the other half is sent off for assay.

The cuttings from the rock annulus bored out may also be recovered and sent for assay as a separate check on values.

HOLE ORIENTATION

A large orebody can obviously be drilled with vertical holes at regular intervals. However, many orebodies are of the type shown in Figure 3.4. It can be seen that if inclined holes are drilled then the chances of intersecting several high grade orebodies are increased. It is also clear that by drilling in from the footwall side a longer length of core from within the mineralised zones will be obtained than if the approach was made from the hanging wall.

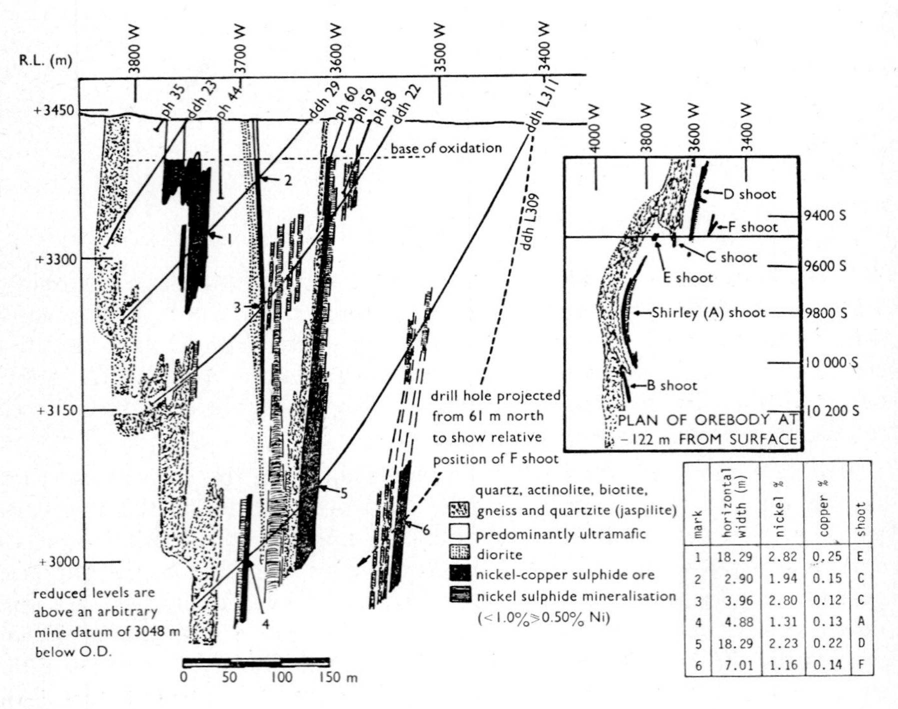

mark	horizontal width (m)	nickel %	copper %	shoot
1	18.29	2.82	0.25	E
2	2.90	1.94	0.15	C
3	3.96	2.80	0.12	C
4	4.88	1.31	0.13	A
5	18.29	2.23	0.22	D
6	7.01	1.16	0.14	F

Fig. 3.4 Nickeliferous orebody drilling (*courtesy of Poseidon N.L.*)

It must be noted however that holes should always be drilled across an orebody and not down its dip. A metalliferous orebody is frequently zoned, as in Figure 1.2, and a down-dip hole could travel in a rich or a poor zone all the way. This would give a misleading assay. A hole across the orebody gives a representative sample of its grade variation.

Figure 3.4 shows that an original surface survey was made by percussion drilling (PH = percussion hole). These holes coincide in depth with the oxidised zone. Such a zone often consists physically of weak crumbling rock in which diamond drilling is difficult, with frequent rod sticking and lost circulation water. The water losses themselves can be serious in arid areas. It is often good economics to contract out the drilling in two stages. The upper part of the hole, say 50 to 60 m, can be percussion drilled at $10-15 per metre, and cased (lined with steel tube) if necessary at $6-10 per metre. Then the rest of the hole is diamond drilled at $25-35 per metre. There is a distinct saving both in direct hole costs and time costs because percussive drilling is two to four times as fast as diamond drilling, say 4 to 5 metres per hour against 1 to 2 m/hr.

BOREHOLE SURVEYING

Another feature of deep drilling is that the boreholes wander off course. They usually flatten in the manner shown in addition to a tendency to corkscrew. For prospecting it is usually sufficient to bore the hole and then survey it afterwards for inclination and compass bearing. A full hole survey is made with a camera photographing at fixed intervals the inclination of a plumb-bob over a compass chart with dip circles marked on it.

Spot readings can be made with instruments such as the Tropari. A double celled tube holds a compass in gelatine, and a phial containing hydrofluoric acid. The tube is rapidly lowered into the hole and left to stand at the desired depth. The gelatine sets to fix the compass bearing while the hydrofluoric acid etches an ellipse on the phial to give the hole inclination.

The spot readings may be necessary when *wedging* a hole. It is rarely necessary in prospecting to straighten a hole but it is frequently deliberately deviated. It is possible to drill for several hundred metres with continual core logging, then if interesting mineralised zones have been intersected the rods can be drawn back and one of a number of wedging devices inserted in the borehole above the ore zone. A new length of core can then be taken by forcing the core barrel out a few degrees from the original line through the orebody. The cost of the shared length of hole per core can then be saved although wedging costs upwards of $2000 per deflection. Several deflections can be made from one borehole.

BOREHOLE LOGGING

Borehole logging is carried out at two stages. Records of penetration rate and downthrust on the drilling bit are kept to show changes in strata and possible cavities, shear planes, etc. When the hole has been bored it is not abandoned immediately. The high cost of drilling is such that for some deep orebodies it may be considered economical to spend several thousand dollars per hole on having a complete geophysical survey made.[3] Such companies as Schlumberger have developed a range of instruments to carry out IP, SP, neutron logging, etc., within the small diameter of a borehole. This is done on a contract basis because specialist personnel are needed to log the hole and interpret the readings. A logged hole may convert a near miss into a possible hit.

HOLE DEPTH AND SPACING

Theoretically it would be good practice to bottom every hole to below the mineralised zone, but that would be very expensive at perhaps $25-50 000 per hole, or even more. A more moderate exploration program would place a few deep holes to determine the structure of the orebody and then put in several more shallow holes to confirm continuity between the deep holes. In this way sufficient ore reserves can be proved to warrant the huge investment to open up a mine. Once the mine is working further holes can be drilled from underground to explore the deeper parts of the orebody. The costs of these can be paid for from working revenue instead of increasing the original capital costs. An indication of normal practice can be taken from North Broken Hill Ltd's Annual report for 1968. The mine produced 486 354 long tons of ore and drilled 4023 m of underground routine holes for ore delineation, plus 1687 m of underground exploratory holes and 11 093 m of surface exploratory holes.

For cost reasons hole spacing can be very much dependent on hole depth. It is also dependent on the geological structure and type of the orebody. Shallow deposits of precious metals frequently have a very patchy ore content and need sampling holes spaced on close centres. Conventionally these are placed on a rectangular grid. On large shallow deposits an arbitrary grid of 400 ft or 200 ft centres was often chosen, this might equally well be replaced by a 125 m or 75 m grid. The high profits that can be made from mining rich gold bodies have paid for a large part of the research on sampling methods, coincidentally it is these bodies that need the most effort in sampling. Details of the current U.S. Bureau of Mines research program on sampling are given in reference 1 and in the Bibliography at the end of the chapter.

Statistical methods generally depend on a random ordering of events. The events in many ore bodies are often far from random in that they can have a regular shape and a graduated but regular zoning of ore within veins or masses, which often have a rich core and low grade edges. Consequently the pattern of drill holes should be designed to determine the shape of the body and to make sure it is sampled from edge to edge.

Deep Bedded Deposits

The feature of these is that by definition they are continuous over large areas. They can be followed in from an outcrop by shallow pitting and percussion boreholes. However they may dip down to several hundred metres and then cores have to be recovered by diamond drilling. If extra cores are needed for analysis then the holes may be wedged above the seam. It is also possible to use extra large rigs which will core at 0.2 to 0.3 m diameter for recovery of bulk samples from a few boreholes.

Because the seam of coal, rock salt, potash deposit, etc., being sampled may be buried as much as 2000 to 3000 m below surface it would be expensive to core all the way to the mineral. Once a few holes have penetrated the overlying strata a change may be made to large rotary drilling rigs of the oil-well type. Toothed or tungsten carbide button bits will be used to penetrate to close above the seam horizon. Rock identification to this level is by collecting a few handfuls of cuttings every one to two metres. Bit pressure and penetration records are kept to show where strata changes may occur. For the final few metres of hole a change is made to a hollow bit and core barrel similar to the diamond drill equipment. The more powerful rig used for rotary drilling enables the hole to be started at about 0.2 to 0.3 m and even if part of the hole has to be cased the diameter at coring can still be about 0.15 m to recover a 0.1 m core. Cores of this size can show dip and rock texture much more easily than EX or AX sizes, and provide more mineral for analysis.

Holes for bedded deposits may step out at 1 to 5 km intervals rather than 100 m for metalliferous deposits. Provided the widely spaced holes show the same general dip of seam and comparable mineral values the seam is assumed to be continuous between the holes, especially if a geophysical survey has previously indicated this. If an irregularity occurs then in-fill holes may be drilled to detect any faults or seam splitting which has occurred.

Exploratory Shafts

With many orebodies or with seams that approach to within a reasonable depth from the surface a final stage of exploration is to put down a small diameter shaft, say 2 to 3 metres diameter into the upper part of the orebody, and then to put out exploratory drifts and crosscuts. This gives absolute proof of the nature and quality of the ore or other mineral. Several tonnes of ore can be sent off for pilot scale primary metallurgical tests. The geologist and the assayer can determine the variation in quality of the samples. The local geological structure of the orebody can be carefully studied and then the whole of the geophysical and borehole evidence is carefully reassessed to conform to the underground exploration data.

Also from the exploratory underground workings and testing of any extra borehole cores the mining engineer will assess the strength and hardness of the wall rocks and ore so that he can design the mine workings.

Underground Sampling

Several publications on mining geology or economics[4] deal with these techniques at great lengths. The important thing to remember is that most mineral deposits have either been laid down horizontally or have been intruded with a hot core and cold sides. Therefore sampling should be taken at right angles to the hanging wall and footwall, or between the roof and floor as appropriate to metal mines or bedded deposits. In massive deposits sampling is across any visible banded structure, otherwise an arbitrary pattern is used.

In coal deposits a special saw can be used to cut out a full column about 10 cm square between roof and floor. This is carefully packed and transported to the laboratory. Alternatively each band of a seam can be sampled separately with careful measuring of the average width of each band.

Similar methods can be used in metal mines. A full column about 10 cm wide and 5 to 10 cm deep can be cut between the hanging and footwalls. In repeat cases a groove sample about 2-3 cm wide and deep is taken. Alternatively, spot samples (chip samples) are taken at regular intervals across the development or stope end to obtain an average value. If bulk samples are required they should be taken from several places across and along a stope, not from one place only.

For orebodies with erratic values and widths the samples are assayed and averaged. Narrow veins of precious metals usually have assays reported as gram-metre, multiplying the weight of metal per metric ton, given in grams, by the width of the vein in metres. Values vary between sampling points and are usually arithmetically meaned to give an average value per metre. If samples are at random intervals or from different widths a weighted mean is used.

Many formulae are in use for weighting assays in precious metals. Truscott[4] and McKinstry[5] discuss their use. McKinstry suggests that an arithmetical average is correct if high-grade values occur in random arrangement of if they occur in a high grade shoot traversing the ore-body, although the shoot could be treated as a separate entity. Where the high values form the core of an oreshoot, or occur as regular shaped bodies, then a weighting method is used. The engineer can choose from:

$$\text{average value} = \frac{\Sigma\, WAF}{\Sigma\, WF} \quad \text{or} \quad \frac{\Sigma\, WAF^2}{\Sigma\, WF^2} \quad \text{or} \quad \frac{\Sigma\, WA^2F^2}{\Sigma\, WAF^2}$$

where W = width of ore body sampled, A = assay value of total sample over that width, F = frequency, i.e. number of times the given value of WA or WA^2 occurs in the population.

The first expression appears to be the most popular, but a final choice would be left until mill recoveries can be compared with underground samples.

The average grade of a block of ore is calculated from the average grades of the sampled openings which form its boundaries, i.e. length of drive x width of drive x average assay is calculated for each of the drives which form the block (or for two drives and two raises), then the total length of the drives is multiplied by their average width, and then average grade of block =

$$\frac{\Sigma\text{ (length x width x assay)}}{\text{(total length x average width)}}$$

For average grade of a volume of ore the prismoidal formula

$$V = (A_1 + 4A_2 + A_3)\ h/6$$

can be used where V = volume, A_1 is the area of the upper surface, A_2 is the area of the mid-section, A_3 is the area of the bottom surface and h is the height of the prism.

For massive bodies, and for deposits whose values do not vary greatly, a geometrical construction will probably be used as in the next section. However for both types of deposit it is likely that large companies will change to a statistical approach using a computer program to work out average values. This is already being done for many open pit mines where each working bench has to be sampled for grade control.

Ore Reserve Calculations

With modern methods of analysis which should include mineragraphy and possibly electron probe microanalysis, the elaborate methods formerly used to detect and avoid 'salting' of samples should be superfluous. In any case no scheme can defeat the person who blatantly distorts recorded analyses, or who misplaces a decimal point somewhere. However, common sense should indicate that the results of a single borehole which indicate a bonanza should be treated with scepticism, or a mad rush to sell shares while they rise in value.

Several boreholes or samples are needed to evaluate the grade and tonnage in an orebody. The calculation is usually made from a pattern of geometric bodies. Patterson[6] gives an approach to this problem on which Figure 3.5 is based. The orebody is divided physically into a series of prisms which are summed to provide the final figures. The prisms may be based on uniformly spaced squares centred on boreholes on a regular grid. They can be unequal volume rectangular blocks centred on random holes or they can have polygonal ends with the polygons constructed by bisecting angles between three holes, or from perpendicular bisectors of lines between adjacent holes. In the S.I. system the volume in cubic metres is converted to weight in kilograms by multiplying by the density (formerly specific gravity).

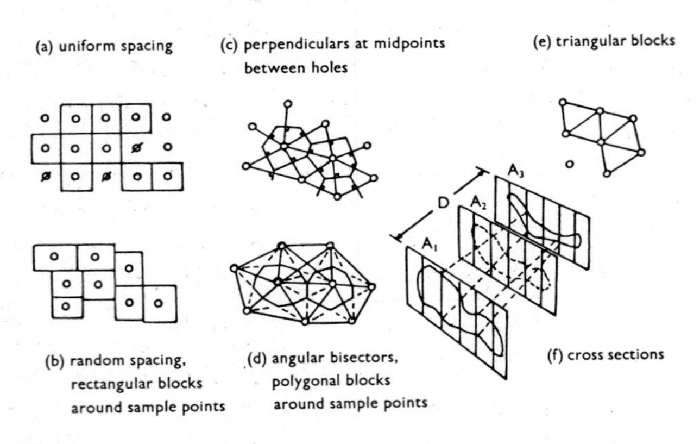

Fig. 3.5 Geometric patterns for reserve calculations

In the rectangular and polygonal methods the grade of ore in one drill hole assay is assumed to extend halfway to any adjacent borehole. This is obviously a very simple assumption which could lead to serious errors. Exceptionally high values on any one hole could influence the results unduly. It is common practice to take the mean value of the high assay and at least two adjacent holes, or a weighted mean of the high assay and the adjacent holes, as the value from which to calculate the ore grade in the suspect block.

The triangular method of estimating the grade of ore is also an approximation because it assumes an exact linear change of grade of ore in direct proportion to the distance between drill holes. The cross-sectional area method also assumes average grades across the area and uses the mid-ordinate (Simpson's) rule to calculate average grades and tonnage within the block. The tedium of these calculations can be eased with computer programs[7, 8]. Within Australia such organisations as AMDEL (Australian Mineral Development Laboratories) have commercially available programs for ore reserve calculations. The more recent of these divide ore bodies up into uniform blocks. For open pit mines the block height is the proposed bench level interval as this simplifies the mine planning and overburden stripping ratio calculations.

For most mining purposes the grade approximations made in geometric models have to be accepted. They are probably within the same limits of accuracy as most of the other assumptions made in the mine feasibility studies although mistakes are occasionally made.

EXTRACTION, DILUTION, AND RECOVERY RATES

The calculated reserves for a mine, whether metalliferous or coal, must be adjusted to allow for the mining and beneficiation processes. It is rare for all of a mineral to be recovered.

The effect of a cut-off grade is dealt with in Chapter 4. In essence an orebody or a coal seam is not always mined to its limit. An arbitrary line is drawn between mineral which can be profitably extracted and that which is unprofitable at current market prices. This can cause some loss of reserves.

In the mining process itself it may be necessary to leave mineral unmined to support the mine roof or walls. Chapters 6 and 7 will give some values for each method, but the *extraction ratio* of mineral mined to mineral present can be as low as 40 per cent if subsidence must be prevented. For pillar mining in coal or other bedded deposits the final amount extracted may be only 60 per cent unless a mine is carefully designed. Even then it would be wise to plan for an extraction ratio of 75-80 per cent. Even longwall mining rarely recovers more than 90 per cent of the coal in situ, and all these figures are for a proportion of the planned extraction height. If roof or floor coal is to be left then the extraction ratio will be decreased.

Dilution will also occur in the mining process. Waste rock in a metal mine, or roof and floor rock in a coal mine, will become mixed with the mineral and reduce its value. In a coal mine, for a given rate of extraction, all the mineral handling system should be designed for about 25 per cent excess to allow for handling dirt with the coal. Part of this will be combined with the coal in the seam, and cannot be left behind; part will be roof and floor rock. For example a coal mine that wishes to sell 1 000 000 tonnes of clean coal a year may have to mine and handle 1 250 000 tonnes of mineral and dirt out of the mine (*raised and weighed*, or *run of mine coal*). Similarly in a metal mine the process of recovering ore in stopes may produce 10-30 per cent dilution. Either the margins of an orebody must be left, or this dilution has to be accepted.

The *recovery rate* is the proportion of mineral mined that is actually passed on for sale. In a coal washery clean coal is separated from dirt. However some *middlings*, a mixture of coal and dirt, always exists and an arbitrary distinction is made between dirty coal and 'coaley' dirt. Of all the coal sent out of a mine about 85-90 per cent will be recovered in the washery for sale, the rest will go on to the waste heap.

Recovery of metal concentrates from ores is more difficult. The metalliferous particles are finely intermixed with rock with perhaps 1-5 per cent metal present. The rock is crushed in a mill (metallurgical plant) and a chemical or mechanical process is used to separate the metallic compounds from the rock. Some recovery ranges are given in

Appendix I. Recovery of tin values can be as low as 40 to 60 per cent of the tin present as metal in the ore, i.e. if the tin is present as 1 per cent metal in the ore, then 0.4 to 0.6 per cent metal is recovered and 0.6 to 0.4 per cent goes into the tailings. Recoveries of other metals are usually higher, but never 100 per cent.

Abandoned Boreholes

For want of a better place this mining note is put here. Any drill hole from the surface to an orebody which is to be mined by underground methods should be carefully plugged. The plug should be waterproof through any aquifer and in any lake bed. Holes are frequently only plugged for short distances above an orebody but the waterproof plug should extend for at least one hundred metres above the orebody to allow for the caving action which breaks the roof beds.

Although a major disaster may not result from unplugged holes a mine can be faced with heavy pumping costs because a borehole usually enlarges if water pours through it. The staff and men who were called from a Christmas party at Kambalda to plug an open hole to the salt lake from below did not appreciate the change from beer to brine!

References

1. KOCH, G. S. *Statistical analysis of assay data from the Round Mountain Silver Prospect, Custer County, Colo.* United States Bureau of Mines Report of Investigations, R.I.7486, 1971.
2. STANDARDS ASSOCIATION OF AUSTRALIA *S.A.A. Standards Code AS CK14, 1967. Sampling coal seams in situ by boring.* Sydney.
3. LYNCH, E. J. *Formation Evaluation.* Harper and Row, New York, 1964.
4. TRUSCOTT, S. J. *Mine Economics* (3rd Ed., revised by J. Russell). Mining Publications, London, 1962.
5. MCKINSTRY, H. E. *Mining Geology.* Prentice-Hall, New York, 1948.
6. PATTERSON, J. A. *Estimating ore reserves following logical steps.* Engineering and Mining Journal, **160**, p. 111, September 1959.
7. HEWLETT, R. R. *Computing ore reserves by the polygonal method using a medium-size digital computer.* U.S. Bureau of Mines, R.I.5952, 1962.
8. HEWLETT, R. R. *Computing ore reserves by the triangular method using a medium-size digital computer.* U.S. Bureau of Mines, R.I.6176, 1963.

Bibliography

Extractive Metallurgy

GILCHRIST, J. D. *Extraction Metallurgy.* Pergamon Press, London, 1967.

DENNIS, W. H. *Extractive Metallurgy.* Pitman, London, 1965.

BOLDT, J. R. and QUENEAU, P. *The Winning of Nickel.* Methuen, London, 1967.

Drilling

COCKING, J. R., CROWE, J. G. D., MCCOGGAN, G. A. and MITCHELL, A. *Deep wireline drilling at the southern leases of New Broken Hill Consolidated Ltd, Broken Hill.* Aust. I.M.M. Monograph No. 3, 'Broken Hill Mines 1968', p. 57. (Australasian Institute of Mining and Metallurgy, Melbourne, 1968.)

CUMMING, J. D. *Diamond Drill Handbook* (2nd Ed.) J. K. Smit and Sons, Ontario, 1962.

McGREGOR, K. *The Drilling of Rock*. CR Books, London, 1967.

MORGANS, B. H. *Recent developments in coredrilling*. The Mining Engineer, No. 123, pp. 163-77, December 1970.

Sampling

HAZEN, S. W. *Some statistical techniques for analysis of mine and mineral deposit sample and assay data.* U.S. Bureau of Mines Bulletin No. 621, 1967.

KOCH, G. S. and LINK, R. F. *Sampling gold ore by diamond drilling in the Homestead Mine, Lead, S. Dakota.* U.S. Bureau of Mines, R.I.7508, 1971.

LINK, R. F. *et al. Statistical analysis of gold assay and other trace element data.* U.S. Bureau of Mines, R.I.7495, 1971.

Ore Reserves

AUSTRALIAN INSTITUTE OF MINING AND METALLURGY and AUSTRALIAN MINING INDUSTRY COUNCIL *Report by Joint Committee on Ore Reserves.* Supplement to The Aust. I.M.M. Bulletin No. 355, July 1972.

Chapter 4. Feasibility Studies

By the time a company starts its feasibility studies on whether to mine a mineral deposit it may already have spent from $500 000 to $10 000 000 on exploration. However, it must still be realistic about writing-off even large sums as a loss and not waste further money on an uneconomic mine. Usually, money spent on exploration receives special tax concessions because of the high risk rate.

Griffis[1] quotes details of Cominco Ltd exploration projects: 1000 properties were explored in 40 years. Only 78 warranted major work. No ore bodies were developed at 60 of these. Eighteen mines were brought into production· but eleven did not return the investment. Only seven out of the original 1000 prospects became profitable mines. The mines in question include precious metals (e.g. gold and silver), base metals (the cheaper non-ferrous metals), potash, coal, natural gas, oil and fluorspar, so that the company was not selective, just careful.

It should be appreciated that a feasibility study is not a slow and steady plod with a pause between each event. It is often a harassed rush to recover money already spent and may be conducted in several overlapping or separate stages, often called Stage 1 planning, Stage 2 planning, etc. Stage 1 is usually a quick assessment in outline to see if there is a chance of a profit, or of improved proceeds from an existing mine. Stage 2 is the more detailed approach to work out a profit per ton, or a return on money invested. Stage 3 would be detailed engineering planning and ordering. Figure 4.1, adapted from a paper by Hope[2], shows the interrelationship between money already invested and various planning stages on the Bougainville Copper project started by C.R.A. About $20 million had been spent before the evaluation report was accepted and about $30 million by the time the report was completed. Three years and $350 million later the mine was ready to start up.

Scale of Operations

Most of the foregoing material is based on the supposition that a large mining company is working a large deposit. Exploration losses are offset against profitable mines. Near-surface orebodies with a mineral content of around half per cent of copper (for instance) can be worked

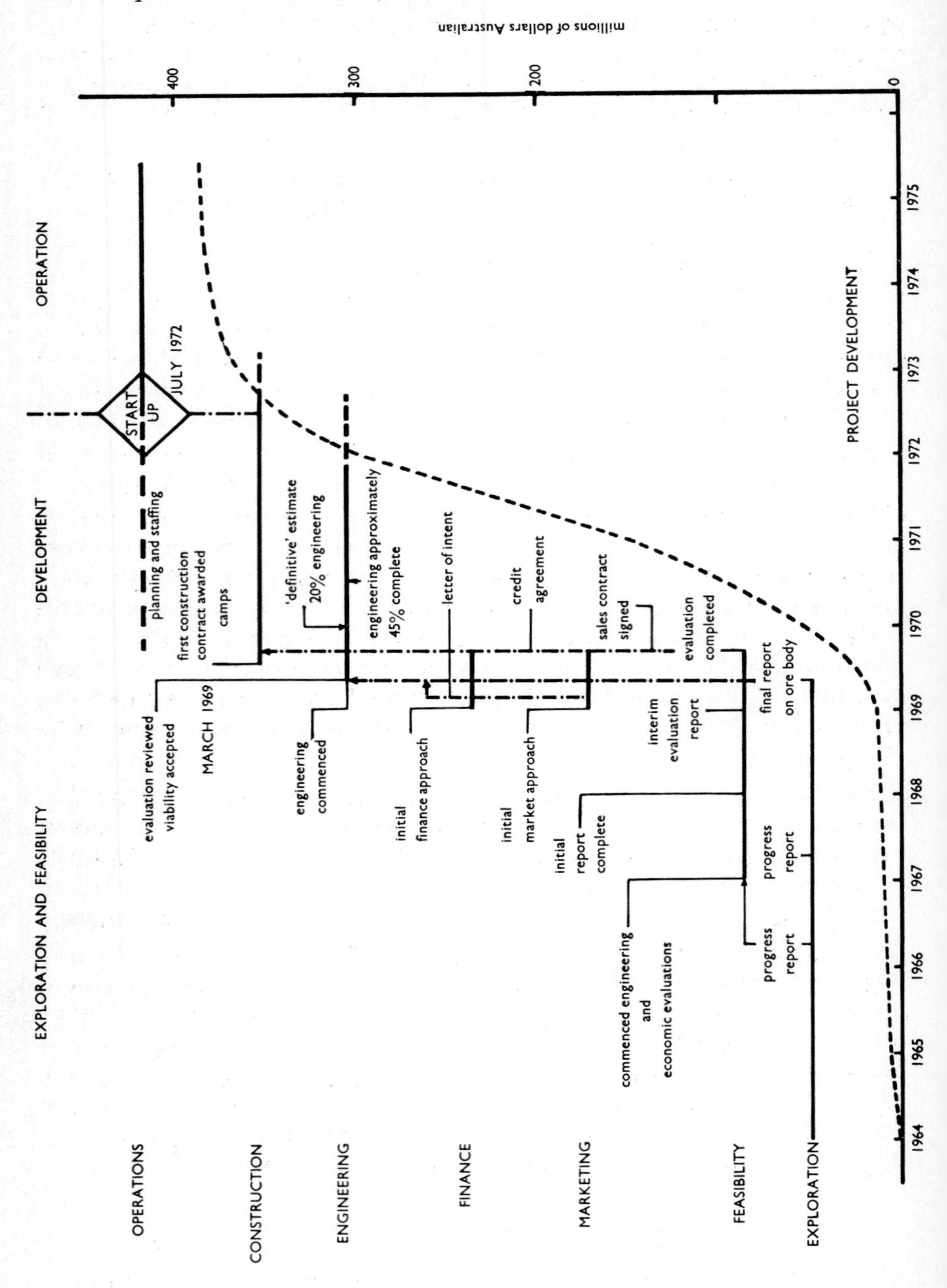

Fig. 4.1 Relationship between planning stages and expenditure on Bougainville project

by open cut methods producing about 20 000 to 70 000 tonnes per day of mill feed. Each tonne of ore gives about 4 or 5 kg of pure copper at about $1 per kg selling price. When the costs of transport, insurance and refining are removed, then the value of the ore fed into the mill (beneficiation plant) is about $2.50 per tonne on 1970-1972 prices. The open pit mining costs would be about 25c per tonne on this scale for breaking and removing ore and rock. For a ratio of 2 tonne of overburden to 1 tonne of ore, plus mill costs of about $1.00 tonne, and overheads of 25c per tonne, the total mine costs are $2.00 per tonne to show a profit of 50c per tonne. Notes later on sensitivity analysis will show that this could as easily be a loss of 50c per tonne, so that a 40 000 t/day operation could have a profit or a loss of $20 000 per day. Since the capital cost was probably around $300 000 000 the returns are not high for the risk involved.

For underground mining the costs increase from caving at about $1.50/t to narrow stoping in large mines at about $4.00/t. As the scale of operations decreases to around 4000 t/day the overheads rise to about $1.00/t, mill costs to about $4-5/t, and the total cost per ton of ore is about $8 to $10. The selling price of metal is still the same as for the large scale operation although the capital investment would be less. The only action that can be taken is to work a higher grade ore which gives three to four times the metal output per ton of ore handled. For copper one would need a cut-off grade of 1½ to 2 per cent or more for underground mining and even then it would have to be a large mine.

For any mine there is always a compromise between capital and revenue costs. The cheapest system to run may be the most expensive to buy. If capital is not easy to raise then the mine may have to settle for higher operating costs per ton.

At the bottom end of the scale is the small mine working orebodies of 500 000-5 000 000 t at (say) 250 to 1000 t/day, or even less for a mine with 10 to 20 men in a small deposit. They have their own problems of finance in that large companies are not interested. Although they may only need $500 000 to go into full production the owners may never raise this much capital and high grade ore will be worked selectively, ('picking the eyes' out of the deposit) and the rest is left as an unwanted relic until metal prices warrant re-opening the mine. Frohling[3] gives an account of how to finance a small mine.

Factors Affecting Mine Planning and Operation

These can be grouped into physical factors external to the mine operation; factors which exist because of the physical position and state of the orebody; and the effects of economic, social, and governmental controls.

EXTERNAL FACTORS

These are the effects of the geographical location of the orebody. The controls which exist are

(a) transport of minerals to market and supplies to the mine and mill

(b) availability of labour, housing, educational and recreational facilities

(c) environmental climate in the mining area with associated benefits or disadvantages.

The obvious place to start a mine, if one was allowed to, is right in the centre of a large city so that one has housing, transport, water, power and market all on the doorstep. A mine owner can be occasionally lucky. Balmain colliery worked under the site of the present coal-loading wharves in the centre of Sydney, N.S.W.

Many coal mines in Australia are in fact quite close to major cities. The New South Wales coal deposits are spread out along the urban conglomeration of Newcastle, Sydney and Wollongong so that there is a ready availability of labour. Existing roads and railways give easy access to ports, and to the steel works and non-ferrous smelters in Newcastle and Wollongong (Port Kembla). Power stations to supply the high concentration of people living in this area can be directly connected to the mines by conveyor belt (as at Munmorah and Liddell). In return the power stations provide relatively low cost electricity for mine operations. Water is also cheap and plentiful. All these cost advantages enable the newer and larger New South Wales underground coal mines to compete with the large scale opencut mines in Queensland which have a two hundred kilometre haul over the coastal range to the ports of Rockhampton and Gladstone.

The opposite end of desirability in distance and environment is found in North-Western Australia in the development of the iron-ore fields of the Pilbara region. It was known for many years that there were large deposits of iron ore in this semi-desert region at 400 to 500 km from the sea, with no communications, no towns, nor port, no surface water, and a summer daytime temperature of over 40°C. While the iron content was assumed to be only about 30 per cent in limonite no-one could afford to work it. In addition, there was an embargo on exports of iron ore from Australia, so that there was not much incentive to explore for rich iron ore because there were sufficient supplies in more accessible areas for the home market.

Eventually the persistence of a few people, and some hard routine exploration over the banded iron formations of the Hamersley and Opthalmia Ranges found haematite with over 60 per cent iron. One block of it measures 6.5 x 1.6 km and holds 550 000 000 tonnes of ore. With the chance of spreading overheads over large scale production of high grade ore the mining companies began work on a search for customers. A market for 40-60 000 tonnes per day was needed. Long term contracts were essential and these were eventually forthcoming from Japan. On the strength of these contracts finance was raised to meet the enormous cost of mine and infrastructure (houses, schools, communications, etc.).

It was only at high production levels that ports and railways could be intensively utilised. The method of pit operation or the grade of ore did not dominate strategic planning. It was the length of railway track to be constructed, the towns to be provided, the port and harbour dredging, and the provision and reticulation of power and water. By the end of 1970, Hamersley Iron had invested over $330 million in providing facilities for ore and pellet production, and the rail connection from the mine at Tom Price to the port at Dampier. Only about $60 million of this was spent in the development of the mine proper. At this stage, ore production was 17.5 million tonnes and pellet manufacture 2.5 million tonnes a year.

For every dollar of capital spent on mine development, and the pellet plant at Dampier, approximately two and a half dollars was spent by Hamersley on items of a 'service' nature. The rail road, for example cost $63 million, and another $19 million was necessary for rolling stock. Housing at Tom Price and Dampier cost $40 million. Port development to cater for ore carriers up to 100 000 d.w.t. and to provide all necessary services, including power and water, cost $83.2 million.

Had the railway, port and townships at Tom Price and Dampier been financed by the government, Hamersley's capital investment would have been reduced by $225 million. Hamersley had to generate f.o.b. (free-on-board ship) revenue of more than $50 million a year, or almost $3 on every tonne of ore and pellets sold, to recover the $225 million spent on infrastructure.

Apart from provision of support facilities, mining companies in the North-West have another major problem, that of finding and retaining enough labour to operate the industry.

While the Hamersley project was under construction up to 3000 workers were employed, but virtually none of the construction workers remained to join the permanent work-force. It was difficult to estimate the extra cost of a transitory labour force compared with a stable one. However, the 100 per cent annual turnover of the Hamersley work-force of 1300 would necessitate an expensive recruitment and training program.

The turnover in the workforce at Hamersley and other remote mines remains high despite additional incentives including free transport from town sites to work areas, subsidised board and house rentals and high incomes. These workforces have clubs, swimming pools, and air conditioned homes which cost about $37 000 each when built. Food is generally subsidised to city prices in the supermarkets. The company pays the bill for all these benefits.

The mine companies do all this to encourage recruitment of a high proportion of married families as these form a more stable workforce. Commonly bonuses are paid after six months, and a large annual bonus to keep workers at a mine. Annual holidays may be subsidised by paying fares to the capital city. In some cases fees for high school and tertiary education are subsidised to make up for the remoteness of mining camps.

The Hamersley development history was repeated for the iron ore mine at Mt Newman with its 425 km rail road to the new port built at Port Hedland. One can say of course that once the high initial capital costs are recovered the companies have a bonus in the low grade ore that then becomes economic.

There are other extremes of nature. In Alaskan or Sub-Arctic Canadian mines both equipment and homes have to be heated from –40°C in winter, rather than cooled in summer. There are still no communications, and avoidance of melting permafrost replaces the problem of bulldust and drought. There are flies and mosquitoes in both Canada and Australia to make a migratory miner feel at home.

MINING FACTORS

Both surface and underground factors affect mine planning. On the surface it is necessary to have space for workshops, offices, beneficiation plant, railway sidings, stores, etc. It may also be necessary to own and clear land for housing. Savage River Mines in Tasmania (iron ore) have levelled the top of a hill to build a town. The Bougainville project is in a mountainous area. Two towns had to be built; Panguna, near the mine, which is restricted to company personnel on critical or standby duties, and Arawa on the coast for everyone else. Panguna will hold 2000 people and Arawa may reach 12 000 people.

The local topography at the mine influences the siting of shafts or entrance tunnels, or haul roads into open pits. It also influences the positioning of rail or road load-out points, siting of dams for tailings, and the location of waste rock dumps.

The following chapters give details of how to extract orebodies, but it is convenient to list here the main factors to be considered. These are:

(a) size, shape and type of orebody (e.g. massive or vein)
(b) type and thickness of overburden or rock cover
(c) amount of water above the deposit or any aquifers in superjacent rock beds

(d) sharpness of mineral cut off: it may be necessary or desirable to keep access to low grade deposits for (say) block caving after initial selective stoping of high grade ore
(e) strength of the surrounding rocks and orebody so that stable openings can be designed for the appropriate time span
(f) hardness and abrasivity of rock and ore for design of breaking by explosives and mechanical handling
(g) behaviour of the mineral when mined; e.g. liability to spontaneous combustion, deterioration of milling properties with time
(h) chemical properties of ore or rock with respect to corrosion of metal components or health hazards
(i) temperature gradients and general heat flow in the mine, plus need to heat or cool incoming air
(j) complexity of treatment plant needed for ores or coal.

SOCIO-ECONOMIC CONTROLS

The economic aspects of scale of operations has already been mentioned. Generally large companies are needed to work large low-grade deposits on a vast scale. The richer the grade and the smaller the deposit then the smaller the company that can work it. Modern methods of financing usually involve obtaining forward contracts and raising the maximum possible fixed interest debt from banks and mining finance companies, with the minimum equity (shares) component. This is a high loan/equity *gearing*. Modern thought is that a 60/40 to 70/30 ratio is desirable. The larger and more reputable the company then the more debt finance it can raise. A small company may have to sell out, go into parnership, or work on an intermittent basis of selling spot quantities of ore or concentrate on an open market.

The more technical aspects of finance are beyond the scope of this book, but can be obtained from the bibliography at the end of this chapter.

The social aspects of mining can be considered under two headings although both are greatly affected by the hysterical outbursts of emotion whipped up by anti-foreign investment and anti-pollution groups.

Political stability in under-developed countries is always a problem. The inflow of foreign investment that indirectly pays for increased social benefits is often hated because the companies investing capital obviously require some dividends or interest, and some control on expenditure. It is all too easy for the local population to look on the interest as 'our money leaving the country', and seats on a local company board as 'foreign control'. They forget the income derived for the country from local taxation and royalties on overseas sales, and the fact that the investment company may not be able to repatriate its capital. This feeling is sometimes apparent in more sophisticated economies such as Australia. It must be admitted that a few mining companies in the past may not have been as scrupulous as one could wish.

The problem of living down a bad past also applies to pollution. Approaches for a mining lease may be greeted with objections to possible blasting vibrations, dust, noise, atmospheric pollution, water pollution and increased traffic density. It is not always realised that vibrations, dust, noise and pollution can be greatly limited by modern mining methods. Indeed legislation now exists in Australia, Great Britain and the U.S.A. to confine mining to approved methods. Land restored after mining can be more fertile farmland or new recreational areas, or be returned to original within very few years. Some Australian mineral sands companies are spending $1250-2500 per hectare ($500 to $1000 an acre) to restore land worth less than $12.50 per hectare ($5 per acre). An increased traffic density may even be an advantage because the mine company will have to widen, strengthen and resurface the roads which then benefit local traffic. Most governments require a financial bond to be deposited before mining can start to cover the anticipated costs of making good after mining.

It is unlikely that a mine would be forced to work underground instead of open cut methods to avoid public nuisance but it is perhaps fortunate that most large opencut mines are remote from all but their own workforce and tourists, who seem to enjoy the guided visits that enlightened companies provide. Some mines have a noticeable operational budget item in public relations booklets and free samples which, with staff wages, can reach tens of thousands of dollars per year.

GOVERNMENT AIDS AND CONTROLS

Apart from environmental control, which is often administered by non-mining government departments, there is usually a rigid control of mining by a state or federal Mines Department with its inspectors. There is also a tax man, who in the early years of a project can give as well as take away.

In the field of environment mounting public agitation has caused most countries or states to enact legislation to control the effects of mining. Great Britain passed its Town and Country Planning Act many years ago and this controls the general use of any land for any purpose, including mining. All surface works have to be suitably screened or of aesthetic appearance. Heights of waste tips and of building structures are restricted. Pollution in Britain may be controlled by several acts at once so that a succession of inspectors may call on a mine and care must be taken to consult local government authorities as well as national bodies. America too has recently passed Federal laws to control mining and most of its states are currently passing local laws or have already passed them to control water and air pollution. In some cases past laxity seems to have been superseded by present harshness, and many good mining companies are paying for the sins of a few.

In Australia there may be an overlap between Federal and State laws. It is too soon to be certain but air and water pollution may well be controlled by Federal laws whereas mining is governed by State acts. In New South Wales the Mining Act No. 49, 1906, which governs general mining activities, was amended in July 1971 to provide a new Regulation, 59A, which prohibits uncontrolled drainage, or pollution of water, and states that all open cut mines must be left in a safe condition. It also lays down that the land must be revegetated to prevent erosion if the Minister (for mines) so directs. Previously to this mining pollution was controlled in that applications for mining leases would only be granted subject to controls on methods of working, and land restoration if necessary.

Some aspects of damage to public property are controlled by other acts. New South Wales has a Mine Subsidence Compensation Act 1961-1967 which regulates that aspect of working mines of coal and shale. It is difficult to summarise an act in a few words but, roughly, if mine workings damage property already built, then compensation must be paid to the owner of that property. If however anyone builds on land already proclaimed as a mining area, then that person will not be compensated for subsequent damage. Common law rights may also exist in relation to some operations.

The best advice to offer to would-be miners is to hire good legal advisers to scour the statute book for appropriate legislation. At the same time the legal advisers may well be sorting out a tangle of lease claims. Old mining areas have frequently been previously pegged for mineral searches and whereas some claims may have lapsed others are still valid. Several claims may overlap, especially if they were only granted for a few minerals. It is not unknown for a company to be conducting litigation on claim leases while actually mining the disputed area, so some financial provision may well have to be made for disputed claims.

In addition to meeting the costs of environmental controls a mining company also has to conform with legislation which controls working practices. Similar legislation exists throughout Australia. The state Mining Acts usually control prospecting and granting of leases, and then there will be an Act, or Regulations, to control the actual working practice at the mines. Thus in New South Wales, in addition to the Mining Act 1906, there is the Mines Inspection Act 1901-1968 which deals with working all mines except those of coal and shale, and the Coal Mines Regulation Act, No. 37, 1912, which covers coal and shale workings. As if this were not enough there is a Scaffolding and Lifts

Act, 1912-1960. If an engineer is engaged in tunnelling work other than mining, or in deepening his harbour, then the Regulations of this Act will control his work under a most inappropriate title. The practising mining engineer should know his local mining laws by heart and these may include Explosives Acts, Mines Rescue Acts, Workers Compensation Acts, and so on.

Many of these extra Acts will cost the company money in compulsory insurance against workers claims, provision for long service leave, sickness benefits, etc., and a project in a new state or country can well lead to unexpected extras.

Mining acts and their administration need money. Consequently every tonne of mineral mined is likely to have royalties and levies imposed on it. There will be a royalty of a few cents per tonne to go to the owner of the minerals, either the crown or private. In addition there may be levies for subsidence compensation, mines rescue stations, miners' welfare, trade union benefits, etc. All this is in addition to taxes and other general government impositions.

However some royalties and taxes can be semi-returnable. Governments make initial prospecting and marketing surveys. They may also pay prospecting subsidies or offset all exploration money against tax. For small miners New South Wales hires out exploration equipment at low rates, and will make current interest loans to enable small companies to buy essential equipment to start up a mine. In some types of mining there may be initial tax offset advantages for development of prescribed minerals. The tax laws have to be studied carefully and can have advantages as well as disadvantages.

TRADE UNION PRACTICES

It is a foolish company that does not consult the local trade unions before starting new mining projects. It is not only good industrial relations, but also recognition of the fact that a union dispute can stop a mine as effectively as a machinery breakdown. Even worse, a continual battle of strikes and petty disputes over working practices can make a mine completely unprofitable.

Sometimes a previous history of restrictive practices in an area means that low grade ore may not be workable unless a union will give a cast-iron legally enforceable guarantee to abandon old ideas of how many men it takes to do a particular job. Even then it may pay the mining engineer to employ methods completely new to the area to ensure that preconceived ideas are not imported into a mine by local workforces.

Sensitivity Studies

The previous part of this Chapter covers a field which can be the subject matter for books on mine law, mine economics and industrial relations. There is obviously a wide field for errors and a fourth book on statistics would be a useful addition to the engineer's shelves.

No person is infallible and errors in estimating costs and revenue can have a surprising impact as the following simple examples show.

TABLE VI.

EFFECT OF CASH CALCULATION ERRORS

	Revenue	Cost	Surplus	Surplus Error	Actual Surplus, % of original
Estimated	100	60	40		
10% less revenue	90	60	30	25% less	75%
20% less revenue	80	60	20	50% less	50%
10% more costs	100	66	34	15% less	85%
20% more costs	100	72	28	30% less	70%
10% less revenue and 10% more costs	90	66	24	40% less	60%

Suppose the revenue of a mine has been estimated at $100, costs at $60, then the surplus is 40 per cent. If the revenue is only $90 then the surplus is reduced to $30; an error of 10 per cent in revenue gives a 25 per cent error in surplus. Table VI gives some other examples.

This simple table is in fact treated in a more complex manner. A modern company may have a computer program available to do the job using Monte Carlo (stochastic) techniques to randomise the cost and revenue figures for each phase of the operations for several projected annual tonnages and mine lives. The final figures will then be examined to see which gives the most profitable scale of operations. Tax incentives and depreciation allowances have to be fed in as do projections of future market trends and mineral prices. An interesting paper on these risk evaluation techniques is given by Brown[4] but space precludes its reproduction here. A paper by Miskelly and Willsteed[5] gives the more financial aspects of determining the benefit of cash flows in future years and describes methods of costing and of assessing taxes.

Case Study—Porphyrion Mineral N/L

To illustrate the task of a mining engineer in a feasibility study, a shortened version of a design project for final year undergraduates is given. The study was based on published figures for the nickel reserves of Poseidon N/L at Windarra with some distortion to suit the design purposes. I should say here that any resemblance between Poseidon N/L plans and the students' plans is coincidental and the results of a logical approach. I would however like to thank Samin Limited for their interest and some information on the Poseidon rock strengths. To distinguish between the two studies we will call our mine Porphyrion (the dark blue moon-man, who has some connection with rocks and lust so must have been a miner).

The following factors have to be determined:

1. tonnage of mineral, both mineable and marketable
2. value of marketable product
3. cost of equipment and development for a given rate of working
4. time required to reach production for a given rate of working
5. cost of production at a given rate of working, including both the operating costs and the fixed overheads: this also includes setting up housing and transport facilities.

The general shape of the orebody is given in Figure 3.4 and the location map is given in Figure 4.2. The first stage when looking at an orebody such as this is a crude determination of reserves and price. Very often it is an intuitive decision. In this case an intelligent estimate from early drilling that the orebody contains at least 9-10 000 000 tonnes of nickel at better than 1 per cent grade indicates that a mine will almost certainly be profitable despite the geographical handicaps.

The next step is to increase the drilling program along the line of the orebodies and put in one or two deeper diamond drill holes to check continuation in depth. Figure 3.4 shows the orebody as two grades; better than 1 per cent cut-off and a $\frac{1}{2}$ to 1 per cent zone. The 1 per cent cut-off will give a 2 per cent average grade for ore mined. It should be remembered that this will be a deep mine and therefore higher grades are needed to make a profit. Market fluctuations may make it worthwhile to mine the lower grade ore during periods of high prices.

It is appropriate here to consider *marginal ore.* Mine costing must be done on profitable ore which will cover the development costs and overheads. However once access is profitably gained to an ore body there may be a halo of marginal ore. The overheads have been covered and any of this marginal ore has only to show a slight profit over the cost of mining and milling. To meet peak demands or to satisfy an urgent order the mine may be willing to mine at absolute break-even costs.

If machines and men would otherwise be idle, perhaps because of trouble in another stope, then marginal ore may be mined and stockpiled on the surface. Later, when there is an imbalance between production and milling the low grade ore can be put through the mill which would otherwise be idle. Occasionally the low grade stockpile can be used to dilute a rich pocket of ore to keep the mill recovery balanced.

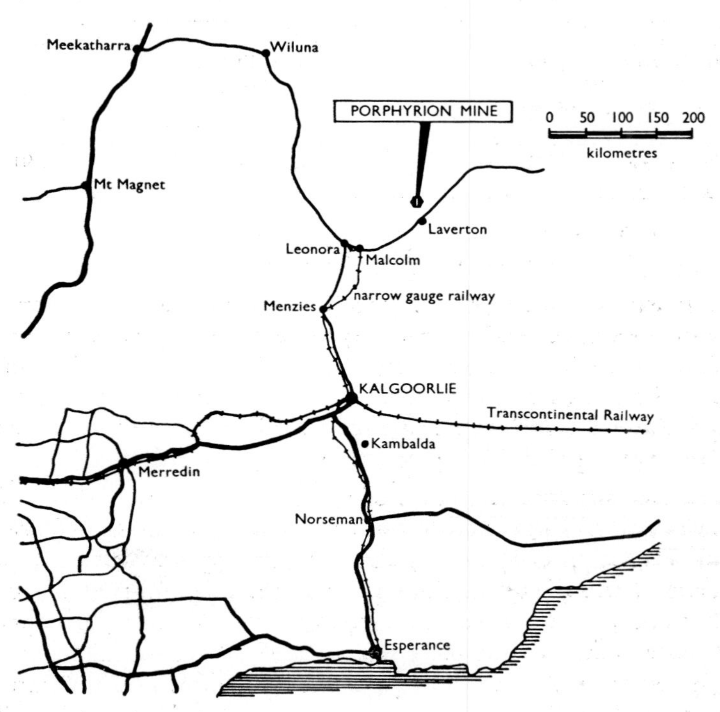

Fig. 4.2 Location map for Porphyrion case study

STEP 1. CHECK RESERVES

The mining engineer engaged with the study will consult with the geologist and examine probable and possible ores and authorise further drilling and sampling to upgrade reserves as necessary. Examination of the several shoots indicates that shoot A is the largest near the surface with about 3-4 000 000 tonnes of $2\frac{1}{2}$ per cent nickel and mining will probably begin here. An exploratory winze is sunk, bulk samples are taken, and the characteristics of the ore body are measured.

The preliminary estimate of ore reserves for Porphyrion is shown in Table VII.

Indications are that at least some of the ore shoots extend down to 800 m. However there is enough proved and probable ore to start mining operations and orebodies in depth will be proved from underground workings.

TABLE VII.

PROBABLE ORE RESERVES FOR PORPHYRION N/L

Ore shoot	million tonnes	Average Grade of nickel, %
A down to 300 metres	3.03	1.3
B down to 300 metres	0.40	2.0
C+E down to 450 metres	0.15	2.7
D down to 500 metres	11.96	2.2
F	Marginal ore only	
Total probable reserves	15.54	1.91

STEP 2. MARKET VALUE

Marketing experts study world production and consumption figures and begin informal and formal approaches to prospective buyers. Marketing can be based on two structures. There is the cost per tonne or per kg of the raw metal which involves costing the mine, mill and a smelter. Alternatively the mine and mill are costed back from the smelter to give a price for concentrate from the mill.

For instance suppose the price of nickel is $2500 per tonne of metal from the smelter. The smelter buys the concentrate c.i.f. (cost, insurance and freight paid by seller) at the smelter gates and only pays for 70 per cent of the nickel in the concentrate; the other 30 per cent has to pay for charges and metal lost in smelting. A tonne of concentrate at the smelter gate is thus only worth $192.50. If the concentrate has to be shipped from Australia to Japan then freight and insurance costs may be about $17.50 per tonne so that the ex-mill value is $175. For convenience in calculation this is quoted as a *cost per tonne unit,* i.e. the value of 11 per cent concentrate is $16 per tonne unit (1 unit = 1 per cent of metal). However the mill does not recover all the nickel in the ore; recoveries vary between 70 and 95 per cent. Assuming a recovery of 85 per cent the value of the ore in the mine is $13.60 per tonne unit. This in turn means that if the *head grade* (i.e. assay value of the mixed rich and poor grade ore leaving the mine) is 2.0 per cent nickel the mine is earning $27.20 per tonne of ore mined from which to pay all its operating costs and overheads for the mine and smelter.

STEP 3. WORKING COSTS

Preliminary costing can often be done by finding a similar mine of about the same projected output as the case study and using its manpower and equipment figures as a guide. There is not space in this book for a detailed scheme so some abstracts of a full design will be given and relevant totals abstracted.

The first decision to make is a method of working. These long, narrow and steep orebodies in fairly competent rock suggest either a mechanised cut and fill or a mechanised open stoping method (see Chapter 6) for minimum costs. The orebody shape and ore strength are wrong for caving methods. Inter-related studies for the mill suggest that sand-fill cannot be obtained from there for hydraulic fill, so surface fill will have to be found. We rule out long-hole open stoping because we think some of the stope walls may have significant weaknesses so that cut and fill will be safer.

Because the orebodies come close to the surface we also decide to use an access decline to obtain a quick start and defer a decision on a shaft for one or two years. Since we are now sure that we are going to mine at some rate we buy out first front-end loaders and drill rigs and start driving the access portals while we finish detailed planning.

The obvious start to mining is in shoot A because it is the largest near the surface. We will put down an access decline at each end of shoot A and drive a development heading from each direction to meet near the centre. The headings will be silled out to develop the stope. Shoot A is low grade and we need a 'sweetener', so we drive out to the other large shoot, D, as soon as possible to boost the head grade. The two declines to A will give through ventilation and a small shaft will be raise bored from D to ventilate that stope.

Some rules-of-thumb have been adopted to start planning. Firstly we make the usual assumption of a mine life of 10 years for the most probable ores, say 10 000 000 tonnes in 10 years; i.e. a production rate of 1 000 000 tonnes of ore per year. Secondly this has to be compatible with a vertical mining rate of about 30 m per year to allow time for further development to keep up with production. For metallurgical reasons mining must start under the oxidised zone. This gives an initial mining horizon of 120 m for shoot A. Some supergene enrichment will keep up the head grade until we reach shoot D. Our initial production rate is then as in Table VIII (p. 78). Shoots C and E are not added into the totals but are held in reserve until their structure is more fully known. It can also be seen that production is not immediate but builds up to the required total.

We now have a starting point from which to plan the mine development and order some of the equipment needed. Deliveries may be from one to two years from date of order so major items of equipment are frequently ordered before detailed planning is finished.

Detailed planning and production scheduling is repeated for several rates of working about this mean and sensitivity studies are carried out to determine the optimum working rate, but this will not alter the basic planning.

STEP 4. TIME TO REACH PRODUCTION

Engineers then plan a step by step progress into production. It is best shown by an arrow diagram (Figure 4.3) that will later be turned into a detailed critical path analysis. For explanations of C.P.A., or C.P.M. (critical path method) a more advanced textbook on engineering management should be consulted. Basically it is a technique of adjusting development program interactions until the optimum time (not

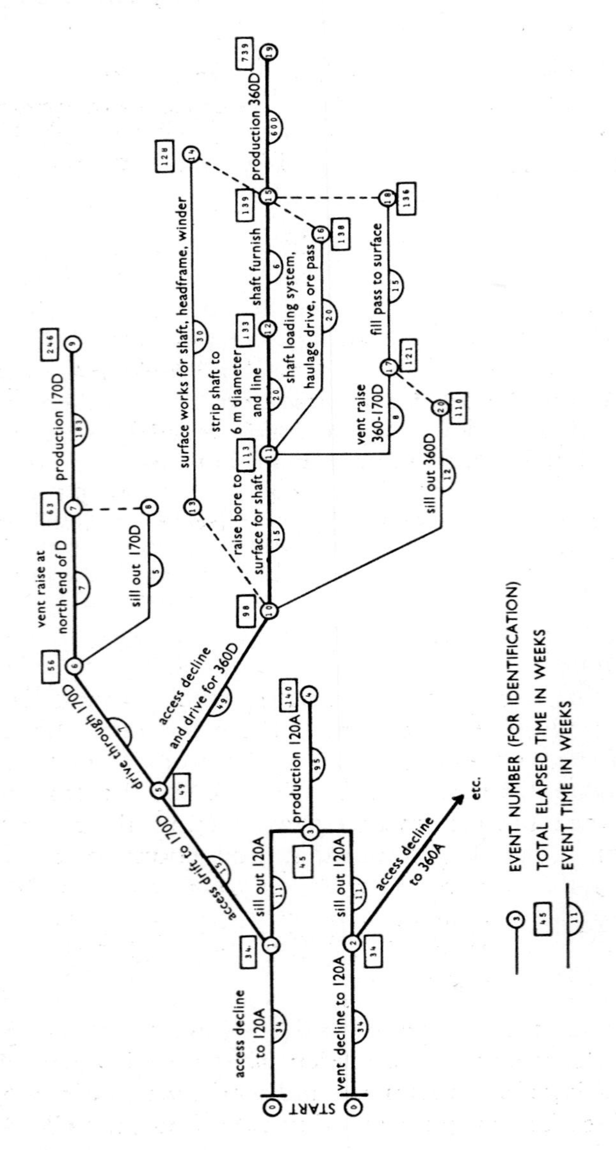

Fig. 4.3 Arrow diagram for development

necessarily the shortest) is derived. Figure 4.4 shows an outline of early workings. In the preliminary stages of planning rough drafts such as this are produced and destroyed several times a day.

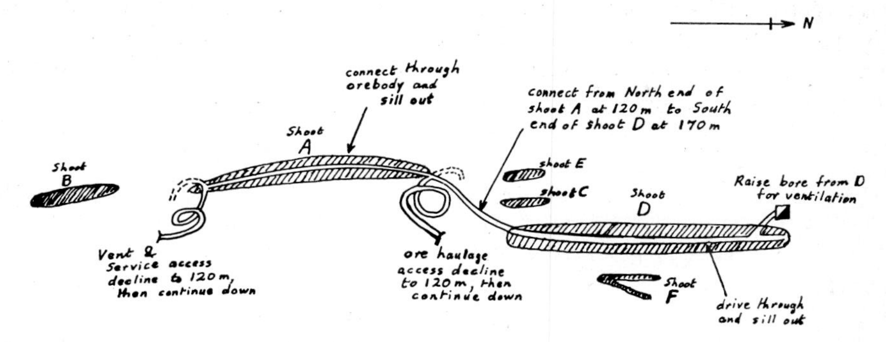

Fig. 4.4 Development sketch for Porphyrion

For an arrow diagram the engineer draws on his experience for probable times for jobs. Drivage rates for declines and development ends were reckoned at 8 m per week. A working week is based on 5 days for planning with a 6 day week normally worked. This gives a contingency margin for production hold-ups, strikes, etc. The diagram allows equipment moves to be checked and scheduled. For instance events 10-11, 11-17, and 17-18 on Figure 4.3 take place in different parts of the mine but are sequential because they will all use the same raise-boring machine.

STEP 5a. TOTAL COSTS FOR A NORMAL MINE

The mine planner now has his stopes planned methodically. He knows the equipment he will need and his labour force. This is not the exact number to fill each job, but a value increased by about 8 per cent to allow for holidays plus another increase for absenteeism. In a small mining community absenteeism would be about 5 per cent, but in a mine close to a large city the average absentee rate could be as high as 15 per cent with about 25 per cent absent on Mondays and Fridays and nearly full attendance mid-week. It may also be necessary to either add an overtime allowance to the wage costing or add still more men to cover emergencies.

The beneficiation plant will be designed contemporaneously with the mine. Mill layout will be based on previous practice and metallurgical tests on bulk specimens. The first guess from other mines is that a 1 000 00 tonne a year mill will cost about $30 000 000. Then we work out a detailed cost of the machinery and the area of building required. All surface buildings will be costed at about $110/m^2$ ($1000/100ft^2$) until a definite quote is obtained from contractors.

When the manpower and equipment for the mine and mill are known then surface workshops, offices, change houses, etc., are planned. Housing and recreational facilities locally available are assessed before a new community is planned.

STEP 5b. TOTAL COSTS FOR PORPHYRION MINERAL N/L

At this stage the arrangements for Porphyrion had to be looked at very closely. There are three major problems; water, housing, and transport. Normally a mill is situated as close to a mine as possible to save transporting the waste rock mixed with the ore. There is also a tax benefit in Australia if the mill is on the mine lease. However a mill for concentrating nickel needs about 1.25 tonnes water per tonne of ore processed. The water supply close to Porphyrion is limited to inadequately explored underground sources. An old flooded mine and non-potable aquifers can supply the mine but not the mill.

Searches for an alternative show that there is an old mill with attendant water supply at Kalgoorlie (Figure 4.2) which we can buy. We must reluctantly transport 4000 tonnes of ore a day instead of 350-400 tonnes of concentrate. There is an offset from saving on housing costs of $40 000 a married unit or $5000 a single man's unit at Laverton where new housing must be built for all mine personnel.

Transport is costed for upgrading the existing railway from Kalgoorlie to Malcolm. This will probably be necessary because bulk haulage by truck over long distances is likely to be uneconomical. From Malcolm to Porphyrion there is a choice of building a new railway or of regrading and bitumen sealing the dirt road to withstand truck haulage. The railway gives a higher capital cost but lower operating costs and a saving on housing because of lower personnel requirements. Truck haulage from Porphyrion to Malcolm gives lower capital but higher operating costs per tonne-kilometre than a railway. Government subsidies towards creation of public assets of this nature will be sought. Porphyrion hoped to raise the capital and chose rail transport.

In practice, and in consultation with W.A. State Government and other mining companies, the rail is to be upgraded to Malcolm for collective use and Poseidon N/L will share with the State of the cost of sealing the road from Malcolm to Laverton. Poseidon has to pay the upgrading cost for the railway and for its own rolling stock.

To summarise, at this stage, Porphyrion's operating costs were $14.70 per tonne made up of

	$ per tonne
MINING (to crusher, including production development)	4.00
RAIL TO KALGOORLIE	4.20
MILLING	4.40
ADMINISTRATION	1.00
TOWNSHIP	0.60
EXPLORATION (extensions to existing orebody)	0.50
	14.70

The value of ore, remember, was $13.60 per tonne-unit, so a 1.08 per cent average grade of ore would break even if there are no mistakes in planning or costing.

The accountants must now deal with raising finance and the mesh of taxation arrangements. Revenue from sales must cover the operating costs, interest and repayment on fixed-interest loans, provision for further exploration for new orebodies, and a return on equity capital to investors. Such factors as sales and tax payment lags complicate affairs and will not be dealt with here. Table IX is included for interest to show a trial cash flow analysis. It can be seen that $51 million has to be borrowed to cover the first two years expenditure before an income is earned. In addition early expenditure on exploration and feasibility studies has already had to be covered and this money borrowed via equity and loans must be serviced.

It is appropriate to leave the case study at this point to start on the mining methods, with the comment that a feasibility study must be carried out in outline by any potential investor in a mining company's shares. An investor can hopefully rely on a large stockbroker to do it for him. However the time necessary to study a prospect thoroughly is such that the lucky gambler can make a profit, or the unlucky one a loss, by instant investment. Institutional investors are more likely to rely on buying at a sound market price to benefit from growth.

Feasibility Studies for Coal Mines

Coal mines are planned in the same way as for metal mines. The coal quality is assessed; the properties sought are described in Chapter 7. One concept that differs is that of an average grade. Although coals can be blended, and some resultant properties such as the ash content can be calculated directly, other properties such as coking and swelling values have to be determined by experiment for each different blend of coal.

A coal seam consists of successive layers, each with its own properties, and the seam is examined to see whether some parts of it should not be mined. The most convenient portions to leave in place are the

TABLE VIII.

POSSIBLE PRODUCTION RATE FOR PORPHYRION

Shoot	A		A		D		D		B		C+E		Total	
Sill Level	120 m		360 m		170 m		360 m		To be decided					
	Tonnage	Grade	Tonnage	Grade	Tonnage	Grade	Tonnage	Grade	Tonnage	Grade	Tonnage	Grade	Tonnage	Grade
Year 1	76 000	1.3	—		—		—		—		—		76 000	1.31
2	390 000	1.3	28 000	1.3	300 000	2.2	—		—		Possible		718 000	1.82
3	320 000	1.3	220 000	1.3	396 000	2.2	105 000	2.2	50 000	2.0	20 000	2.7	1 094 000	1.79
4	Finished		250 000	1.3	396 000	2.2	390 000	2.2	80 000	2.0	20 000	2.7	1 116 000	2.02
5			250 000	1.3	370 000	2.2	425 000	2.2	80 000	2.0	49 000	2.7	1 125 000	2.04
6			250 000	1.3			425 000	2.2	80 000	2.0	20 000	2.7	1 115 000	1.89
7			250 000	1.3			425 000	2.2	80 000	2.0	20 000	2.7	1 115 000	1.89

TABLE IX.

CASH FLOW ANALYSIS—PORPHYRION MINERAL N/L

(Figures are $000, and rounded up)

Year	Revenue	Operating Costs	Working Profit	Tax Deductions	Taxable Income	Tax 38%	Profit After Tax	Capital Expenditure	Net Cash Flow
1	—	3 723	– 3 723	9881	–13 604	—	–3 723	45 800	–49 523
2	27 260	16 165	11 095	9218	1 877	—	11 095	12 550	–1 455
3	36 632	20 530	16 102	9128	6 974	—	16 102	3 895	12 207
4	42 146	19 423	22 723	7683	15 040	3909*	18 814	504	18 310
5	42 944	18 841	24 103	7953	16 150	6137	17 966	791	17 175
6	40 838	19 185	21 653	4593	17 060	6482	15 171	800	14 371
7	40 838	19 185	21 653	1978	19 675	7476	14 177	385	13 792
8	40 838	19 185	21 653	1198	20 455	7773	13 880	145	13 735
9	40 838	19 185	21 653	954	20 699	7866	13 787	25	13 762
10	40 838	19 185	21 653	800	20 853	7924	13 729	—	13 729

roof or the floor coal. These may have a higher ash content and a poorer coking value than the centre of the seam.

A market for the coal is sought, either for electricity generation, industrial steam raising or coking. Occasionally a special impurity, such as germanium, may be present, and the coal ash becomes a useful by-product. After a market has been found then a production rate is fixed and a mining method chosen. Surface and underground equipment is chosen, services are designed and then costs are calculated.

Payments will be necessary for lease rents, wayleaves, royalties, special levies, (e.g. subsidence, miners pension funds, mines rescue, long service leave, workers compensation), community costs, local government rates, water, electricity, telephone, fuel oil and greases, and other consumable stores.

Surface buildings such as the washery, workshops, stores, offices and changehouses must be built. Roads, car parks and rail sidings are necessary. A remote but large mine may even warrant a private aeroplane landing strip.

Underground costs will include not only the mining machinery but the capital values of main conveyors, rail track, locomotives or rope haulage, and rolling stock. Reticulation of electricity, water, compressed air and mine drainage could cost over $100 000 per kilometre. Useful information on production costs can be obtained from a paper by Nielsen[6].

References

1. GRIFFIS, A. T. *Exploration: changing techniques and new theories will find new mines.* World Mining (Catalog, Survey and Directory Number), p. 54, 1971.
2. HOPE, R. B. *Engineering management of the Bougainville project.* Civil Engineering Transactions, p. 45, April 1971. (The Institution of Engineers, Australia.)
3. FROHLING, E. S. *What kind of financing you need to get your small mine into production.* Engineering and Mining Journal, p. 62, December 1970.
4. BROWN, G. A. *The evaluation of risk in mining ventures.* Canadian Institute of Mining, Trans. LXXIII, pp. 298-304, 1970 (C.I.M. Bull. p. 1165, October 1970.)
5. MISKELLY, N. and WILLSTEED, T. V. *The evaluation of mining projects.* Proc. Annual Conf. Aust. I.M.M., Sydney 1969. (Aust. I.M.M., Melbourne 1969.)
6. NIELSEN, M. N. *The control of colliery production costs.* Mine and Quarry Mechanisation, p. 110, 1969.

Bibliography

MERRETT, A. J. and SYKES, A. *The finance and analysis of capital projects.* (See especially Ch. 13, 'Evaluating an Overseas Mining Project'.) Longman, London, 1965.

HEATH, K. C. G. and KALKOV, G. *Graphical valuation methods for use in prospecting and exploration.* Trans. Inst. Min. and Met., Section A, **80**, p. A45, 1971.

SECTION B. MINERAL EXTRACTION

Chapter 5. Surface Mining

Surface mining can be divided broadly into working loose unconsolidated deposits such as alluvial gravel or mineral sands, mining bedded deposits such as coal seams where the mine travels laterally, and mining massive deposits where the mine extends downwards and outwards as it is worked. For convenience some non-entry methods of mining, such as augering and leaching, will also be covered in this chapter. Essentially all these mining methods have not changed greatly over the years, but the advance from horse and cart, to steam shovels, and then to electric shovels and draglines, has enabled larger scale mining to take place.

Attendant on large scale mining are large scale nuisances of dust, noise and disfiguration of the countryside. A brief mention is made of how to reduce these nuisances, but detailed methods will be left to a more specialised textbook.

Alluvial Mining

In the early days this generally meant gold and tin mining and rather primitive methods were employed on small deposits. The Otago district of New Zealand is credited for the major advance of the invention of the spoon dredge in about 1860-1870, although it was both invented, and then developed into the bucket dredge, by two miners from California. Hence it was named the Californian dredge. It was used on a world wide basis and in the early 1900s some 200 Otago dredges were afloat. Only one is still working in New Zealand and the last revival of a gold dredge in 1968-70 in Australia proved a failure. Dredges are still used for tin and mineral sands mining and will be described later.

PLACER DEPOSITS

Heavy stable minerals weathered from rock are deposited in stream beds where the mechanical sorting action of water flow tends to layer them below the fine sand and gravel. Generally only the heavier minerals will be sorted and left. For example,

	Specific Gravity
Gold	15.5 to 19.3
Platinum	14 — 22
Cassiterite (tin)	6.6 to 7.1
Diamond	3.2 — 3.52
Garnet	3.15 to 4.3
Monazite	4.9 — 5.3
Magnetite	5.16 to 5.18
Zircon	4.2 — 4.7
Rutile	4.2
Ilmenite	4.5 — 5.0
N.B. Silica (sand and gravel)	2.6

It is perhaps a rather crude approximation to say that the heaviest minerals are deposited first—hence gold is found in the foothills, and zircon, rutile and ilmenite on the beach—but it does tend to occur that way in practice. *Eluvial* deposits are those close to the parent rock. *Alluvial* deposits are those carried away to river valleys, lakes, etc. The minerals may be deposited in joints in weathered rock. For example, in Malaysia, solution joints in limestone trapped the cassiterite as it rolled along the river beds. Otherwise the heavy minerals drop out on the inside of river bends or in deeper sections of streams and rivers where the water velocity decreases.

This property of the rate of fall of a heavy particle in water being greater than that of a light particle is useful for natural concentration of minerals by river or wave action, and enables the metallurgist to concentrate the mineral after it has been won. However it is a nuisance during mining operations because each time the mixture of sand, gravel, and mineral is disturbed the heavy mineral tends to drop to the bottom of the sequence and get left behind.

In the mining methods given below panning for gold, or assisted panning with a rocker or 'long tom' will be omitted. These methods, although interesting, belong to the era of the lone prospector. Details can be found in earlier books on metalliferous mining. The most recent of these is the book by Griffith[1]. It must be realised that in the time of 200 to 2000 years or more of surface exploration most of the easily accessible deposits have been found and worked out.

HYDRAULIC MINING

As its name implies this method requires a plentiful supply of water so that its use on the Australian mainland is very restricted. In high rainfall areas, such as Bougainville, part of the technique is extremely useful. The soft volcanic ashes and tuffs which form the overburden of the large copper orebody developed by Bougainville Mining were washed away with *monitors,* or *giants.* These are water nozzles constructed with a curved inlet and a swivel bearing in such a way that the reaction thrust of the nozzle is taken by the pipeline anchor. This enables the monitor operator to swivel the jet and aim it at the base of the bank to be undercut.

Either a natural or an excavated slope is a prerequisite of hydraulic mining. The water jets are used to break up the gravel and mineral-bearing alluvium and wash it to a sump or treatment plant. A head of water of around 50 to 60 m (600 kPa) is needed to operate the monitors. Excavation is usually carried out by undercutting the bank and washing the fallen material away. Unless monitors are used to complete the removal of material then sufficient volume of water must be available to provide the velocities given below. This water flow will transport the gravel to a collecting point.

Velocity		Material Moved
m/sec	ft/min	
0.08	15	Wears away fine clay
0.15	30	Lifts fine sand
0.30	60	Moves fine gravel
0.61	120	Moves 25 mm (1 in.) pebbles
1.52	300	Moves 75-100 mm (3-4 in.) stones
2.03	400	Moves 150-200 mm (6-8 in.) stones
3.04	600	Moves 300-450 mm (12-18 in.) diameter boulders

The simplest style of hydraulicking is to wash the gravel straight into *sluices.* These are wooden or metal troughs of from 0.3 m to 6 m width and up to 90 m in length: smaller sizes are used for recovering gold and the larger sizes for tin. The sluices are made up in 3 to 4 m lengths on a slope of about 1 in 20 to 1 in 24 and may have drops between lengths to break up clay balls.

The gravel is broken up in a stream of water 0.15 to 0.3 m deep by riffles. These can be from 25 mm square cross section for gold to about 150 mm square cross section for tin, and from a few centimetres to 6 m apart. The riffles cause a boiling action with eddy currents and separate the heavy minerals leaving them behind. Periodically the water is stopped and the sluices are cleaned out. Closely spaced riffles are usually held in racks for easy removal. In some cases mercury may be put in the lower riffles to catch fine specks of gold. This type of work is labour intensive but cheap in capital investment.

In more modern treatment plants where tin is currently worked by *gravel pit* mining (such as in Malaya) the gravel is washed through a grizzly into a sump, and a centrifugal pump with a specially lined casing and a hardened impeller is used to deliver it to the recovery plant for concentration on shaking tables. These tables are virtually a shaking sluice where a sideways mechanical stroke is used to improve the separation rate of the downward water flow. The gravel will be screened and broken in a rotating trommel before tabling.

It is worth noting that at the Gibsonvale (N.S.W.) operations of Metals Exploration N.L. a *deep lead* (old buried river bed) deposit of alluvial tin is being worked. This is sampled with large diameter auger-bucket type drills on nominal 30 m centres. A plan of the lead is prepared and a contractor strips the 25-30 m of overburden with bowl scrapers aided by rippers. The 2-3 m layer of *wash* containing a few kilograms of cassiterite per cubic metre is then scraped out and deposited on a stockpile which holds about one month's production.

The stockpile is broken down by a monitor jet and washed into the suction of a gravel pump. Slurried gravel and cassiterite is passed through a trommel into a series of cyclones and jigs to recover the tin oxide. Final scavenging is through a sluice but most of the recovery is completed in the jigs.

DREDGING

Dredging is also wet mining but can be carried out in relatively dry areas because only sufficient water is needed to maintain a small lagoon. As mentioned earlier, dredges are not in widespread use. There are two basic types, the bucket dredge and the suction dredge. The latter will be described under mineral sands mining. The current use of bucket dredges for mining, as distinct from harbour and river dredging, would appear to be in Malaysia. There may be as many as one hundred still mining the alluvial tin deposits in the river deltas but as these are gradually exhausted the dredges will have to move out to sea. Consequently they have to be larger and more expensive, and fewer are used for the same output. A large tin bucket dredge may handle around 7 000 000 m^3 of gravel per year: i.e. around 1500-2000 tonnes per hour. Combined operating costs and overheads were about 7-20c per cubic metre in 1971.

The general construction of a bucket dredge is as for the suction dredge shown in Figure 5.1 except that the suction boom is replaced by a bucket ladder. In addition the larger dredges are built as one unit so that the weight of the ladder is counterbalanced by the weight of the concentrating plant. The bucket dredge is preferred in Malaysia because the endless chain of buckets can scrape away the weathered surface of the bedrock to give up to 90 per cent tin recovery. Atlas Copco have suggested that their overburden drilling (OD) method could be used to bore down through the gravel and blast the surface of the bedrock to increase recovery. However the heavy cassiterite can still drop through the broken rock as soon as it is disturbed.

A bucket dredge can dig at a rate of 20 to 30 buckets a minute with a bucket capacity of 0.07 to 0.56 m^3. Mineral concentration on the dredge is more likely to be by tables, jigs or cyclones than by spirals as for beach sands. The bucket dredge can operate down to 20 to 30 m but beyond that the weight of the bucket chain is too much for the buoyancy of an economical sized pontoon. Deeper offshore mining will probably be confined to suction dredging with a submerged centrifugal pump mounted on the suction boom[2]. It must be realised that a centrifugal pump mounted in the dredge will not have an efficient vertical suction lift of more than about 4 metres.

This should not be confused with deep ocean mining where it is envisaged that relatively few buckets will be fastened to a wire rope loop and suspended from a small freighter to mine at over 3000 m.

Mineral Sands Mining

The minerals desired are those known colloquially as heavy minerals, or beach sands but more properly as mineral sands. These are chiefly rutile (mainly titanium dioxide) used as the raw material for making a white pigment (titanium white) after treatment by the chloride process or as a source for titanium metal; zircon ($ZrO_2.SiO_2$) used as a moulding sand, or a glaze, or as a source of the metal; ilmenite ($FeTiO_3$) used as a raw material for pigments by the sulphate process, also as a source of titanium metal and as an additive for metallurgical alloys; and monazite [nominally (Ce La Dy) PO_4 $(ThSi)O_2$], obviously used as a source of the rare earth elements. Minor minerals are garnet, used for abrasive paper and sand blasting, and leucoxene and tourmaline. Ilmenite has 31.6 per cent titanium, and rutile has 60 per cent titanium. At present the market in ilmenite is not good and development work is being carried out on commercial processes to upgrade it as a source of titanium metal to obtain a higher price. Some of the Australian ilmenite has a high chrome content and this carries a price penalty. Currently, upgrading processes are being developed to remove iron impurities and the chromium content to make a 'synthetic rutile'.

The present known reserves of mineral sands in Australia occur mainly on the east coast from just north of Sydney to the mid-coastal area of Queensland, with other deposits in the south west of Western Australia. On the east coast there is a variation in mineral content. Around the Wyong area it is typically 48 per cent rutile, 12 per cent ilmenite, 30-35 per cent zircon and 0.5 per cent monazite. Further north up the coast to Queensland the high-priced rutile content drops to as low as 10 per cent and the low priced ilmenite content increases. This means that large scale mining of sands containing 99 per cent steriles and one per cent heavy mineral can take place at Wyong, with a cut-off grade of about 0.4 per cent H.M., whereas in the Southport area the mining and cut off grades need to be much higher. More data on exploration and evaluation can be found in an Australian government publication[3].

Australia is the world's largest producer of mineral sands and uses both wet and dry methods to produce the minerals. The largest company is Associated Minerals Consolidated Ltd, and it has the world's largest suction dredge, a 1500 tonnes per hour dredge and concentrator, currently on North Stradbroke Island after a period of about three years on South Stradbroke Island. The dredge cost originally about $3m. The second largest dredge, capacity 900 to 1500 tonnes per hour, is owned by Mineral Deposits Ltd, coincidentally Australia's second largest producer. That plant is currently working south of Forster, N.S.W.

WET MINING

Wet mining for mineral sands is principally by suction dredge. These dredges consist basically of four units, which may be housed on one pontoon, or separately. Generally, the smaller plants are all under one roof and the larger plants are split into two parts coupled by a floating hose. Figures 5.1 and 5.2 show the construction and layout of a medium-sized plant.

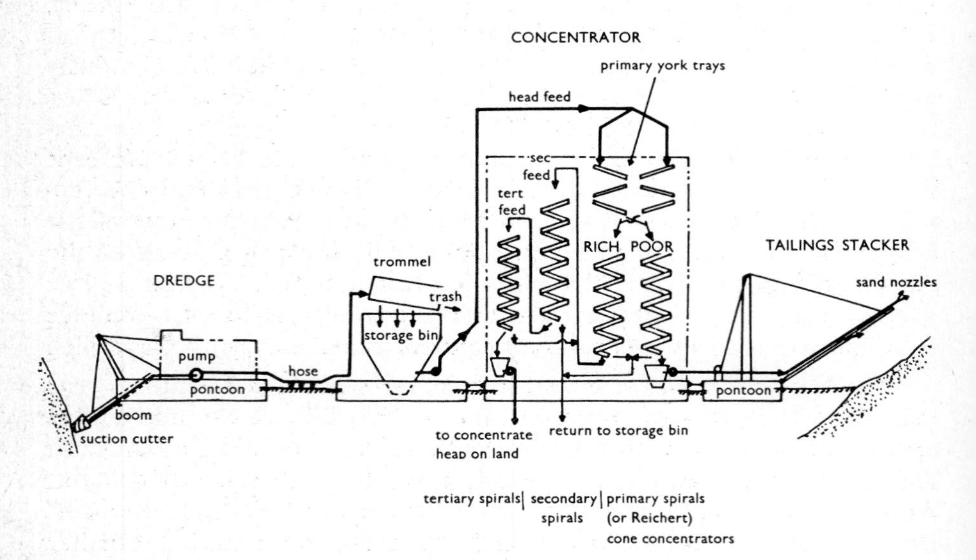

Fig. 5.1 Medium size suction dredge

1500 t/hr suction dredge and concentrator
(*Courtesy Associated Minerals Consolidated Ltd*)

For this 5-600 t/hr plant the dredge would be about 18 m x 6 m x 1.5 m draught and would carry a 260 kw centrifugal sand pump and a 93 kw cutter drive motor. The concentrator plant would be mounted on a base of interconnecting pontoons to give a raft about 36 m x 15 m and 1.2 m draught.

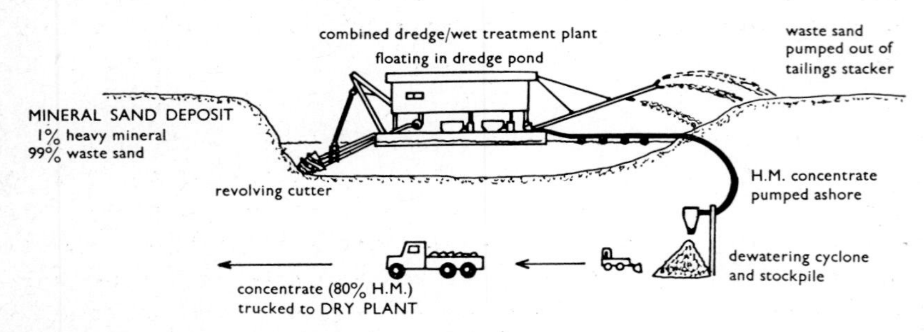

Fig. 5.2 Wet mineral sands mining

The sand is broken by a rotating cutter head. It should be noted that although surface sand is usually loose once the vegetation has been cleared, the lower part of the bed may frequently be indurated (cemented together) by organic material into large hard irregular masses. The conical cutter head is fitted with spiral blades, sometimes toothed, to break the sand or grind off irregular bedrock. Inside the cutter head, at the lower edge, is the suction inlet of a powerful centrifugal pump which sucks the sand up and feeds it through a rotating trommel to remove trash such as old timber and clay balls, and break the sand down further into discrete grains.

Normally the sand bank caves away when undercut and slumps into the pool. If it consistently hangs up a water monitor may be mounted on the dredge to attack the bank.

The underflow from the trommel gravitates to a storage bin in the base of the concentrator section. This bin controls feed surges and also dewaters the dredge feed which has about 20-25 per cent pulp density to a 60 per cent pulp density feed which is pumped to the top of the concentrator. The mixture of heavy minerals and sand is fed down through a series of gravity type concentrating units. In the primary and secondary stages these can be York pinch sluice trays, Reichert cone concentrators or Humphrey spirals. In the tertiary stages they are generally fibreglass Humphrey spirals.

The silica sand tailings are pumped out behind the concentrating plant through nozzles mounted on a long boom. Conservation generally requires that the sand dune must be restored to its pre-mining shape. Consequently the nozzles are arranged to deposit the waste sand as close to that profile as possible. This can cause problems. Generally the water is required to drain back into the dredger pond, which is artificially maintained. When shaping a high dune, if the output from the nozzles is deposited too close together the water content of the sand may be above its liquid limit. This can cause serious slumps of sand into the pond, with disturbance to dredge operation. On occasion damage to the manoeuvering rope anchorages can result. It is usually avoided by extending the tailings stacker pipe and spacing the nozzles out so that the water can drain through a larger area.

The concentrates from the dredge are pumped ashore at about 60 per cent solids and dewatered in a cyclone. They are dropped onto a stockpile from which they are collected by lorry as necessary. It must be remembered that the heavy minerals are in concentrations of only a few per cent. The dredge at Wyong operates in about 1 per cent grade with a cut-off of 0.4 per cent. Consequently a 500 t/hr plant is only producing about 3 to 5 t/hr of concentrate. These are taken to a central mill for separation on shaking tables followed by electrostatic and electromagnetic concentration (Figure 5.3). Most sales contracts have rigid impurity clauses and the individual minerals must be 95 to 99 per cent pure.

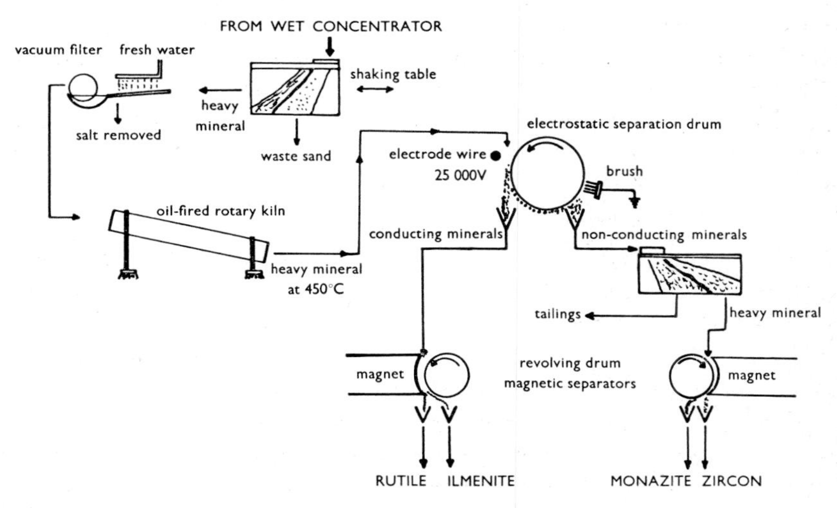

Fig. 5.3 Treatment plant for mineral sands separation

In operation the suction cutter head is traversed laterally across the face of the appropriate section of dune, breaking it down and engorging it. Traversing is carried out by four mooring ropes coupled to movable anchors (or deadmen). The two stern ropes are brought forward at about 45° to provide a crowding and swinging action. The bow ropes are spread at about 90° to provide the lateral traverse. Where the suction head is separate from the concentrator then that unit also has to be equipped with mooring ropes. It may also be equipped with steel spuds. These are heavy steel posts which can be lowered into the bed of the pond to anchor the concentrator or to help swing it to follow the pump hose movement.

The first operation in mining is to sample previously located deposits on a close grid. This can be a square or rectangular spacing of from 45 m x 45 m to 150 m x 8 m. Sampling to the water table can be by open auger. The outer skin of sample is wiped off to clear contamination and the measured grade of the centre core of the sample may be corrected by an operator factor based on experience. Tube samplers can also be used. Below the water table the hole must be cased and a sludge baler used.

The mining plan is then drawn up; a 500 t/hr plant in an average thickness bed will mine about 0.8 to 1.2 hectares (2-3 acres) per month in a series of paddocks. The dredge advances on a 150 m wide front for the width of the mineable deposit and then swings round to retrace its steps (Figure 5.4). In areas of below cut-off grade values it dredges a narrow channel to move itself through to a new mining zone. Where a deposit is fairly rich, because the heavy minerals drop so quickly, it may be economically feasible to use a second suction head to scavenge the pond bottom immediately behind the first head.

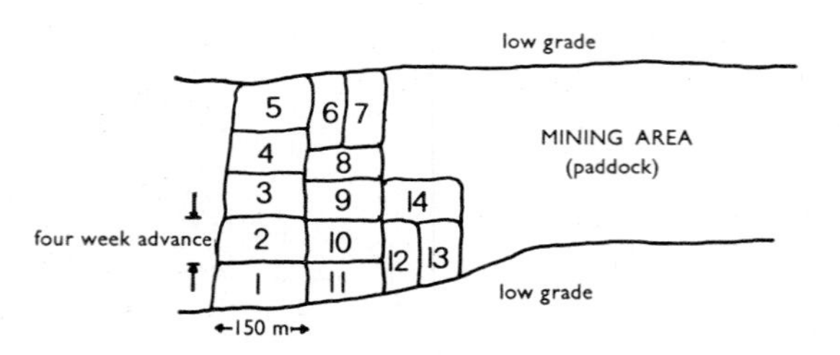

Fig. 5.4 Path for mining dredge

One of the problems of wet dredging in Australia is the supply of water for the dredging pond. This pond is usually above the natural water table and has to be maintained with artificial or mined banks. Water is lost with the concentrates and tailings, plus evaporation. A rich deposit can pay for working in salt water plus its attendant corrosion problems. If the water is pumped from the surf a combination vacuum and centrifugal type pump, such as the Sykes, may be necessary to handle the aerated water without loss of suction head.

With low grade deposits it is fairly essential to use fresh water, consequently wet mining methods may be restricted to areas where fresh or mildly brackish water is available throughout the year.

A dredge operator cannot see the bottom of the sand deposit and has to rely on a depth gauge on the suction boom. Consequently survey pegs are placed at the edge of the pond and water make-up is regulated to keep the dredge on a constant reference horizon. The pond depth is restricted to 3-4 metres to keep the pump suction operative, so the water level is adjusted from the bottom of the deposit.

DRY MINING

Dry Mining methods may be essential, particularly on high dune deposits on raised beaches where water is not available.

Buried Loaders

Draglines are usually not selective enough for mineral sands mining although they can be used. The smaller sizes have a low throughput and cannot load a lorry efficiently, a hopper must be used. Larger sizes would cause a lot of unnecessary dilution, or loss of payable grades. Because these sands are soft material it has been found that bulldozers and front-end loaders can strip the vegetation, top soil and overburden, and then push the heavy mineral sands into a receiving hopper situated on the return end (boot) of a belt conveyor. This is the buried loader. The material is then conveyed to a loading point for trucks, or direct to a concentrator.

Bucket Wheel Excavators

The machine is described in a later section. It is an old machine but a new method for the mineral sands industry. Sands on a beach have been sorted by old or recent wave action and frequently have a banded structure of heavy minerals and light silica sands that is mined as a mixture by suction dredges or most dry methods.

In 1969 Associated Minerals purchased a small bucket wheel excavator from the Victorian Electricity Commission brown coal workings. It was set up at Jerusalem Creek near Evans Head in 1970 to mine rutile deposits about a kilometre from the sea front with no water available locally. The deposit had mixed beds of high and low grade values. To separate these sands effectively it is preferable to keep the range of concentrator feed grades relatively narrow, and it was considered necessary to mine the sands alternately as either high grade or low grade. (Incidentally, the centre of this deposit had already been mined for gold, which gave a low grade mix of non-bedded sands down the middle of the deposits.)

The bucket wheel excavator mines the mineral sands selectively and feeds them as dry material into a hopper, where this is slurried with water from the pond and pumped into the concentrator bin (Figure 5.5). The concentrator floats in a pond for convenience of movement and the pond also acts as a reservoir for treatment water. Alternate feeds of 0.5-4 per cent, and 4-14 per cent heavy mineral, are produced by the excavator. When operating on low grade feed five Reichert cone concentrators act as primary units, each producing a final tailings. When the feed is high grade a number of valves are altered so that three concentrators act as primary units and their tailings are scavenged by the other two units. This gives a high recovery at a lower throughput.

The first bucket wheel excavator has shown that the technique can be successful and a second excavator is being built for AMA with an integrated plant. Mineral Deposits Ltd, are also purchasing a bucket wheel excavator (Sept. 1971). Further details of the Jerusalem Creek venture and details of the machine are given in a paper by Rodgers and Bunting.[4]

Other Methods

Small deposits of mineral sands, or industrial sand and gravel, can be worked with a variety of equipment such as bulldozers, front-end loaders, draglines, etc., but the economy of large scale working is lost and cut-off grades must be higher.

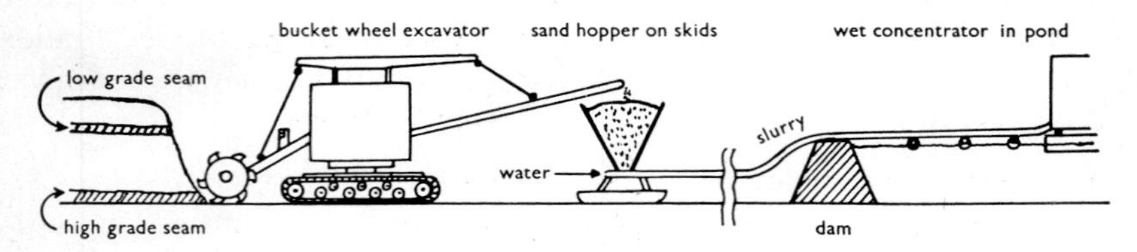

Fig. 5.5 Dry mining with bucket wheel excavator

CONSERVATION

Restoration of mined areas is an essential part of mineral sands mining, and the companies have to post a financial bond to cover restoration before the State will grant a mining lease. Costs of restoration can vary from about $200 per hectare ($500 an acre) to $400 per hectare for bad areas that may need double treatment. Figure 5.6 shows the restoration plan.

The general treatment of the mining area is similar for all types of surface mining. Topsoil should be removed and stored for later replacement before any mining takes place. This is essential because it contains the humus and a reserve of seeds needed to regenerate the area.

Revegetation of sand dunes is difficult because the light topsoil is particularly susceptible to wind erosion. The first area to be mined has its vegetation removed and the topsoil bulldozed to one side. After mining has progressed the tailings are contoured and the topsoil spread evenly across the area ready for revegetation. The areas should be re-seeded as soon as possible although the time interval between stripping and seeding is around 6 months because the sand must be allowed to settle before the topsoil is replaced and cereals sown. This necessitates an area of about 150 m by 450 m being open for mining and tailings consolidation.

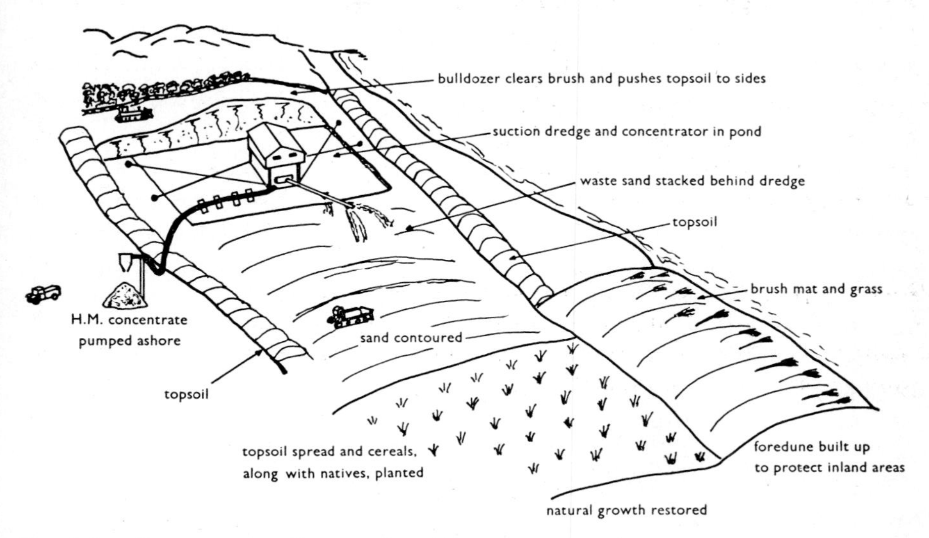

Fig. 5.6 Restoration of mineral sands mining area

When the area is ready for restoration it is divided into three zones. The frontal dune area (not always present) is exposed to wind and salt spray and spreads about 40 to 50 m up the beach. It must be stabilised with a brush mat which gives a wind cover to newly planted spinifex or marram grass. These grasses throw out runners which bind the sand together. With fertilizers and a proper rainy season the grass will bind the soil before the mat decays. The brush is obtained mainly from scrub cut from the area to be mined.

The second zone is the hind dune immediately behind the frontal crest. It is slightly protected but may occasionally have to be treated as frontal dunes. However it is usually possible to sow with cereals and to brush mat. The brush mat drops seed pods, and the company plants other seeds and shrubs. Horsetail oaks, coastal wattle and bitou bush are planted for permanent growth. Pig face and flannel flowers may be added. Cereals and legumes such as rye or sorghum, and West Australian lupins, are planted either in advance of or with the natives, and fertilized. These provide shelter for the young natives and extra humus when they rot down. They are not harvested.

The third zone is the heath area which is wind and salt free. When its own topsoil is replaced and fertilized the native generation is usually satisfactory. The area can be developed as pasture, or quick growing native seedlings may be added to speed restoration.

Extra expense is caused when dry periods hold up restoration, or kill off partly restored areas and necessitate replanting. It may also be necessary to fence areas and leave proper tracks to the beach to prevent indiscriminate leisure use from damaging young-shooted areas. It is perhaps worth noting that tracks on sand dunes, whether for mining access or post-mining leisure, should run along the crests of dunes. The wind then keeps them clear instead of drifting sand onto them. Plastic netting pegged out as windbreaks can help slow down drifting sand.

Governmental concern is now such that advice from agronomists is easily available for any mining restoration and the large mineral sands mining companies also employ their own specialist staff. Companies such as Associated Minerals have research facilities for revegetation and maintain their own nurseries. The latest AMA innovation is to use an elevating bowl scraper to remove topsoil and transfer it directly to the area to be restored. This preserves the essential bacterial life and native seeds to accelerate restoration.

General Open Pit Mining

The largest open pit in the Western world is the low grade porphyry copper mine at Bingham Canyon, Utah which in 1965 produced 29 000 000 tonnes (32 000 000 short tons) of ore and 76 000 000 tonnes (84 000 000 short tons) of waste. The mine was a mountain and is now a pit 3.2 km long, 2.4 km wide and 0.8 km deep with more than 60 benches.

Where a mineral is near the surface then an open pit is an automatic choice, the only question being that of the economic cut-off limit where a change may be made to underground mining if the deposit warrants it. The advantages of open pits are:

1. higher productivity
2. greater concentration of operations
3. higher output
4. lower capital and operating costs per tonne mined
5. greater geological certainty and easier exploration
6. less limitation on size and weight of machines
7. better recovery of mineral
8. simplified engineering and planning
9. greater safety.

Most of these are a consequence of being in the open air.

The reason for choosing open pit mining must be a matter of overall economics determined by a feasibility study. Factors affecting the choice are:

1. relative thickness of overburden and orebody
2. size of orebody
3. relative costs of mining by open cut and an applicable underground method
4. relative dilution and loss of mineral in both cases
5. relative costs of development
6. climate (e.g. rain, snowfall, etc.)
7. topography
8. continuity of projected operations
9. availability of skilled labour
10. capital available.

The profit margin on open-pit mining can often be predicted to closer limits than for underground working and sophisticated computer programs are available for calculations.

COMPUTER PROCESSING OF ECONOMIC DATA

It will be readily appreciated that opencast mine planning involves considerable and tedious calculation of volumes and costs. This task can be greatly simplified by automatic data processing. Borehole information can be fed into a computer store and updated as further holes are drilled. The computer can then be programmed to select the required boreholes from the store and to process this information. This may take the form of calculating the depth of overburden, or the discard or dilution at the interfaces between overburden and mineral, dilution due to thin bands of sterile inclusions, average thickness of recoverable mineral, volumes within designated areas, grades and yields of mineral.

Statistical methods permit the calculation of confidence limits from the data available, or alternatively further data can be called up to obtain results within specified confidence limits (e.g. the borehole grid spacing can be reduced or intermediate holes drilled).

The volume to be mined is then split up into a three-dimensional grid of roughly equal sized blocks. A convenient size is 15 m edges, because this is a normal bench interval. If the pit is to have shallower benches then the cube size should be adjusted. Each block is graded for ore values and ease of working, and can be assigned a market value; Figure 5.7 shows a greatly simplified block plan. Grade of blocks is automatically lowered when air is included at the surface. The final mining profile has to include sterile blocks to enable a safe slope (approx. 45°) to be maintained. Consequently the computer program must repeat calculations for each block of ore to be mined in the pit to ascertain how many superimposed blocks must be taken, and their total market value.

At any time during the life of the mine the program may be re-run to see how changes in market values or in mining costs have affected the optimum pattern of extraction. If the selling price goes up, then the stripping ratio can increase, and more overburden can be removed to recover more mineral. Ore may have to be left in the bottom of the mine if too much overburden would have to be removed to mine it. Alternatively the rest of the ore body may be mined by underground methods if it extends in depth.

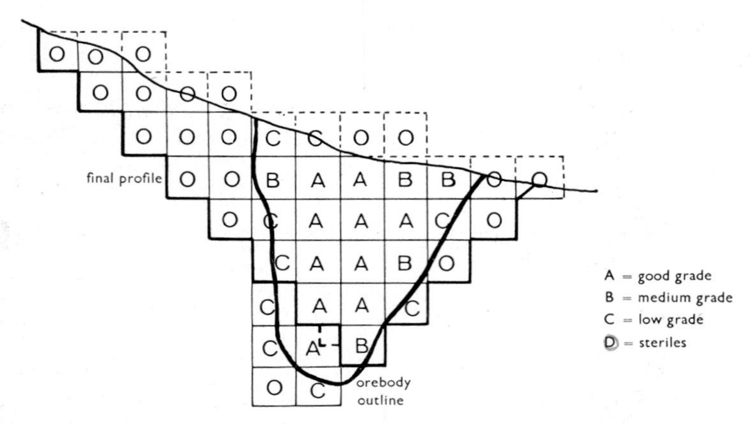

Fig. 5.7 Cross-section of computer plan grid for open pit

OPERATING COSTS AND CUT-OFF CURVES

The *stripping ratio* of a pit is the ratio of overburden actually removed to mineral mined. The *overburden ratio* is the ratio of the vertical thickness of overburden to the thickness of mineral to be mined and should be used for stratified deposits. However the terms tend to be used rather loosely.

If C_o is the total operating cost per unit volume of mineral at the pit limits, then

$$C_o = C_m + SC_w$$

where C_m is the cost of excavating unit volume of material and transporting it to the pit limits, C_w is the cost of excavating unit volume of waste and S is the stripping ratio.

The capital cost of equipment and office overheads must be paid for, so if A is the total fixed charges per unit volume of material and C_p is the total production cost per unit volume at the pit limits, then

$$C_p = C_o + A$$

The larger the pit then the smaller the value of A proportionally.

The cost of the saleable product is increased by the beneficiation processes and by transport so that,

$$C_s = \frac{1}{RG}\left(\frac{C_p}{D} + B\right)$$

where C_s = cost per unit weight of saleable product at the point of sale in percentage grade units,

R = recovery of mineral per unit, as a ratio (i.e. 0.8 not 80%),

G = grade of mineral in percentage points (i.e. 30 not 30/100),

D = density of mineral, weight per unit volume, and

B = cost of transport and preparation per unit weight of mill feed.

C_s for instance would be in dollars/tonne/grade unit.

The calculation for C_o through to C_s has to be performed for each block in the pit and is obviously computer material. For each arrangement of machinery and each mining method a series of cut-off profiles can be obtained as in Figure 5.8A. Their general relationship with depth and overburden ratio is shown in Figure 5.8B. It can be seen that bedded deposits are much more amenable to open pit mining than are vertical pipes of ore. Figure 5.8B is derived from a paper by Brealey and Atkinson[5].

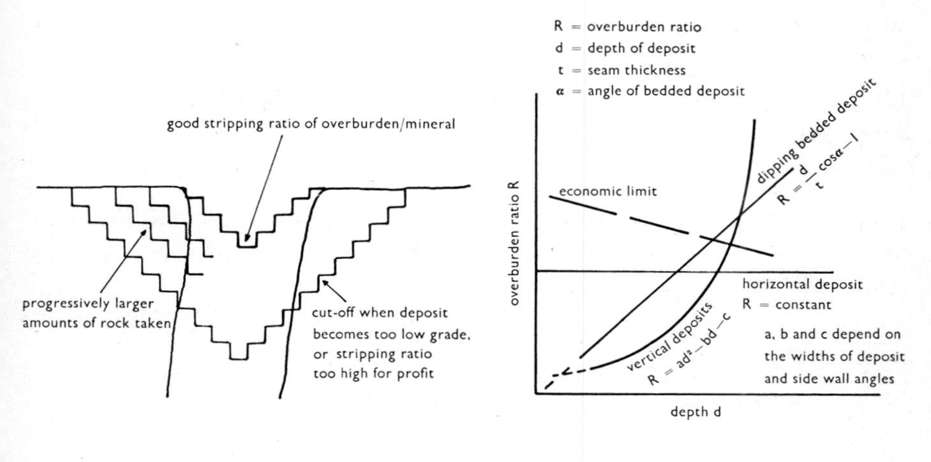

Fig. 5.8 Evaluation of deposits for open pit mining

OPEN PIT PLANNING PROCEDURE

In open pit mining, as in underground mining, planning is of considerable importance, because failure to maintain adequate 'blocked-out' reserves of mineral will result in a slow insidious reduction in the mine's ability to maintain production, which ultimately necessitates an extensive re-development program to restore the position. In bad cases, loss of production is severe, and in all cases capital is required for the re-development operation; a consequence of the earlier practice of 'living on capital'. This situation usually comes about by failure to strip off overburden fast enough, particularly where the overburden ratio is increasing rapidly (e.g. when advancing into a hill).

The concentration of output in a few production units can also present problems if large variations result from lack of forward planning. The correct balance must be maintained between benches which must advance in synchronism, and also the construction of access roads and auxiliary works must keep pace with this advance. Planning must also ensure that expensive machine groups are fully utilized in production and not subjected to unnecessary movement.

Short-term planning can never be static and must operate flexibly within the framework of a long-term plan, in order to allow for new geological information, revised volumetric control data, meteorological conditions, changes in production requirements, changes in ore grade, etc. The general procedures adopted can be summarized as follows:

1. long-term planning; exploiting a deposit profitably to obtain extraction of the whole of the reserves or up to the cut-off point
2. medium-term planning; programs which are in greater detail and are related to annual production requirements
3. short-term planning; detailed control of day-to-day production.

PLANNING INFORMATION

The following maps, sections and data are maintained for planning purposes.

Maps

1. Topographical plans, with surface contours at close intervals
2. Borehole plans, showing altitudes, mineral thicknesses and values
3. Isopachytes (for stratified deposits) of deposit thickness, overburden thickness, overburden ratio, and mineral grade
4. Hydrological plans, showing actual levels of free ground-water tables and piestic levels of confined ground-water
5. Contours of the base of deposit, and the base of overburden
6. Plans of available reserves
7. Mining plans showing blocks to be worked and machine moves

Sections

8. Geological sections (particularly for non-stratified deposits)
9. Bench levels
10. Mineral grade
11. Exposed reserves
12. Ground-water tables

Data

13. Volumetric control
14. Machine output
15. Transport capacity
16. Meteorological records
17. Management statistics.

FACTORS AFFECTING MINE LAYOUT

Brealey and Atkinson[5] classified mineral deposits as in Figure 5.9. This makes the influence of the shape of a deposit more apparent for design work. In general, stratified deposits are in softer younger sedimentary rocks and non-stratified are harder, older, igneous rocks. The shape of a mineral deposit fundamentally determines the shape of the mine, and the physical characteristics of the mineral determine the choice of machines. The major factors affecting the layout are as follows:

1. the shape and depth of the deposit. This is of particular importance since the selection of a transport system is restricted by the distance and vertical height through which overburden and mineral must be moved, i.e. routes and gradients.
2. the properties of the mineral and overburden:
   (i) slope angle limitations, both of working faces and of sidewalls, and of spoil heaps
   (ii) degree of hardness and abrasiveness, which influence the selection of the excavating and transport equipment
   (iii) variations in grade which may require selective mining to be practised and affect the shape of the pit due to the cut-off grade limits
   (iv) competence of the various rocks to support the equipment.
3. the geometry of the excavating machinery, particularly digging height or depth, dumping height, and reach
4. the drainage requirements, particularly the maintenance of drainage slopes to sumps in such a way as to safeguard working faces, benches, spoil heaps and access roads at all times during the life of the mine from the influence of ground water. Saturated slopes are much more likely to collapse than are well drained slopes.

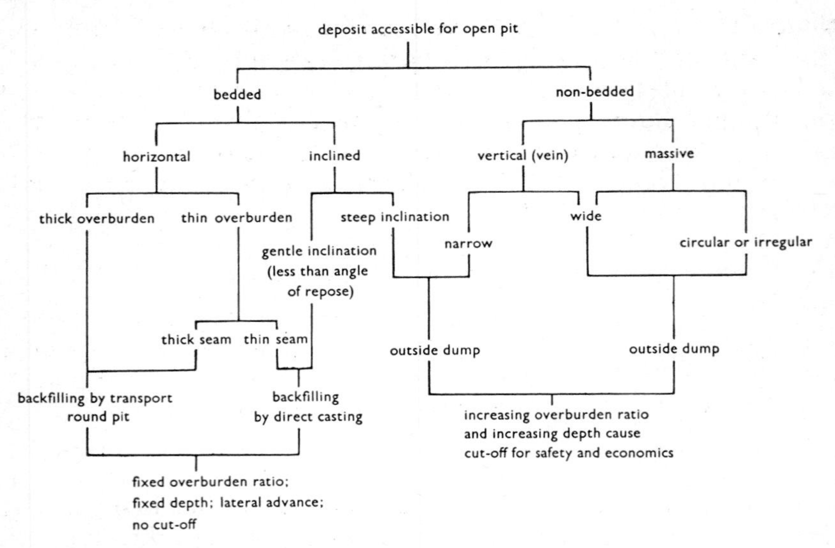

Fig. 5.9 Classification of deposits for open pit mining

BACK FILLING

When a stratified deposit dips at an angle less than the angle of repose of spoil heaps, plus a factor of safety, it is possible to dump spoil inside the excavation, i.e. to back-fill. This is desirable because it:—

1. is usually the shortest distance for horizontal movement,
2. gives minimum height of lift, and
3. permits economic reclamation of the worked out area.

Where possible the overburden spoil is direct back-cast across the cut with a dragline or stripping shovel (described later). As the overburden becomes thicker there is a limit to the reach of the dragline or shovel and the overburden has to go round the pit and be filled in from behind.

Extra thick or steep deposits cannot be back filled or else the mineral would be covered up. At a later stage, with suitable precautions, waste from an underground mine may be dumped into the abandoned pit. Without reclamation laws, it is unlikely that the overburden would be put back in. Present legislation usually only requires slopes to be made safe, and secure fencing to be erected.

STRIPPING RATIO

Depending on richness of ore, nature of overburden, and extent of deposit, the stripping ratio may go up to about 20 to 1 or occasionally more. This usually needs an extensive shallow deposit and large machinery. On a steep deposit the mine may extend very little beyond a ratio of 1 to 1 to 3 to 1, before going underground. In this case the open pit may be worked by hired machinery or by contractors while the underground mine is being developed. A limit may be set for the open pit at the bottom of oxidised ores. The sulphides are then taken by a change to deep mining concurrent with a change in milling methods.

Surface Mining Machinery

There is an overlap with civil engineering practice in the field of open pit mining. However the large open pit mines have a longer life, and more material to move, than in most civil engineering projects, so that equipment tends to be bigger and stronger.

The minerals worked in big open pits are mainly coal, iron ore, and low grade copper ores. Other minerals may be worked as surface outcrops but the size of the outcrops usually implies a smaller operation with smaller sizes of machinery, although techniques of use are the same.

The largest items of equipment, particularly draglines, are generally found in the large opencast coal mines where the coal and overburden rocks tend to be softer and to have a lower density; about 1 to 2.5 t/m^3. In the metalliferous pits, which are multi-benched, machines are smaller, and trucks smaller, both to cope with the limited space and the higher rock density. Copper ores are around 2.5 to 3.2 t/m^3 and iron ores range from 2.6 to 4.7 t/m^3. Both ores are usually more abrasive than coal and shales.

INSURANCE

It is worth noting that large items of mining equipment are insured until their depreciated value becomes less than the premium. In addition, lost production is also insured, although there may be a limit, such as 30 days, on payments. Recently when the boom of a dragline collapsed in Queensland heavy replacement metal parts were air-freighted from the United States to rebuild the boom. Production loss cover was such that the insurance company were anxious to have a quick repair.

POWER SHOVELS

The larger sizes of shovels and draglines are electrically powered and generally use Ward-Leonard control for hoist and swing motors to give accurate position and speed control. An a.c. motor generator set provides the d.c. current for the motors. Modern solid-state devices (thyristors) are replacing d.c. generators in smaller new equipment to give improved performance. A surface mine may have to reticulate electrical power at 11 kV, or even up to 35 kV, to the working areas through carefully protected rubber-sheathed trailing cables, and plan its day to day activities around the cables. On-site power generation is avoided where possible because of higher costs. The larger machines can carry diesel generators if necessary.

Smaller sizes of shovels may have a.c., d.c., diesel, diesel-hydraulic or diesel-electric transmissions.

Capacities of shovels and draglines are measured in terms of bucket volume. Regular sizes of power shovel bucket vary from about 0.4 m^3 to 19 m^3 but popular sizes are

	m^3	yd^3
Truck loading in benched metalliferous mines or quarries	3.7-4.6	5-6
Truck loading in opencast coal mines, large iron ore mines, and large copper mines.	10.5-11.5	14-15
Overburden stripping in opencast coal mines. Not used for truck loading.	Made to order up to 138 m^3 (180 yd^3) for one-off specials.	

These are nominal bucket volumes, actual loading capacities are about 65 per cent of nominal in rock. Small buckets in soft material can load to near capacity. The two most popular bucket sizes are related to the size of haul truck used in mines. A rule-of-thumb is that a shovel should fill a truck in 4 to 5 passes. A smaller number of passes can result in spillage, more passes waste time. The smaller bucket fills a 45-50 tonne truck and the larger fills a 100-120 tonne truck. The largest sized shovel mentioned above strips overburden and its 138 m^3 bucket is mounted on a 65 m boom. The a.c. installed load is 22500 kW.

Figure 5.10 shows the operating principle of a power shovel. There are three motions on the shovel.

1. Dig, by pulling on the rope that goes over the end of the boom to the bucket (or dipper in American terms), thus lifting it through the rockpile.
2. Crowd, by rolling the dipper down past the boom. This forces the bucket into the rockpile as it moves forward.
3. Swing, to take the bucket round and empty it through its back door.

On the older machines pulling and the boom lifting were rope operated and a boom-mounted motor provided the crowding action. The use of hydraulic operation is spreading and several smaller shovels now have hydraulic rams to provide boom lifting, a hydraulically operated knuckle-joint dipper stick, and a hydraulic bucket tilting action. All these combine to give the dig and crowd action. Shovels up to about 23 m³ are now available with a hydraulic boom-dipper knuckle and a rope hoist for the boom.

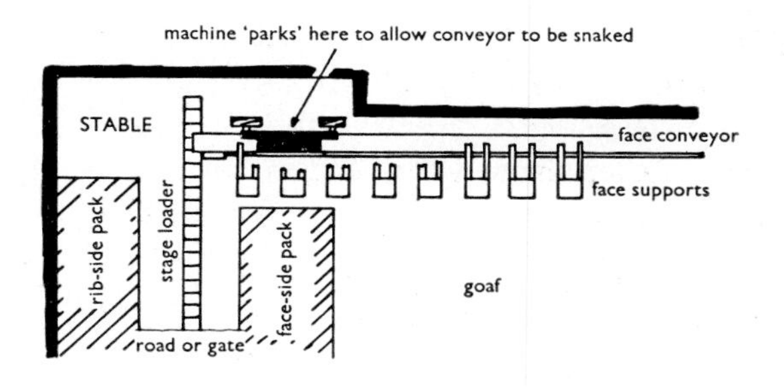

Fig. 5.10 Power shovel principles

The crowding action of the dipper stick and its heavy, stiffly-mounted bucket, give a shovel a much stronger loading breakout force than either a dragline or a front-end loader. For instance a 4.6 m³ power shovel has a breakout force of about 90 tonnes, whereas a 7.6 m³ loader has a breakout force of 35 tonnes. Consequently a shovel can break out soft shales and coals without blasting, and also give a smoother bench surface.

A shovel will dig down below grade (i.e. below track level) to about half its dipper stick length and up to about two-thirds its boom height. The hydraulic shovels can go down to about full dipper length and up to above boom height; a manufacturer's table will give the exact dimensions.

Normally a shovel runs on the floor level it digs for itself and works only from that height up. A bulldozer of size appropriate to the shovel is kept for cleanup duties and floor grading. The shovel is not sufficiently mobile to perform its own cleanup, and ground bearing pressures have to be kept to a minimum by providing a smooth path under the crawler tracks.

BACKHOES

A backhoe is not often used for mining but can be used for ancillary operations such as ditching. It is effectively a shovel with the bucket reversed and it loads by pulling the bucket towards the machine body. It is usually all-hydraulic and intended for working below grade.

DRAGLINES

The smaller draglines will generally run on crawler tracks but the larger walking draglines sit on their tub (base of the body) for operation. This is to keep the ground bearing pressure to a minimum. For tracked operation of shovels or draglines the crawlers are kept as broad as possible and may be twinned for operation on soft ground. Ground bearing pressures are usually limited to 60-150 kPa.

Opencast draglines normally work on top of a highwall that has been drilled and blasted. The broken ground is bulldozed flat to avoid large rocks buckling the tub plates. The whole machine can rotate through 360° relative to its tub (Figure 5.11).

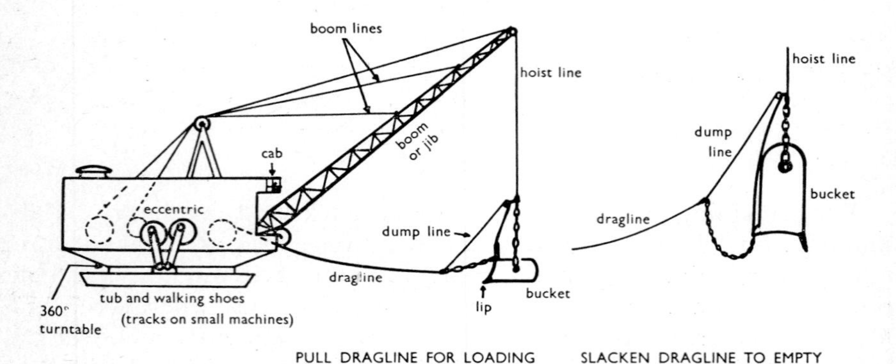

Fig. 5.11 Walking dragline principles

It is interesting to note that on the large machines with a boom length of 80 to 95 m the average speed of the bucket, which may weigh over 100 tonnes, is 48 km/hr, for a normal swing cycle.

Large draglines move by walking on a pair of shoes: an eccentric or an hydraulic mechanism lifts the tub up and propels the dragline forward between the shoes, and then cycles the shoes forward for another lift. Direction changes are made by swinging on the tub with the shoes in the air.

Because the ground will be either soft enough for unaided digging, or will have been drilled and blasted, it is essential that the dragline should have a firm base to prevent it sliding into the pit, or sinking into the ground. If the layer of overburden immediately below the topsoil is not sufficiently strong to take the enormous working weight, then the dragline may have to stand on a lower stratum. This is not desirable because a dragline is designed to work below grade, filling its bucket by pulling it in with the main dragline. When filled the bucket is lifted with the hoist rope, keeping some tension on the dragline. On releasing the dragline the bucket will drop and empty. If the dragline has to work above grade (called chopping-down) its efficiency is reduced by about 20 per cent.

When working normally a dragline can dig below grade to a depth equivalent to two-thirds its boom length and reach up for dumping to about half its boom length. To dump higher would cause the spoil to run back around the dragline base, and risks buckling the boom by accidental impact from the bucket.

An idea of the enormous investment in a dragline may be obtained from the machine at the Thiess-Peabody-Mitsui mine at Moura, Queensland. This Marion 8900 dragline has a 100 m^3 bucket on an 84 m boom. The service weight is 6600 t. It takes 195 t of overburden in one bucket load with a cycle time of one minute. Hence it moves about 6500 tonnes per hour, and it works 24 hours a day, seven days a week. Its initial cost was $12 million and total cost to date about $20 million. It is now only the second largest dragline in the world! Bucyrus Erie have built 'Big Muskie' to work in Ohio: service weight 13 500 tons, bucket load 325 t per one minute cycle which is nearly half a million tonnes per day.

CHOICE BETWEEN SHOVEL AND DRAGLINE

The choice is usually confined to dealing with overburden, where methods other than trucking are used. Trucks are normally loaded with a shovel or front-end loader.

For dealing with overburden a dragline has a greater reach and dumping radius, so that deeper overburdens can be tackled, but a shovel has about a 30 per cent greater loading efficiency so that it clears material faster. A dragline is more versatile and easily manoeuvred. It is usually located on the ground surface which eliminates problems from slides of the spoil heap or the highwall, and from water. In the initial stages it has the advantage of being able to open up a box cut more easily and efficiently, and at the final cut the dragline can cast spoil both into the cut and on to the bank.

However the location of the machine on the original land surface can have its disadvantages; for example the topography may require preparation of a working bench for the machine, which can be expensive. Also it is not unknown for a dragline to fall off the highwall.

The dipper on a shovel is fastened by a direct link to the driving mechanism, so that the digging action is more positive. This higher loading factor becomes increasingly important as the material becomes harder, because less expense is involved in blasting. The loading cycle is shorter for a shovel, especially if the dragline is involved in chopping down from a higher level. A shovel usually has a ready-made working surface because it is loading from the ore or mineral surface.

The final choice must depend on circumstances but in most cases draglines are used, especially for the softer deeper overburdens. Face shovels are better where the cover contains competent beds such as limestone or sandstone, or the overburden is thinner.

Draglines are also used as auxiliary equipment located on the spoil bank, pulling back from the main machine so that the spoil bank is kept below about 16 to 17 m which is about the maximum stable height. This clearing action helps to level out the hill and dale construction of the spoil heap and is an aid to final restoration. When the overburden becomes deeper and a larger dragline must be bought the smaller one can be moved back on to the spoil heap.

Another use for draglines has been developed. They are occasionally used as skip hoists to lift mineral from the cut and avoid maintaining a haul road. Derrick cranes have also been used.

The bucket wheel excavator is not usually considered as an opponent of the shovel or dragline. Its particular field is the continuous excavation of soft bedded deposits. It is particularly good for mining these selectively as it has a precise control of horizon level.

There are occasions for mineral loading when a crawler dragline may be preferred to a power shovel. A crawler dragline has the inherent disadvantages of poor spotting ability for truck or hopper loading, a slower cycle time, and a less positive digging action. However the dragline can work from any horizon so a competent bed can be selected. Consequently for wet pit operations, such as sand, gravel, chalk, etc., a dragline is used. It can also recover remnants from pit floors and dig sumps and box cuts. If the mineral bed is very disturbed a shovel cannot be used on the resultant steep inclines and a dragline will operate from an artificial level surface to dig out the mineral.

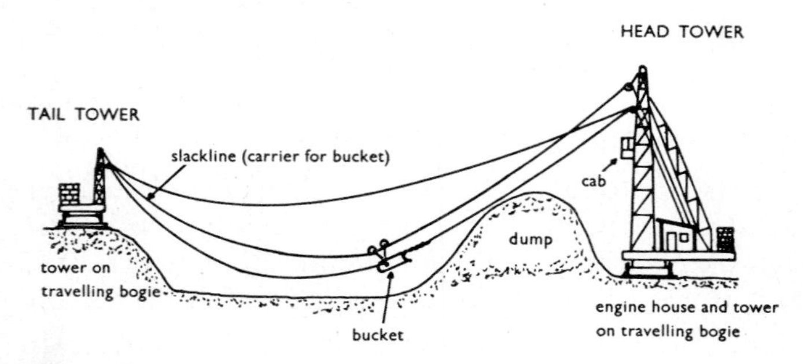

Fig. 5.12 Tower dragline

TOWER DRAGLINES

A tower dragline is sketched in Figure 5.12. Their use is now mainly restricted to sand and gravel operations where they are particularly useful for recovering material from river beds or lakes.

BUCKET WHEEL EXCAVATORS AND BUCKET CHAIN EXCAVATORS

These are often abbreviated to BWE and BCE. Their main use is on soft coals or overburdens, and on stockyard reclamation in bulk materials handling. They are essentially continuous feed machines and are designed to give a high output in soft materials. Both originally worked with rail truck loading but on new installations it is more usual to feed directly on to a conveyor. A typical installation is shown in Figure 5.13.

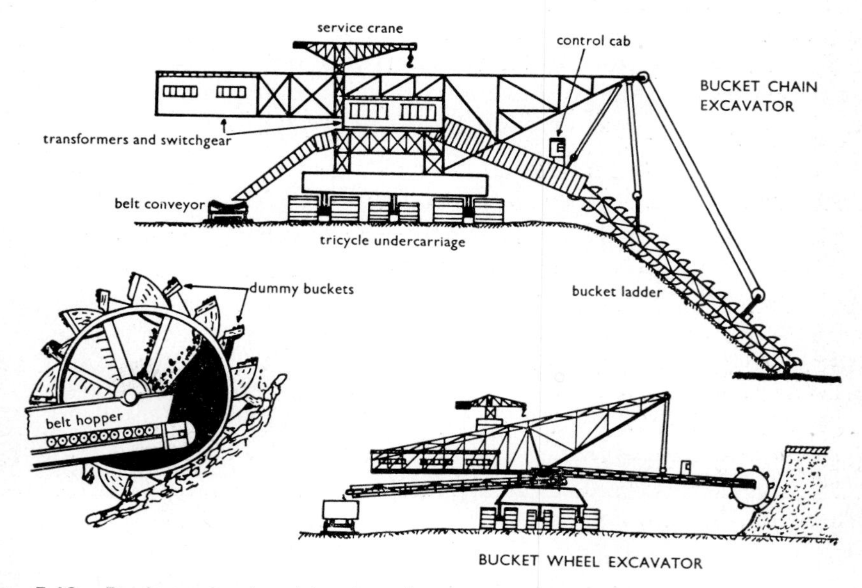

Fig. 5.13 Bucket wheel and bucket chain excavators

The BWE consists essentially of a large cutting wheel with buckets mounted on the periphery. Digging buckets open to the interior of the wheel but in harder ground alternate buckets are dummies, consisting of cutting teeth only with no bottoms. Their function is to even-out the cutting stresses on the periphery. The bucket wheels can be celled, with individual chutes to the transfer point supplying the boom conveyor, or half-celled, or cell-less. A cell-less wheel is better for sticky material. The brown coal at Yallourn and Moe (State Electricity Commission of Victoria) tends to stick badly and the S.E.C.V. have mounted plastic conveyor belt linings on to the wheel centre to improve clearance. Material from the bucket wheel interior slides onto the boom conveyor and is transferred onto a tail conveyor for main conveyor or rail truck loading. The BWE usually cuts material by taking sickle-shaped swings whilst its tail conveyor remains stationary. It then crawls forward and takes another series of dropping swings. Alternatively it can take dropping cuts from each position of the bucket wheel.

The BCE uses a chain of buckets instead of a cutting wheel and is essentially designed for digging down in soft uniform deposits. Its efficiency is greatly reduced if used above grade. The throughput of these machines in Victoria is up to 1750 t/hr, although machines with a throughput of 10 000 m^3 have been built.

Both BWE and BCE are limited to operating gradients of about 1 in 20 and travelling gradients of 1 in 10. They are large and cumbersome and even though fitted with tricycle crawlers are not very manoeuverable. Despite their size they have a low power consumption, partly because they work in softer materials and partly because they are slower moving and do not have to cope with high peak stresses. Atkinson[6] gives some approximate figures, these are reproduced in Table X.

TABLE X.

APPROXIMATE POWER CONSUMPTION OF DIGGING MACHINES

Machine	Power consumption kWh/m^3 excavated
Loading shovel	0.45–0.71
Opencast stripping shovel	0.52–0.91
Walking dragline	0.88–1.21
Bucket chain excavator	0.41–0.60
Bucket wheel excavator	0.30–0.50

Since the second world war the BWE has been gradually replacing the BCE. It has several big advantages: a BWE can mine selectively, taking minerals and steriles separately, and can mine thin beds. It can also dig relatively hard ground and tackle occasional boulders. When digging in the same conditions maintenance, and wear and tear, on the BWE is less than for the BCE. A bucket wheel excavator can also handle stickier materials without build-up in the buckets.

The main advantage of the BCE is its downward digging ability so that it is used for the bottom cut in wet pits, or for digging down from the travelling level of an overburden bridge. It can also be used on undulating mineral beds.

The last two continuous excavators bought by the State Electricity Commission of Victoria are both similar bucket wheel excavators, nos. 11 and 12. No. 11 was commissioned in 1970. No. 12 was ordered early in 1972 at a cost of $4 million for delivery in 1974. These excavators will dig the wet, woody brown coal at 1800 t/hr and the sandy overburden at 2300 t/hr, their service weight will be about 1600 tons. Further operational factors of BWEs and BCEs can be found in Atkinson's paper[6]. For electrical and mechanical design notes, and for working details, the paper by Rodgers and Bunting[4] on bucket wheel excavators is extremely useful.

FRONT-END LOADERS

A front-end loader is virtually a tractor with a loading bucket on the front (Figure 5.14). Some of the latest models have been fitted with a bucket at each end to improve their carrying capacity, but the rear bucket is restricted in height of lift. Front-end loaders are usually rubber-tyred, but can operate on crawler tracks. A new type of tyre has been produced with a flexible steel track fitted around a pneumatic tyre centre. This could be an improvement on a tyre fitted with protective chains.

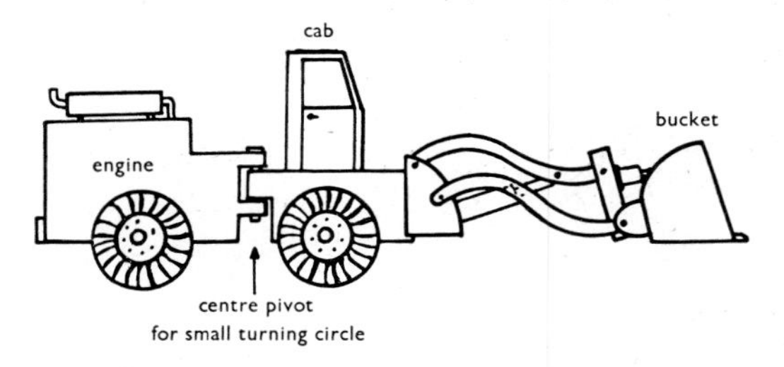

Fig. 5.14 Front-end loader

Bucket sizes on loaders are now up to 17-23 m^3. It is accepted as a rule-of-thumb that to replace a shovel on permanent loading duty a front-end loader should have about double the bucket capacity of the shovel to give equal output. A shovel swings to load but the front-end loader has to reverse out of the pile and turn before loading.

Loaders above 1.5-2 m^3 are generally articulated about a centre pivot and have four-wheel steer to keep shunting time to a minimum. On small operations, or with low benches, or thin coal seams, the mobility and flexibility of the loader may make it nearly equal to the shovel.

The front-end loader is popular for its versatility: it can do its own clean-up and grading, and because of its higher speed it can be used for blending from two different loading points. It doesn't have the ruggedness of a shovel, but it is useful as a spare loading unit, a trouble-shooter, and for stockpile dumping and reclaim duties.

CHOICE BETWEEN SHOVELS AND FRONT-END LOADERS

Many short papers have been published on the advantages of front-end loaders over shovels, particularly by the loader manufacturers. The shovel manufacturers have retaliated in kind, but it would appear from present practice that both machines are needed at open pit mines. The shovels are scheduled for normal production loading and front-end loaders are kept to help meet peak demands and as spare loading units to cover breakdowns and routine maintenance. Opencast coal mines and small multi-benched pits with variable grade ore may prefer front-end loaders to cut down on shovel travel time. However the newer hydraulic shovels in small bucket sizes have had their travelling speeds increased to compete with the loaders. In conditions of uncertainty (political or economic) the smaller amounts of capital invested in loaders is attractive.

For interest an article from each side of the fence is summarised below.

(1) By a loader supporter

FRONT-END LOADER *v.* SHOVEL (World Mining, September 1969). Follows a two-to-one bucket size ratio, assumes depreciation = cost/life, and ignores cost of extras or of equipment because of possible large variations.

(a) Cost of 4.6 m^3 shovel = \$300 000
9.2 m^3 loader = \$111 000

(b) Shovel life 20-30 years, loader life 5-6 years

(c) Shovel output 789 t/hr, loader 890 t/hr
(Note—both these figures appear to be theoretical values for easy digging conditions and 100 per cent efficiency. In harder rock the loader output would drop more than the shovel output. L.J.T.)

(d) Hourly costs including labour, fuel, etc., \$35 for shovel, \$29 for loader. Cost/tonne of material:—electric shovel 4.2c, loader 3.5c. Shovel also needs a share of a bulldozer for clean-up operations at \$12 per hour.
(Note—these costs presumably only allow a low cost for tyre wear and maintenance on the loaders and would probably apply to aggregate quarries or to soft coal, not hard rock work. L.J.T.)

(e) Easier to train a loader operator than a shovel operator.

(f) Shorter life of loader gives a chance to keep up with technological improvements.

(g) Not so much capital tied up with a loader as with a shovel.

(h) Front-end loader more versatile and can be used for odd jobs such as clearing stockpiles, road making and hauling.
(Note—the bucket type may have to be changed between loading mineral and clearing stockpiles. The teeth for the former job can cause deterioration of a processed stockpile. This reduces versatility. L.J.T.)

(i) Shovel life 40 000 hours, loader life 12 000 hours.
(These figures are conservative for a heavy duty shovel and optimistic for a loader under moderate conditions. L.J.T.)

(j) Shovel salvage value 15 per cent, loader 10 per cent.

(k) Repair parts and labour—shovel 90 per cent of depreciation, loader 60 per cent of depreciation plus tyres 40 per cent of depreciation.

Both loader and shovel were assumed to operate 2000 hours per year, i.e. one shift only. In fact most mine operators would operate at least 12 hours per day and even operate three full production shifts for six or seven days a week. The heavy duty shovel then shows to a greater advantage.

(2) By a shovel manufacturer (From a publicity pamphlet)

SHOVEL	LOADER
Cast dipper built for heavy rock work	Light welded plate bucket breaks up
High breakout force (crowds well) fills dipper better	Cannot crowd effectively, wheels slip. May need bulldozer to push
Digging smooth, operator more comfortable	Machine jerks. Operator shaken about
Can dig a down grade or up, and slope a bank	Works best on level, won't go down in hard ground or trim a high bank
Loads above grade, and loads over end of truck	Bucket too wide and insufficient reach for loading over end of a truck
Digs in close quarters. Can swing over rubble	Needs room to back and turn
Can blend off vertical height of face	Insufficient reach
Handles coarser materials	Needs more intensive shotfiring
Safe against a rock face	Operator may be too close to wall for safety

Crawlers built for rock work	Short tyre life, needs recaps and/or chains
Converts readily to other attachments (e.g. dragline, crane pile driver)	Only a loader/dozer
Longer productive life with better availability	Less than $\frac{1}{3}$ shovel life. May be obsolescent lowering resale value
Wider resale market, even 40 to 50% after 10 years	Narrower market, resale 30% (?) at 5 years
	May require dozer to push material to it Floor needs padding with fines

A final point in using shovels or loaders is that the rock or ore must be better fragmented and wider spread for good front-end loading. For shovel loading the rockpile is better when it is in a compact mass from a high bench as this avoids having to move the shovel. It is suggested that inclined hole blasting should be used to spread the rock for front-end loading. Vertical blast holes are still favoured for shovel loading in mines and quarries.

SCRAPERS

These are sometimes called tractor-scrapers or bowl scrapers. A sketch is shown in Figure 5.15. The bowl of the scraper fills by a planing action and is emptied by an internal blade pushing forward. Because of their loading action they can only be used on soft ground or ground which fragments well after ripping or blasting. In mining operations they are usually restricted to overburden stripping, although this can be done on a large scale. The Twin Buttes copper mine in Arizona, U.S.A. spent two years stripping alluvial overburden at 225 000 tonnes per day with 52 bowl scrapers before it started mining ore at about 30 000 tonnes per day.

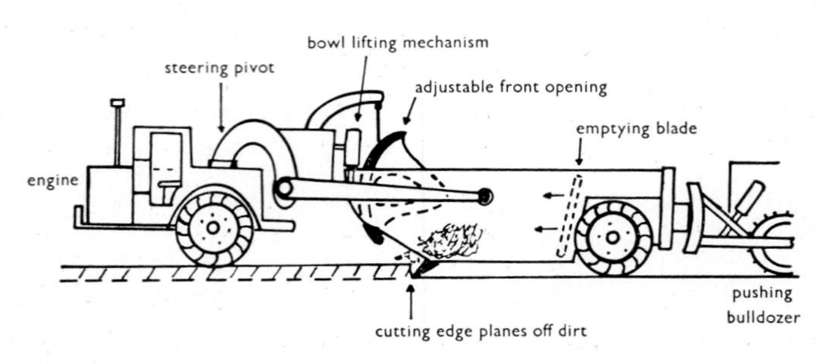

Fig. 5.15 Scraper

Many sizes of scraper are available, perhaps the most popular mining size is in the 23-38 m³ capacity range. Scrapers are not just excavators, they are load-haul-dump machines which complete a full cycle of rock removal. Because of the time they spend on travelling it would be uneconomical to provide an engine and transmission strong enough to break out hard ground. Consequently the usual practice is to provide sufficient power for haulage and to provide a pushing bulldozer for loading assistance. One bulldozer can assist several scrapers in turn in each loading group. When planning a stripping operation the effect of queuing delays on scraper time cycles must be considered.

There are three main types of scraper available. The *single engined* conventional scraper is the widest used but does not perform well on steep adverse grades, or with high rolling resistances, poor floor conditions, or on short haul distances. It has to be boosted up to travelling speed when loaded.

The *tandem-powered* machine has an engine mounted at each end of the scraper, one driving each pair of wheels. It can tackle wet and sticky conditions and work on uncompacted spoil heaps. It can partially self-load on the level or self-load fully on downhill runs, but it normally needs assistance from a pusher. A pair of tandem scrapers can give mutual assistance for push-loading each other and a bulldozer may not be needed.

Elevating scrapers are available that will self-load with flights. They are good for short or medium haul operations and do not need pushing. They cannot however handle sticky materials or boulders above 200 mm. The machine also will not handle adverse grades or high rolling resistances such as uncompacted or wet and sticky ground.

BULLDOZER-RIPPER COMBINATIONS

When soils and the weaker rocks become too hard for scraping, and before an operator resorts to blasting, he may use a ripper mounted on a bulldozer (Figure 5.16) to break up the rock. Ripping is generally confined to weak rock that will ultimately be loaded by a scraper or front-end loader, although the bulldozer carrying the ripper blade can push the broken ground to a shovel. Ripping may be extended beyond normal economic limits in residential areas where blasting would be unwelcome. It can also be used for cutting back rock in potential slide areas of an open pit to avoid a blast which could trigger off the slide.

To make a ripper one to three shanks, or tines, are mounted in a frame pivoted on the back of a large bulldozer. Downward pressure is provided by hydraulic rams. The most suitable rippers for mining operations are the radial type where depth of penetration is easily controlled. The ripper shanks swing in an arc and the mechanism is simple and robust. Most manufacturers will provide ample information on ripping equipment and techniques. Atkinson[7] in an earlier paper than that quoted previously covered the subject comprehensively.

Rippers can be conveniently mounted in any case on bulldozers used with scrapers. The extra cost is small and the bulldozers can rip tough spots whilst pushing the scraper. Before a decision is made to buy extra bulldozers specially for ripping the overburden should be carefully evaluated. It should be weak and blocky in any case.

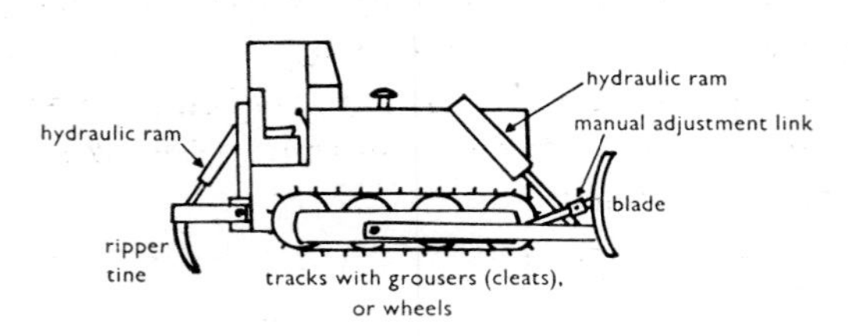

Fig. 5.16 Bulldozer-ripper

The most recently suggested criterion for a rapid evaluation is to use the velocity of sound propagation through the rock. Tests with a seismograph will show whether the ground is weak and blocky because it will have a lower sound velocity for a given rock than if the rock were solid. In general, rock with a sound velocity up to about 1800 m/sec are considered to be amenable to ripping with possible extensions on the heaviest equipment to rock of around 2000 m/sec sound velocity. Laminated rocks obviously rip more easily than massive rocks.

GRADERS

A grader is used for the final surfacing and repairing of haul roads. Unlike the bulldozer, whose blade cross-angle can only be varied manually, the blade of a grader can be varied in tilt and cross-angle whilst the machine is in motion. Wheel cambers can also be altered so that a road can be cross-graded for drainage, or a bank can be profiled. The grader can be fitted with a scarifier (like a rake) to break up the road surface in front of its blades, or can have ripper shanks fitted at the rear.

ON-ROAD TRUCKS

Most people are familiar with road haulage trucks. If they are used on sealed roads, particularly public ones, they need low axle loadings, hence they have a multiplicity of wheels. The smaller tyres used for on-road trucks do not generate as much heat from internal friction so that speeds of about 45-50 km/hr can be sustained. To keep the width of vehicle down and increase manoeuvrability a road truck consists of a prime-mover and one or more trailers as permitted by State legislation. A paper by Peters[8] gives details of operating conditions and costs for trucks used in the Tennant Creek area of Northern Territories, Australia.

These figures indicate that for truck haulage of several kilometres it is well worth while to seal the road surface and this is borne out by other operators. In practice a 6 m wide road is used with 4 m of seal offset to one side to favour loaded trucks. Unloaded trucks drive on to the hard shoulder to pass. Other experience indicates that in planning truck haulage it is best to work road vehicles for not more than one shift of 12 hours per day otherwise operating costs and maintenance charges appear to rise very steeply, probably through a breakdown of proper maintenance schemes.

Rough figures used as a guide for truck haulage costs prior to costing a specific installation are: cost of prime mover and two trailers for 70 tonne capacity, $110 000; hourly operating cost about $10, hourly ownership cost about $14 based on 2000 hours per year over 5 years.

Special problems may arise when operating a large truck fleet in proximity to suburban areas although there can be some compensations. All trucks must be inspected before leaving the site, loads should be covered with tarpaulins if dusty and must be struck off below the side-boards. Mud must be hosed off the truck to prevent fouling roads, stones should be prised out from between double tyres. Speed limits must be enforced and the company may have to employ roving inspectors to enforce company rules such as no overtaking another fleet lorry. If a mining company annoys local residents by obstructing or fouling roads there can be endless trouble. The company may also have to keep public roads in good repair. The only compensation is that the company may be able to build up an owner-contractor fleet from local people instead of buying all their own trucks.

OFF-ROAD TRUCKS

These have rigid bodies and are usually end-dumpers. For many years they only had four wheels which were of large diameter to cope with muddy conditions and loose rocks. Sizes stayed around 35-50 tonne capacity because this suited the 3.7-4.6 m^3 shovel normally used. Trucks for civil engineering construction are often quoted in body cubic capacity but for mining a large range of minerals of variable density it is better to quote load capacity in tonnes. A truck that could carry 40 m^3 of coal might well break down if loaded with 40 m^3 of iron ore.

When 11 m^3 shovels were introduced for large operations then it became feasible to use 100-120 tonne trucks to be loaded with them; these trucks tend to have two wheels in parallel at each corner. Also larger mines meant longer hauls so that although larger trucks travel more slowly it was worth developing even bigger trucks to increase tonne-kilometre rates. One hesitates to nominate a largest size of truck in normal use as it could be out of date by the time this book is printed, but 200 tonne trucks are still on probation.

These off-the-road trucks have to be enormously strong both in chassis and body to stand up to rough travel and loading. A 100 t capacity truck may have an empty weight of 75 t; its engine power would need to be about 750 kw. This kind of power would tear a conventional gear and shaft transmission apart. Trucks up to about 75 tonne are diesel engined with conventional transmission. Above this the diesel prime mover drives a d.c. generator. An electric motor is mounted in place of the normal axle and drives the wheel it is attached to (an 'electric wheel'). A 200 tonne truck capable of 20 000 tonne per day has a 1230 kw locomotive diesel engine and a traction motor for each of its four rear wheels. An even larger monster for a 230 tonne payload has eight wheels of which six are powered by electric motors driven from a 1940 kw 12 cylinder locomotive diesel engine. However the only real future of these large trucks appears to lie in the development of gas turbine drives to give a better power to weight ratio[9].

The electric wheel truck has led to the development of the trolley truck, e.g. by Le Tourneau. Power requirements usually limit operating grades of large trucks to about 1 in 12 but a new type of truck has been developed to pick up electric power from an overhead wire by trolley pole and boost the electric motors for a steep climb out of the pit at, say, 1 in 6.

Truck bodies (load trays, or boxes) need careful attention to minimise wear. Rubber linings may be economical and another useful idea has been to install transverse strips across the floor. These retain a layer of fines to cushion the impact of rocks dropping in and sliding out. Non-proprietary tailgates can be fitted to prevent spillage on haul roads. In wet or freezing climates the engine exhaust gases are vented through the channel sections of the truck body to keep it dried out and prevent fines sticking or freezing in to reduce capacity. Care must be taken to check that carbon build up in the channels does not increase back pressure on the engine.

Haul roads for trucks must be kept in good condition. They should have a proper road base and be of ample width for passing. Empty trucks always keep to the outside of a haul road on a bench to reduce the chance of a loaded truck breaking the lip off.

A haul road should not have a gradient greater than 1 in 9 even for a deep pit operating small trucks. Good surface maintenance and strict speed limits on down grades reduce truck maintenance and tyre costs. The gradients should be as smooth as possible to avoid frequent gear changing which is bad for both trucks and drivers. Haul roads need constant cleaning and regular watering to keep dust down. Even in wet areas, such as the west coast of Tasmania, after 3 to 4 hours sunshine the dust on an unwatered haul road is dense enough to reduce visibility to practically nil behind a truck.

Operating costs and methods of determining off-road truck requirements can be found in an article by Morgan[10].

TRAINS AND CONVEYOR BELTS

The flexibility of haulage trucks, and improvements in their design and reliability, has enabled them to largely replace trains and conveyor belts, although the latter may still be favoured for large scale operations. Bucket wheel and bucket chain excavators can only be fully utilised with a continuous transport system such as a conveyor belt.

Trains used to be the automatic choice for large pits mining low grade copper or coal. They would comprise rotary tip or bottom dump wagons of about 100 tonne capacity. The locomotives used to be steam but are now almost invariably a.c. trolley installations. The tracks were pushed over bodily from either a pivot point or in parallel advance. These installations were obviously best used in flat deposits of a large extent. The limiting grade was around 1 in 25 for short pitches to climb out of the pit, with about 1 in 80 as an operating grade. Natural topography could dictate the route out of the pit.

Improvements in conveyor belts have caused them to largely supersede trains as continuous transport. The belts can be loaded by BCE or BWE in the pit. Occasionally a stripping dragline may load through a surge bin hopper onto a conveyor belt to remove excess overburden from the cut. The spoil is taken around the pit to the spoil bank. This may be necessary with pitching seams which prevent direct back casting because the extra overburden from each slice would not fit into the previously mined cut.

Conveyor belts were first introduced on a large scale in 1952 and their application has increased ever since. The conveyor supports are mounted as modules on skids, in the same way as railway tracks are mounted on sleepers, to facilitate sideways advance. This has enabled large conveyor belts of up to 3 m wide, running at speeds up to 5.6 m/sec and with capacities of more than 25 000 tonnes/hr, to be used. Conveyor belts require straight faces, but grades can be up to 1 in 2.5. Preferably grades should be less than 1 in 15 to 1 in 20 to keep power requirements to a more economical level. Using a mobile crane or an adapted bulldozer a small crew of men can transfer a 300 m length of these large conveyor belts in to a new track in less than a shift. The belt is moved over intact by sliding it sideways.

Deep pits working metalliferous orebodies have found it an economic proposition to put a drop shaft into the bottom of the pit to below the ore body, install a primary crusher there, and transport the ore clear of the mine by a belt up an inclined drift (Figure 5.17).

This latter method can also be used with skip or rail haulage. Mount Lyell uses an underground trolley loco and drop bottom mine cars to transport the ore which has been hauled up to the rim of the pit, put through a crusher, and dropped into an underground tunnel for a level haul to the mill. The choice is obviously influenced by the general topography.

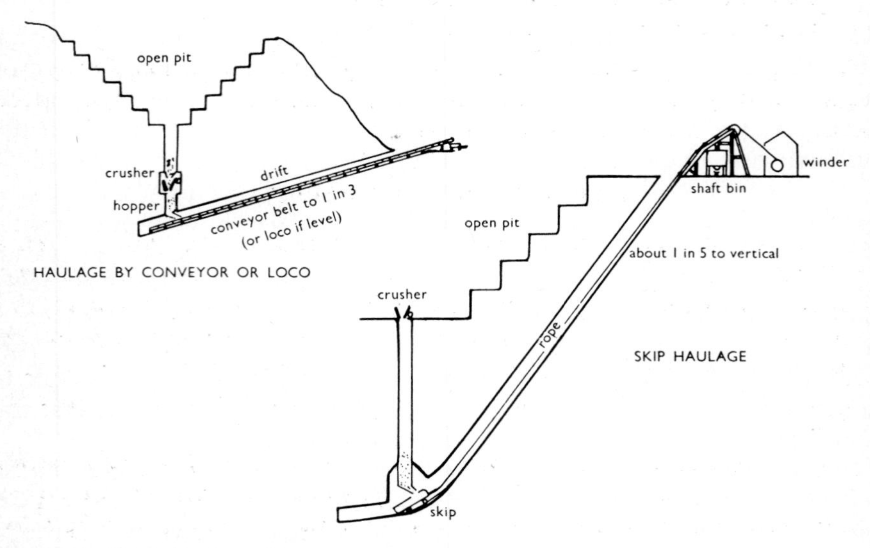

Fig. 5.17 Drift haulage from an open pit mine

ANCILLARY EQUIPMENT

The items of equipment described are major units costing hundreds of thousands of dollars. They must be serviced and properly maintained; electricity, water, and drainage for the mine have to be provided. Drilling machines of various sizes are needed and the largest of these, fitted with 12-26 m masts to drill holes up to 0.3 to 0.38 m diameter can be very expensive.

A list of auxiliary equipment will include items such as the following.

- Bulldozers
- Cable cars (self-motorised drums or mounted on lorries) for laying and reclaiming electric high voltage cables
- Graders and scrapers
- Small face shovels, front-end loaders, ditchers
- Lubrication trucks
- Fuel trucks
- Haul road water sprinkler trucks
- Mobile workshops with compressors, welding sets, etc.
- Service trucks, four-wheel drive vehicles, company officials' transport

Mobile cranes
Explosives trucks
Various size blasthole drills and lump breakers, all mounted on crawlers or rubber tyres
Pump units
Fire truck and ambulance
Two-way radio communication equipment
Explosives mixing plants and trucks.

In addition to these it is also necessary to provide offices; change rooms; maintenance workshops with major overhaul and repair bays; stock piling and load-out equipment; etc.

CHOICE OF EQUIPMENT

The theoretical methods of selecting equipment are beyond the scope of this book. A textbook by Antill[11] deals with the selection of earth moving equipment in terms of performance and use of queuing theory, and shows how Monte Carlo simulation can provide realistic estimates. It is limited in that it deals with small equipment used in civil engineering and it should be read in conjunction with the paper by Atkinson[6].

The easiest choices can be made when the company already owns a similar mine and can obtain realistic performance figures for equipment. Fortunately more material is now being published to help selection. For instance Price and Hortie[12] have published a detailed study of truck selection at a Canadian mine. Each item of cost such as tyres, engine maintenance, transmission maintenance and lubrication was studied in detail. Records were kept of all downtime. The data on engines and transmissions, etc., was used to forecast the likely performance of new trucks before a purchase was made.

Mt Morgan mine in Queensland has used a similar method to check on tyre costs. Careful records were kept of each tyre in use until the company was in a position to negotiate a contract with one supplier for tyres at a guaranteed cost per tonne of ore mined. Savage River Mine in Tasmania has bought its tyres for a guaranteed life in hours, but from a different tyre company. In this case the tyre company keeps a field service engineer on the site to do all tyre changing and maintenance. These are the obvious trends to adopt in equipment buying. It is much better to buy a guaranteed performance than a simple machine.

Another vital component of choice is delivery dates. If equipment is not delivered on time a multi-million dollar project may be held up for months, losing thousands of dollars in interest and wages to idle staff, for the sake of minor savings on an initial quote. Penalty clauses may be worthwhile in a delivery contract.

The essential part of choice of equipment is that an integrated system must be bought. It is becoming more common to buy a turnkey installation from a prime contractor but his quotations must be checked. All equipment must have matching sizes and performance, nothing should cause queues and bottlenecks. It could be superfluous to say that if one has a 1000 t/hr excavator one should also have a 1000 t/hr conveyor belt. But one also has to have a 1000 t/hr treatment plant and a 1000 t/hr disposal system out of the plant which in fact could have to be three 400 t/hr separate systems to cope with variations in quality.

Other items must be considered such as the choice of a crusher. Large iron ore mines have chosen gyratory crushers because a load of ore can be tipped straight in and the crusher will accept a choked feed, but the ore must be well fragmented.

If an ore is slabby and does not fragment well then a jaw crusher will have to be used because for the same throughput it will handle larger pieces, but it will not choke feed. Consequently an apron feeder or a surge bin arrangement will have to be added, or smaller trucks used.

Sizes of equipment must be carefully matched. The shovel size and truck size need to be related but also the lump size of broken mineral has to be planned. Lumps must not wedge in the shovel bucket or hang up in ore bins: chutes from ore bins must be matched to the lump size and blasting practices, and grizzly screens must cope with the feed.

However, overkill must be avoided. There is no point in making everything cope with peak demands. Three 400 tonne/hr shovels could feed indirectly onto a 1000 tonne/hr conveyor because the operator can rely on a diversity factor to smooth the flow. It is no good putting large equipment in a variable grade mine on grounds of economy. Smaller shovels and trucks must be used so that the ore can be blended to obtain the maximum selling price. It is true too that, as for exploration and all mining, if a small company is involved it will have to be satisfied with small equipment and sacrifice economy of operation for economy in capital.

It can be seen that correct planning is only possible after far more experience than this book can offer and mine design can only be done with the mutual cooperation of practising engineers and consultants. The diagrams and method descriptions following will give typical groupings of equipment but the mining engineer must be prepared to tailor methods to mineral deposits and change equipment layouts to suit. He should not be hidebound or follow fashions. When Costain Mining were faced with the removal of peaty overburden from their Westfield opencast coal site in Scotland they imported a suction dredge, flooded the site, and mined the overburden as if it was a mineral sand project.

Open Cast Mining of Bedded Deposits

GOOD NEIGHBOURS

Strip mining of coal or other bedded deposits uses the most spectacular machinery, although it does not make the deepest holes in the ground. As a result some strip mining operators in the past have disfigured large areas of the countryside. Modern operators have become more pollution conscious and go to great lengths to restore mined areas and even to improve the countryside.

There are areas where restoration is not yet compulsory but these are decreasing in number. It is as well to consider all mining in a manner which will not leave the countryside with a permanent scar. Utah Developments at Blackwater in Queensland has had to put up a bond of $20 per hectare to restore land worth only about $2 per hectare. It is estimated the final cost may be as much as $200 per hectare.

The first stage in opencast mining is to open up the ground to get at the coal or other mineral. This first slot is known as the *box cut* and is either made in the shallowest part of a pitching deposit, to mine to the dip, or for a thick deposit may be made near its centre, or at the highest part, so that a series of benches may be taken outwards. Figure 5.18 illustrates the terms used for benches.

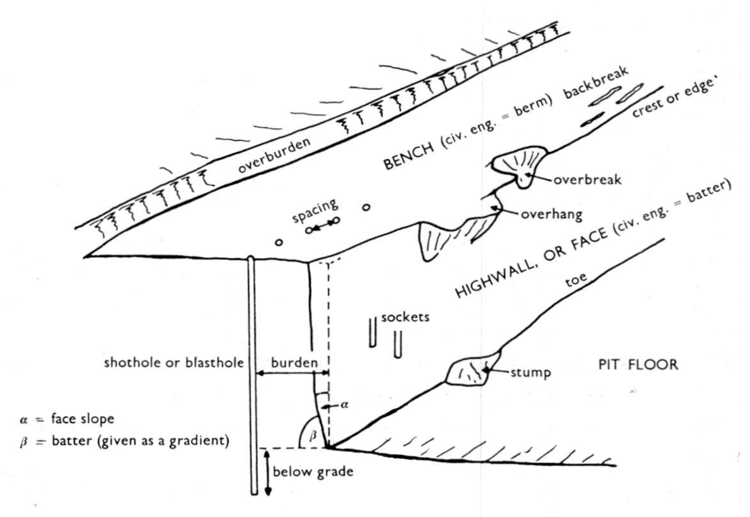

Fig. 5.18 Terms used in open pit mining or quarrying

It is obvious that the overburden from the box cut would be a nuisance if it were dumped on top of an area to be mined subsequently. For restoration the topsoil, subsoil and rock overburden must be removed separately for replacement in the correct order. It is suggested that rather than dump all this material in some random spot it should be used to make a protective embankment for the site, particularly if the mine is close to a main road or a residential area. An embankment a few metres high can be built around an appropriate part of the lease, perhaps terraced for safety or appearance; the topsoil is put on and the whole embankment grassed over. If the site is to be operative for more than (say) eight to ten years, then small saplings can be transplanted onto the embankment.

One must use a sense of proportion and not expect grass and trees to grow freely in arid climates. However, an embankment such as this makes a pleasant screen to what could otherwise be an ugly sight for passers-by. Nevertheless one must not forget the spectators' viewing platform for tourists. An embankment also reduces wind-blown dust nuisance and acts as a baffle to noise from the site. Trees and shrubs are particularly helpful in this respect.

In the previous section a comment was made on prevention of nuisance from trucks. If trucks are to be driven off a site on to the road, and are likely to be muddy, then a concrete wash pad should be laid down at the exit from the pit. Lorry wheels should be hosed down and stones must be removed from the wheels. A walkway should be erected so that the driver can trim his load. The relatively small amounts of money these things cost are more than repaid by good public relations.

A final point of being an efficient operator is that of blasting. Delay blasting of shotholes in sequence is common in any case, but due regard should be paid to the proximity of houses, even if they belong to the company. No workman will remain on a site if his wife continually complains of crockery jumping off the shelf each time the mine blasts. If a large round of shots is to be fired then the amount of explosive in each delay period must be controlled. The amount permitted is proportional to the distance and inversely proportional to the square root of the explosive weight per delay.

The Standards Association of Australia[13] permitted limit for the resultant of the three components of ground vibration is an amplitude of 0.2 mm at a frequency of 15 Hz. The U.S. permitted level is slightly higher and the U.S. Bureau of Mines recommends a scaled charge distance of $D_s = \frac{D}{\sqrt{E}}$

where D_s = scaled distance, D = distance from shothole to measuring point and E = explosive weight per delay.

It can be seen that 1 kg of explosive at a distance of 100 m would have the same effect as 100 kg of explosives fired at 1000 m. In the absence of tests the U.S.B.M. recommended a scaled distance of 50 (for ft and lb imperial units) which is equivalent to 22.6 for metres and kilograms. The scaled distance of 50 is related to a peak particle velocity of 2in/sec (i.e. 22.6 at 51 mm/sec) whereas the S.A.A. value of 0.008 in (0.2 mm) is related to a peak particle velocity of 0.75 in/sec (19 mm/sec).

MINING ADVANCE

After the box cut has been opened by cutting down through the overburden with bulldozers, draglines, or shovels and trucks, etc., the first cut of coal or other mineral is taken out. If it is soft enough it can be dug directly, otherwise it is drilled and blasted. A bench system is then set up and the deposit is progressively mined by advancing the benches horizontally. As space is created lower benches may be opened up in thick deposits. The volume of the box cut will depend on the scale of operations, the dip of the deposit and general economic circumstances.

For bedded deposits there are two principal methods of advance which can be used separately or combined. These are shown in Figure 5.19. The names parallel advance and rotating pit are self-explanatory.

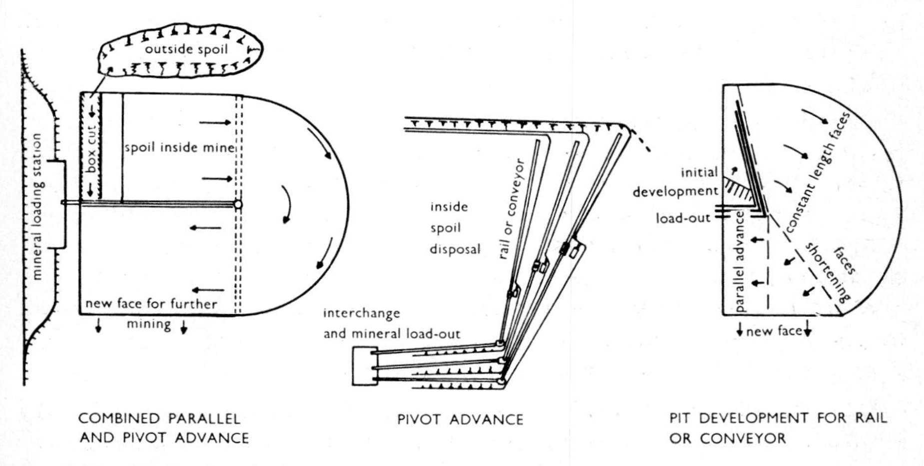

Fig. 5.19 Methods of advance in open pit mining

The main points of parallel advance are that:

(a) access to the pit is facilitated

(b) there are no slope angle problems or space restrictions from bench convergence

(c) transport of mineral out of the mine and overburden to the spoil bank requires equipment capable of being advanced at regular intervals. Trucks of course fulfil these requirements.

The features of a rotating pit are that:

(a) permanent transport arrangements are easily arranged at the pivot point for the mineral, for overburden to the spoil bank, and for interchange of mineral and steriles from the same bench
(b) the volume of material excavated for each block width is only half that obtained for the same block width in parallel advance operation, so that the number of conveyor, rail track or haul road advances is doubled
(c) average bench transport distances are increased by about one third.

Both of these methods are easier to work properly in large stratified deposits where bench lengths are measured in kilometres and face lives are timed in years. There may be some difficulties in keeping an even advance. If the overburden varies in thickness, as in hilly country, short term measures may have to be taken such as (1) create a curved face to cast the overburden on an outside curve to create additional spoil space, this is easier in contour mining; (2) temporarily reduce the pit width to the minimum possible; (3) pre-strip some of the overburden, possibly using outside contractors; (4) drop the horizon of the dragline to let it chop down in advance or run a stripping shovel partly on the spoil bank instead of wholly on the mineral bed.

METHODS OF WORKING

Shallow Deposits

The simplest case is where a single seam is being stripped under an even thickness of overburden, and the strata are fairly flat. In this case the seam can be mined by direct back casting as shown in Figure 5.20. The overburden can be stripped by either dragline or shovel. The decision should be made after a full feasibility study.

If a dragline is used the position will be as in diagram A. If the overburden needs drilling a mast-type rig mounted on crawlers will be used. The mast can be up to 26 m high and carries automatic drill rod changing equipment to drill holes up to 50 m deep with a minimum of rod changes. Blast holes of 0.15 to 0.38 m will probably be drilled on a grid spacing of about 7.5 m x 7.5 m to 15 m x 15 m and loaded with explosives. The spacing is wide because the ground is to be broken, but not displaced. The explosive normally used is ANFO (ammonium nitrate—fuel oil mix) in dry ground with an AN—C slurry waterproof mix for breaking out the toe and in wet holes. The amount of ANFO used will be about 0.1 to 0.7 kg per tonne of rock broken.

Overburden drilling is carried out from a surface that has been bulldozed clear, and the drill moves along pegged-out guidelines. A badly drilled hole will spoil the efficiency of the firing pattern and the dragline will have trouble digging the poorly broken rock. Drilling and blasting must be carried out in advance of dragline progress. The shattered surface of the overburden is bulldozed flat and the dragline works on top of the broken rock.

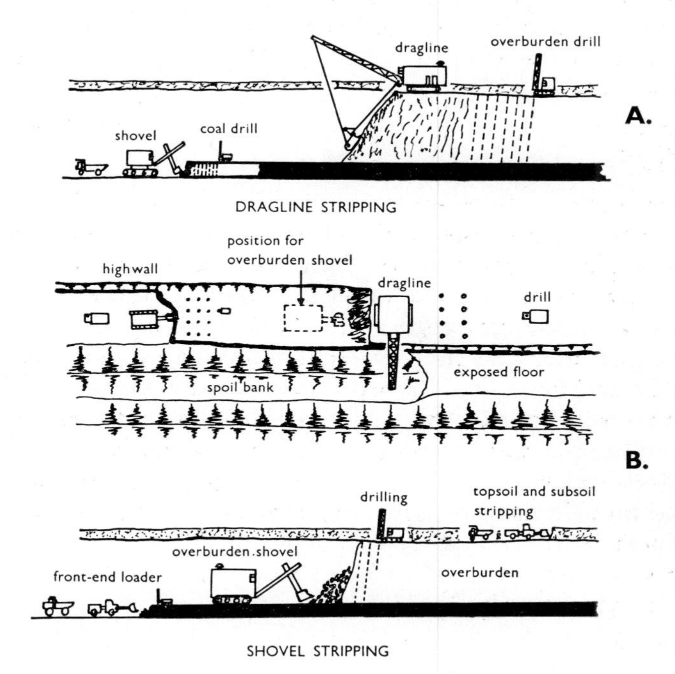

Fig. 5.20 Mining by direct back casting

If the rock is not broken by explosives the highwall should be carefully studied to determine its joint pattern, and to see whether major faults or planes of weakness will cause it to slide into the pit. The weight of the dragline could cause a slip. If the rock is blasted it is usually well drained and reasonably stable but in high rainfall areas, or bad climatic freeze and thaw conditions, the highwall may have to be specially drained or kept at a low angle to prevent collapse.

The overburden is stripped off by dragline, or by shovel as in Figure 5.20B; the excavator swings to dump the spoil into the emptied box cut or previous bench space. When the overburden is clear a smaller drill follows up to drill and blast the coal or mineral if necessary. The drill mast will be tall enough to drill the depth of the seam in one rod length, or a suitable bench interval of 10-15 m, whichever is less. Hole diameter will be about 75-125 mm in coal, with ANFO again being the main explosive used. Hole spacings will be closer than in the rock: the coal must be broken with light charges, not pulverised.

The broken mineral, or unbroken if soft enough, is loaded out into lorries, or possibly through a portable primary crusher onto a belt, by shovels or front-end loaders. The mineral is then sized, possibly with further breaking or crushing, and stockpiled for load-out to trains or beneficiation plant for treatment.

The width of cut opened out depends on the stability of the highwall and of the spoil heap. Highwalls can usually stand a gradient of about 3 in 1 (i.e. about 18° to 20° from the vertical). It is undesirable however to have equipment working against the highwall and deep pits may leave a 6-10 m strip of coal unmined against the highwall on each cut taken. The spoil heap will stand at the normal angle of repose of the loose material. This is about 30-35° for shales and 35-45° for limestones and sandstones; both measured from the horizontal.

There is thus an interaction between the maximum width of cut for a given dragline or shovel, and the size of dragline or shovel boom for a given cut design width. The design width needed depends on the mineral loader and transport system. Where a long boom dragline is used, as at Moura, it is possible to have a wide cut. A haul truck can be reversed in between a pair of shovels which load concurrently from opposite sides. The truck size can thus be doubled over the normal 4 to 1 ratio. When the overburden thickens the cut has to be narrowed and the loading shovel can swing 180° if necessary although that is inefficient. It is difficult to generalise but a coal strip of about 15-75 m would be taken at each cut worked, dependent on the size of the operation.

If there are highwall or spoil heap stability problems then direct back casting may not be used and the highwall will be stripped in advance by benching as for deep deposits.

Deep Deposits

When the mineral is buried below about 45-50 m it is impracticable to strip the overburden in one bench. Figures 5.21 and 5.22 show methods of stripping thick overburden. It has to be benched down and the shovel is the usual machine, operating on benches at about 15 m intervals. The overburden from the upper benches is sent around the pit by haul trucks or by a conveyor system. On a small pit operation trucks will be favoured, but in larger pits two or three shovels can load onto one conveyor for maximum economy. Lower level benches can transfer overburden directly across the cut with a bridge conveyor.

For pitching seams a dragline previously used on a single bench can be kept on the bottom bench for direct back casting. It can also operate for a period by chopping down the extra thickness provided that the spoil is back-handled by a bulldozer or other equipment. The smaller dragline can be transferred to the spoil heap for back casting and a larger dragline mounted on the bench. A study of a decision on equipment selection for thick overburden at Goonyella mine in Queensland is given by Simchuk[14].

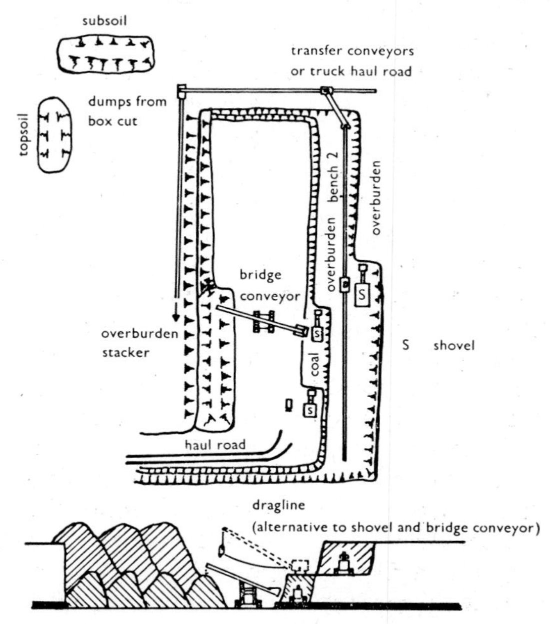

Fig. 5.21 Benching deep overburden

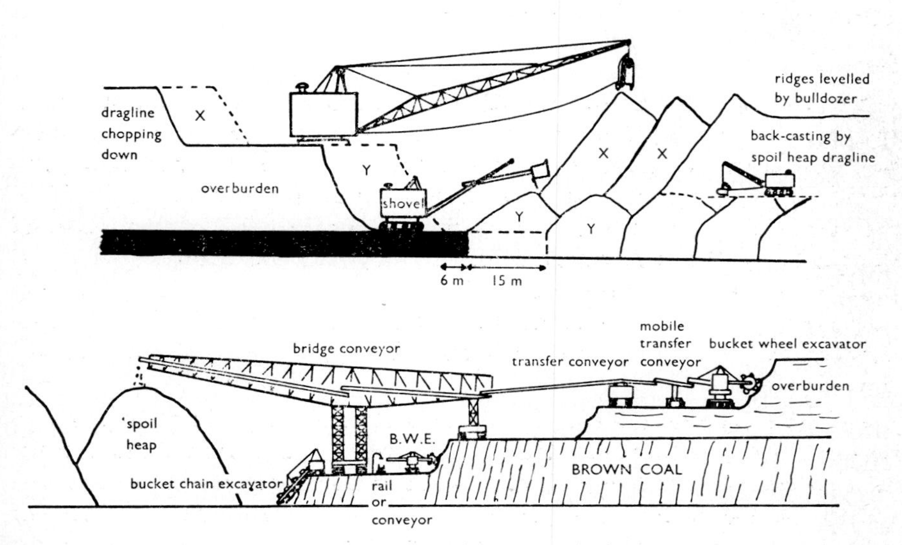

Fig. 5.22 Large volume overburden removal

Thick Deposits

If the deposit as well as the overburden is thick then special methods of mining can be adopted. The Victorian brown coal deposits at Yallourn and Moe are typical. At these deposits overburden is stripped by BWE and sent round the pit by conveyor belt. The coal can then be mined by BWE and BCE and sent by conveyor directly to the power station and briquetting plants.

European brown coal deposits are frequently mined with a system of transfer conveyors and bridge conveyors as in Figure 5.22 to take the overburden directly across the pit. The choice of method depends on the thickness of the deposits. A single conveyor bridge can span up to about 400-500 m and the total distance from the first overburden bench to the spoil heap can be around 700 m for a group of conveyors as shown. To justify these large machines a bench would be a few kilometres in length and the mine would have a life of several years. Bridging boom stackers of 150-200 m would be used in smaller operations.

Multiple Seams

Multiple seams are treated in the same manner as thick overburden. Each layer of overburden is benched down by shovel and truck to a workable coal seam. The seam is extracted and then the next layer of overburden is removed. A pit working several seams has to be carefully planned. The mine operator has to purchase equipment that will not be too large for the smallest seam, or too small for the largest seam. Luckily overburden and coal shovels and trucks can be interchanged with careful planning. That planning has to keep overburden stripping between each seam in sequence and to minimise movement of various sizes of equipment from one job to another.

CASE STUDIES

1. Maesgwyn Cap

George Wimpey as contractors are working an opencast site with four anthracite seams at Maesgwyn Cap in Wales, U.K. The maximum depth is 161 m and the overburden ratio including batters is 35 to 1, probably a world record for coal mining. The seam thicknesses are 0.91 m, 3.0 m, 0.76 m, and 2.1 m in descending order. The 100 m of overburden to the second seam is reduced in 12 m lifts by shovel and truck. Then a 30 m^3 dragline with a 75 m jib excavates the 10.7 m and 19.8 m of overburden between the bottom two seams by direct back casting. The equipment listed below is used to produce about 8000 to 10 000 tonnes of coal a week. The bottom three seams had previously been partially extracted by underground mining which causes some production problems. Total reserves were 6 million tonnes.

Maesgwyn Cap Equipment

Draglines	One Ransome W1800 electric walking dragline, bucket size 30.1 m^3, 75 m jib. Dumping height 30 m. One Marion 7400 electric dragline with 9.4 m^3 bucket and 53 m jib.
Shovels	Three BE190B electric, with 6.9 m^3 buckets. Two 150B electric, with 4.6 m^3 buckets. Two Lima diesel with 4.6 m^3 buckets plus several 1.1 m^3 coal loading shovels.
Dump trucks	21 Euclid 23 m^3 and two Aveling 17.6 m^3 for overburden only. Haul road is max. grade of 1 in 18. All trucks are rear dump.
Coal lorries	On-road type for haulage to washery. Sixteen AEC 6 or 8 wheel type, capacity 7.6 or 10.7 m^3.
Drills	Four Bucyrus 40R with 12 m capacity masts, one 76 mm blasthole drill, IRDM3.

It should be noted that this is a relatively small operation working in difficult conditions.

2. Moura Mines

A contrast to Maesgwyn is provided by Thiess Peabody Mitsui Coal Pty Ltd at Moura in Queensland. This mine has the second largest dragline in the world. The opencast workings produce about 3 100 000 tonnes of coal per year, and another 600 000 tonnes is produced from the associated underground mine.

Opencast reserves are about 35 000 000 tonnes and total reserves are more than 75 000 000 tonnes in an area of 470 km^2.

The overburden of shales and sandstone is stripped down to 48 m depth as a cut-off limit. The equipment includes:

Draglines	One Marion 8900 100 m^3 electric with 84 m boom. One Marion 7900 30.1 m^3 electric with 84 m boom. One BE 480W with a 9.2 m^3 bucket.
Shovels	Two Marion 151M with 10.7 m^3 coal loading buckets as main production units.
Front-end loaders	Two Cat 952 with 10.7 m^3 buckets for coal loading. The mine also has two Hough 400 units for stock-pile duties and a Cat 988 unit for loading and clean-up duties.
Trucks	14 Euclid bottom dump trucks, mainly 100 tonne units with some 50 tonne units. These are only used for coal; no overburden is moved with trucks.

Drills — Two rotary blast hole drills on crawlers for overburden drilling; both BE61R using 0.38 m bits.

There are five seams to be mined but because of their general dip and vertical interval they are not mineable together. Each seam worked is started with its own box cut, the deepest can be mined first. Overburden stripping is performed 24 hours a day, 7 days a week. The overburden is drilled with holes inclined at about 18° from the vertical to give a sloping highwall for stability. About 300 holes are put into an area 450 m x 55 m (on a 9 m x 9 m grid). ANFO explosives are used at about 0.7 kg/tonne.

The coal also has to be drilled and fired before loading out on two 10 hr shifts per day. Maximum haul distances are nearly 13 km to the dump where the coal is then fed through a rotary breaker. Coarse sizes are washed in a magnetite dense medium bath, small sizes are cleaned in a dense medium cyclone and the fines are cleaned by froth flotation. All coal is crushed to minus 32 mm and stockpiled for load-out. Transport is by 3000 tonne load trains to the port at Gladstone 177 km away.

Open Pit Mining of Massive Deposits

The basic operations of a large open pit are summarised in Figure 5.23. They indicate one of the major differences between coal and metalliferous deposits. Coal is fairly uniform in its chemical constitution and is mined for total extraction, apart from weathered outcrops and occasional barren patches from faults or rock inclusions. With metalliferous deposits the grade is often more variable and selective mining may be necessary to give a constant head grade. Areas of rich ore are blended with lower grade ore to meet the selling contract specifications with minimum waste.

BENCHING AND HAUL ROADS

A deposit has to be benched down in successive intervals and the way in which this is done depends on the topography of the deposit. Many metalliferous deposits start out as a hill or a low mound because they are more resistant to weathering than the host rock. Benching must then start near the crest of the hill. The deposit is progressively flattened out and mined downwards.

The way in which a pit is mined will depend on the shape and size of the orebody. Each bench could be taken to the periphery of the orebody before the successive benches are taken, or the pit can be benched downwards at a maximum rate and spread outwards just enough to provide mining room. An economic analysis has to be made of the optimum relationship. If the overburden is thin and the orebody is an inverted cone then the choice is fairly free. If the overburden is thick then the mine may have to be deepened rapidly to pay for overburden removal from revenue rather than capital although the working program must keep the stripping ratio fairly constant. Another important factor may be change of ore grade with depth. If a balanced ore grade is desired over the life of the mine it may be deepened immediately. If maximum earnings are needed in early years then the near surface of the orebody may be mined first if it is higher grade.

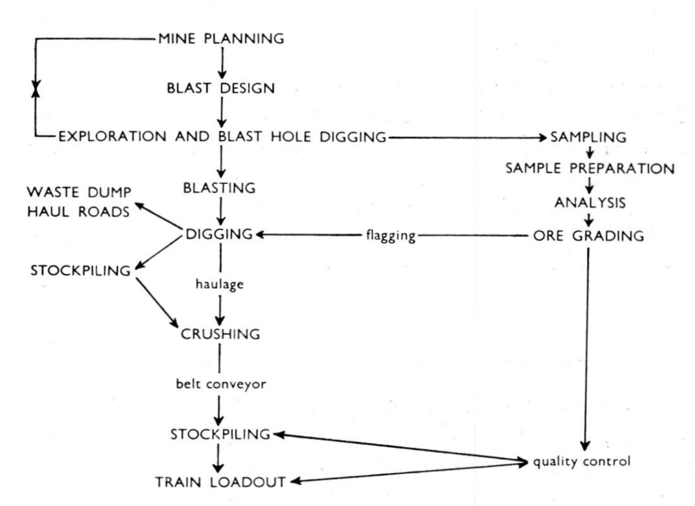

Fig. 5.23 Basic operations for a large open pit mine

For all the period that a pit is mined a haul road must be maintained into the pit. At shallow depths the haul road is straight but as the pit becomes deeper the haul road has to spiral or zig-zag down.

The choice of spiral or zig-zag depends on the shape and size of the pit and on its slope stability. In a small pit a spiral is the only practicable solution other than tunnel or shaft access. In a larger pit the haul road may be established as a zig-zag on a waste wall while the pit expands away from it. If a wall has to be cut back to a low angle to give stability on sloping rock beds, then it is often convenient to run the haul road up this slope to give flatter haulage gradients.

Figure 5.24 shows the general layout of the Mt Newman iron ore project[15] in 1969. The mine was developed by removing 3 million tonnes of overburden which consisted of low-grade ore and waste. Then the first three 15 m high benches were formed with just over 1.5 km of mining faces. The main haul road from the benches was about 3 km long and 18-21 m minimum width with a maximum downhill grade of 7 per cent (1 in 14.3). Loaded trucks keep to the right down the hill, that is on the side nearest the hill.

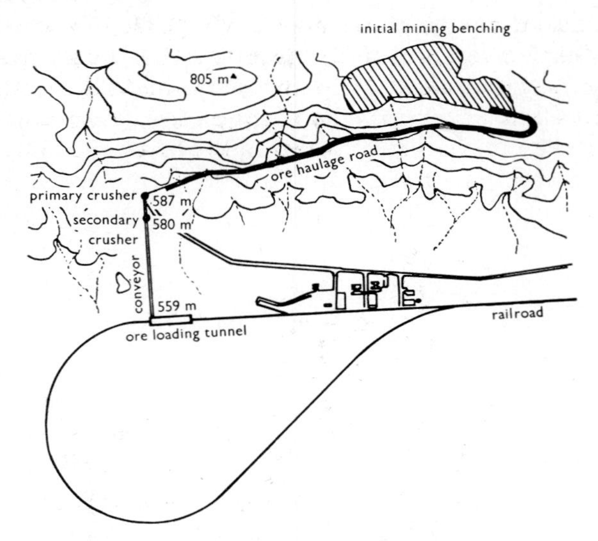

Fig. 5.24 General layout of Mt Newman iron ore mine

The bench layout and shovel positions for the pit in February 1971 are shown in Figure 5.25. It can be seen that some shovels have alternative loading positions to complete the production requirements. Connection from the various benches to the haul road is by a series of ramps. The separate exit for trucks hauling waste to the dump is also shown.

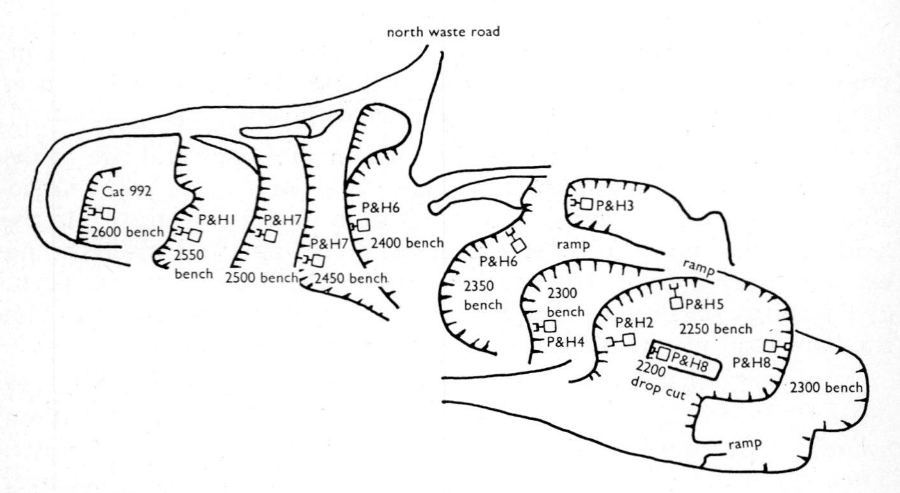

Fig. 5.25 Bench and shovel positions for Mt Newman in February 1971

As a pit becomes older then it may progress into a hole in the ground. Figure 5.26 shows the Mount Lyell open pit in Tasmania after 35 years of mining. It is about 115 m long by 61 m wide by 170 m deep with another 30 m depth of ore remaining. The haul road zig-zags up the South wall of the open cut, runs out to a crushing plant on the western rim, and then drops again to the mine buildings as in Figure 5.27. Ore and waste are brought out of the pit by truck.

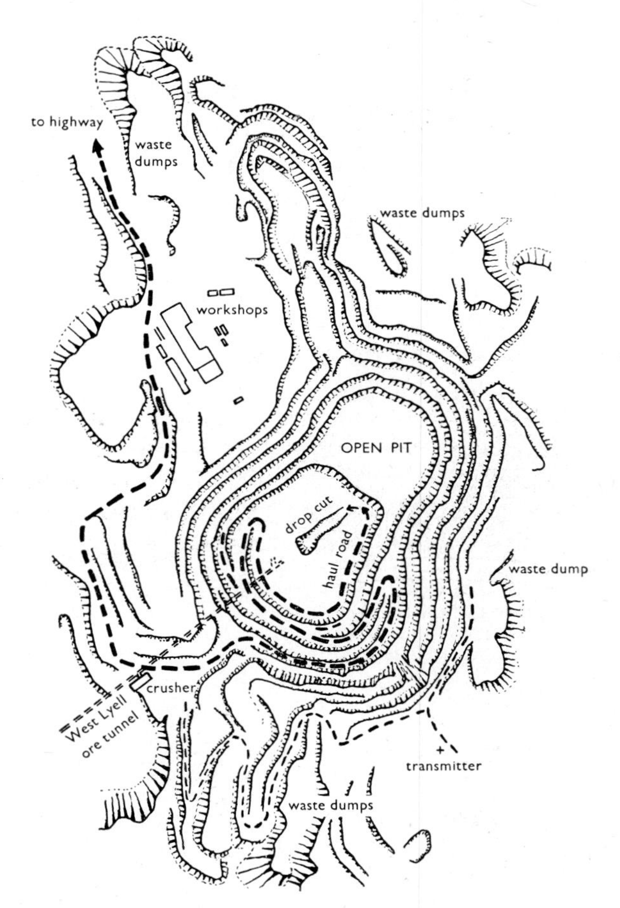

Fig. 5.26 Plan of Mount Lyell open pit

The waste is dumped on nearby hillsides and the copper ore is put through the crusher. It falls from there into a 24 000 tonne bin excavated in the rock and thence by transfer pass to the West Lyell Tunnel loading station for a horizontal haul by trolley locomotive and mine cars to the mill. The total fall from crusher to tunnel is 143 m.

In contrast, Mt Morgan open pit mine in Queensland is a similar size pit to Mt Lyell, but deeper, and uses a spiral haul road to bring ore and waste out of the mine. Their haul road is a uniform 1 in 8.36 and has been treated with crushed dyke screened to minus 26 mm. The Haulpak dump trucks have their sixth gear blanked off to limit the maximum speed to 40 km/hr.

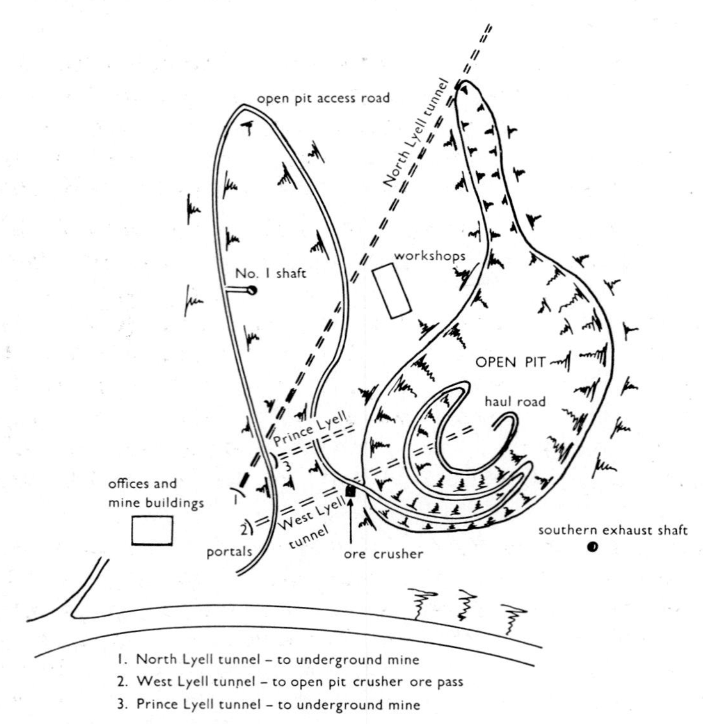

Fig. 5.27 General layout of Mount Lyell mine

PREPARATION AND MINING OF BENCHES

The first bench has to be started by levelling the ground, subsequent benches below it are of course started on the previously mined surface. If the initial bench is on a hillside then it will be cleaned off by bulldozer and levelled by using a small self-propelled drill and small scale blasting. If a wide bench is needed then the ground may be partly filled level with broken rock, but this is difficult to drill. Drill steels may stick and be broken. Care must be taken to recover them because the steels can damage both loaders and crushers.

The equipment used in multi-bench mining varies with the size of the benches. Draglines are rarely used, nor are scrapers after the initial removal of overburden. Loading shovels of 9-12 m³ would be usual for large scale mining with outputs of above 2 000 000 tonnes per year and where bench widths can be kept above about 20 m to accommodate the large shovels and the trucks that must go with them. Where bench widths have to be kept down to about 12-15 m and the consequent narrow, and probably steep, haul roads necessitate small trucks, then shovel sizes would be about 4-5 m^3. Front-end loader bucket sizes would be about the same as for shovels in metalliferous mines loading higher density ores.

Large iron ore mines use the same rotary blast hole drills as for coal mining, but with tougher bits, to drill holes of 0.2 to 0.4 m diameter in the benches. There are some exceptions, as in the tough and abrasive taconite iron ores of the Mesabi Range in the United States, where such exotic methods as flame penetration drills have been used.

In harder ores, or for smaller benches, a down-the-hole drill will generally be used to drill holes of up to 0.15 m diameter. When continuous downward benching in a mineral is necessary there is usually a lack of definite horizontal bedding planes to give a clean break when blasting. Blast holes have to be drilled below grade, that is up to 3.0 m deeper than the new bench horizon, to shatter the bottom of the old bench sufficiently for clean loading and a smooth new surface (see Figure 5.18). It is necessary to drill about 10 per cent of hole depth below grade in any case to allow for cuttings which fall back down the hole when the drill is removed.

On steep narrow benches it may be necessary to drill *toe* holes. These are near-horizontal holes drilled at the base of the bench to help break the ground. Toe-holes are not favoured as small pieces of rock which scale off the face of the bench can injure operators, or may cut explosive firing lines and cause a misfire. A book on drilling and blasting will give more details of these operations. Toe holes are generally smaller diameter, say less than 0.1 m. These smaller size drills are also used to drill large boulders left from the primary blasts so that they can be broken by secondary blasting.

Blast holes are generally charged with ANFO or slurry explosives. However in smaller diameter holes, or in hard or wet rock, then stronger semi-gelatinous or gelatinous explosives of nitroglycerine base type may be used, although nitroglycerine type explosives may be up to three times the cost of ANFO. The frequency of blasting may be from daily to weekly, with several working areas of the benches blasted at the same time. Generally, at least 2-3 days supply of broken ore will be kept blasted in advance.

Broken ore will be loaded out as for opencast mining of coal but operational planning may be more complicated, particularly in deep circular pits, to avoid broken ore or potentially dangerous blasted ore from blocking the haul road or threatening a lower working area. Ore and rock transport in modern pits is generally by trucks because of their flexibility.

ORE GRADE CONTROL

Some grade control information would have been obtained from the original exploration drilling, but this is not sufficiently detailed for good day-to-day planning. Grade control is usually by assay of blast hole cuttings. Each hole bored produces a neat pile of rock chippings in inverse order to the strata bored through. If the hole has penetrated banded deposits then the pile can be cut vertically in half to expose the sequence, and a channel sample is taken. Each band is separated for assay. If the vertical sequence is fairly uniform then the sampler will be satisfied with a shovelful from the centre of the pile. Experience of the relationship between hole sampling and final product sampling will show which approach is satisfactory, and may supply a correction factor for regular use.

Each hole is sampled and analysed separately. The hole is numbered and marked physically both on the bench and on a bench plan. The results of the analysis are also marked on the plan and each hole may be colour coded so that the ore graders can see blocks of approximately the same grade.

The grade of each block is averaged over the holes in the block by using the formula

$$\text{Blast grade} = G = \frac{\sum_{i=1}^{i=n} g_i}{n}$$

where g_i = grade of the hole i

n = number of holes in the pattern

This is sufficiently accurate for most grade control purposes. As a physical check a composite sample is made up from each of the hole samples and is analysed. All this is completed before the blast takes place and leaves two options. Each block of ore of a different grade could be blasted separately if necessary or, after the blast has taken place, the ore graders will mark each block out with coloured streamers.

Knowing the grade of each block the ore graders can direct where the material is to go; it has to be routed to the crushers, waste dumps, low grade stockpile or crusher backfill (used for haul-road construction, etc.). By careful recording and calculation the graders know the average grade of ore in the low grade stockpiles, and the grade of ore passed through the load-out point.

Feed grade for the crushers is calculated from the equation

$$\text{Feed grade} = \frac{\sum_{j=1}^{j=n} T_j G_j}{\sum_{j=1}^{j=n} T_j}$$

Where T_j = tonnes per shift from a given area, j.
G_j = grade of the area.

Assays of the treated ore output will be taken frequently (say at hourly intervals in large pits) to check on the grade calculations. However it must be remembered that this is an historical check and the effect of time delay of a half or whole shift on throughput must be allowed for. Stockpiling before load out, or the storage bin, should be designed so as to mix treated ore to even out minor grade fluctuations.

Ore grade control is obviously dependent on short term mine planning. A general practice is to have an updated monthly pit plan on the wall of the production office. A Perspex (polymethyl methacrylate) sheet is fastened over the plan to allow marking and erasing with coloured pencils without damaging the plan. The plan will be marked with such things as

(a) drill patterns, both planned and completed,
(b) position and grade of broken ore,
(c) shovel or loader positions to give required crusher feed grade and waste removal rates, and
(d) areas requiring dozing and grading.

To allow for breakdowns the grade controller should have written out a schedule of alternative loader positions. If a high or low grade loader is out of action for a significant period (say more than half a shift) another loader must be moved to medium grade ore as appropriate to balance the crusher feed. Short term breakdowns may be countered by using a front-end loader to replace the defective unit.

CASE STUDIES

All these case studies have been chosen to illustrate various points of open pit mining so that some are deliberately incomplete. There is no point in a tedious catalogue of equipment whose specifications can change every year.

1. Mt Newman, Western Australia

The plan for Mt Newman has already been given in Figures 5.24 and 5.25. In 1971 the mine was in the middle of an expansion program and had only been in operation for two years. It is mining haematite iron ore with an iron content of up to 69 per cent. Drilling is with BE 60R rotary drills. These have a 24 m mast and are used to drill 0.25 m diameter holes to a depth of 18 m to clear the toe of the 15 m high benches. Three drills were in use for an output of 14 000 000 tonnes per year and two more were being prepared for future use. Several percussion drills (IRDM3) were also used to drill 0.19 m diameter holes in the waste rock, and smaller drills were available for secondary blasting and development work.

Loading was with eight P & H 1900E electric shovels with 7.6 m^3 buckets into a fleet of sixteen 109 tonne (120 short ton) diesel Haulpak rear dump trucks. Cat 992 and 988 front-end loaders with 6.1 m^3 and 4.6 m^3 buckets were also available for loading duties. Bench clean up and development duties were performed by Cat D9 crawler bulldozers and Cat 824 rubber-tyred bulldozers.

The ore is crushed to minus 10 cm and stockpiled over a train load-out tunnel. Pneumatically-operated self-choking chutes fill a 150 car ore train with 14 000 tonnes of ore in about 2 hours for the 426 km haul to Port Hedland. Four trains a day were despatched in February 1971. Final crushing takes place at Port Hedland.

2. Hamersley Iron, Western Australia

This is another large haematite iron ore mine in the north-west of Western Australia. The output from the mine situated at Tom Price was about 22 million tonnes in 1970 and is scheduled to be increased to 30 million tonnes when markets are available. High grade lump output (–30 mm + 6 mm) is 63.8 per cent iron and the lower grade fines (–6 mm) are 62.0 per cent iron.

Production drilling is with three BE 60R rotary drills, as for Mt Newman, but uses a larger diameter 0.31 m tricone roller bit with a life of about 900 m in ore. About 120 m of hole is drilled in an eight hour shift. Normal hole pattern is a grid on 7.3 m x 7.3 m centres, which is reduced in harder ores. Holes are drilled 3 m below grade. Surveyors mark out the required hole depths on pegs at 30 m intervals and drill holes are checked for depth before blasting. The maximum tolerance is to within 1.5 m of the required depth. Smaller drills are available for development and secondary blasting.

Blasting is with an ammonium nitrate/sodium nitrate/aluminium powder/fine coal/sulphur slurry explosive. The mix is carried as an AN/SN solution and a bin of mixed solids on an explosives lorry and is blended as it is pumped into the blasthole. Holes may be topped off with ANFO. Single blasts of 200 000 tonnes of rock are reasonably common and the largest blast was about 1 200 000 tonnes. Blasting is usually on a daily basis, carried out during the midday crib (lunch) break.

The loading equipment is

4 diesel-electric Marion shovels 9.2 m^3
3 electric 6.6 kV Marion shovels 9.2 m^3
1 diesel-electric Marion shovel 4.6 m^3
1 diesel K.W. Dart front-end loader 7.7 m^3

The capacity of the 9.2 m^3 shovels is about 500 000 tonnes per month each so that there is spare loading capacity at present. All shovels are equipped with two-way radio communication equipment.

The haulage fleet is all rear dump trucks;

7 K.W. Dart, articulated, of 100 tonne capacity
38 K.W. Dart, unit chassis, of 100 tonne capacity
5 WABCO trucks (120 short tons) 109 tonne capacity

Average haul is about 1200 m and a truck carries about 14 loads in an 8 hour shift. Truck availability in dry weather (the usual climate) is about 80 per cent.

The ore is fed through the crushers and sized before stockpiling over the load-out tunnel. Trains haul the ore to Dampier for shipment. Hamersley Iron had to build all its own housing, rail and shipping facilities before it opened the mine. Water is pumped in to the township of Mt Tom Price from the Fortescue River about 113 km away. All electricity (about 23 000 kVA capacity) is generated at the mine for transmission at 33kV.

It may be of interest to note that summer daytime temperatures are frequently in excess of 38°C. The rain comes as a tropical monsoon with a typhoon season. At least the cab of all equipment is air conditioned, this both improves comfort and keeps dust away from workmen and control equipment. Haul roads must be continuously watered because evaporation rates from the surface are high.

3. Savage River Mines, Tasmania

This mine, with four P & H 4.6 m^3 shovels and fourteen 50 tonne Euclid trucks (plus three to allow for maintenance) to move 38 000 tonnes of material per day (50 per cent waste rock, 50 per cent magnetite iron ore) is worth a brief mention. It is in a high rainfall area (2.5 m per year) and pumps its ore through a pipeline 85 km to the coast.

The notable point is that the bench height is kept to 10.7 m instead of the more usual 15 m. This is claimed to have the advantages of better fragmentation of the relatively soft rock at the toe of the bench: broken material is less jagged and easier to clean up and tyre life has increased from 500 to 1500 hours. There is also no overhang on the crest of the bench face so that the safety of the operation is increased.

4. Mount Lyell, Tasmania

The mine is shown in Figures 5.26 and 5.27. West Lyell Open Pit consists of schist impregnated with chalcopyrite copper ore. About 30 per cent of the material mined is waste. Bench height is 13.7 m and the rock is broken with 0.15 to 0.23 m diameter holes drilled below grade. Broken rock is loaded with 3.8 m^3 electric shovels into 30 tonne Euclid rear dump diesel trucks. There is a standby 4.6 m^3 front-end loader. The trucks dump through a crusher into the underground tunnel mentioned previously.

5. Mt Morgan, Queensland

This is an open pit gold mine which also produces substantial amounts of copper. It plans for an output of 1 420 000 tonnes of ore a year with an overburden ratio of between 2 to 1 and 3 to 1. Its depth was about 260 m in 1971 with a probable final depth of about 330 m. To keep the stripping ratio to a minimum the overall slope batter is about 53°. Berms are about 12 m wide and 30 m high with a bench face angle of about 60°. These slopes are known to be above average in steepness and are carefully monitored, especially in the wet season. The mine is in the tropics and has a summer monsoon. Rainfall can be several centimetres within a day, and the bottom of the pit has flooded to a depth of a few metres in exceptional downpours of over 15 cm in 24 hours.

Drilling is with a selection of relatively small crawler mounted percussive drills and toe holes are frequently needed in the high narrow benches. The need is enhanced because of the presence of old underground workings which have caused breaks in the bench and spoil normal blasting techniques. Loading is with one 4.6 m^3 shovel and three 4.6 m^3 front-end loaders with another front-end loader on standby.

DRAINAGE

Drainage is a particularly vital point of most open pit mines. In wet climates pumping capacity to cope with the highest expected seasonal rainfall must be installed if the pit is a natural sump. The Mount Morgan pit is in a hot tropical area where rainfall evaporates readily so that pumps do not have to deal with steady high rainfall. In the wet areas of Tasmania an ore haulage tunnel from the bottom of a pit can double as a water drainage tunnel if the topography is suitable.

Areas with winter snow can have problems. If overburden mixed with snow is removed and stacked in winter then the snow will melt in the spring. The waste pile will be saturated and earth slides can be anticipated. A low angle and shallow depth for the winter rock pile is a necessity. Mention will be made later of dealing with contaminated water drainage.

GLORY HOLING

Glory holing is on the margin of open-pit and underground mining. It is a small scale operation that can be used for placer mining or civil engineering site excavation. Underground drives are put in below the area to be mined and then short raises are driven up into the ore body or excavation site. The raise is widened into a funnel by blasting and the broken ore falls down the raise for subsequent loading. Hopefully any overburden stays intact or does not fall sufficiently to dilute the ore. Glory holes are run into each other laterally to excavate an area. The general method is as for the base of the shrinkage stope in Figure 6.20. Heavy rainfall can be an enormous problem for glory holing, but the method is useful for frozen ground under a snow cover.

Abandoned Pits

HIGHWALL SEALING

When an opencast coal mine is abandoned there is a risk that the seam will catch on fire, because of spontaneous combustion. Consequently mining legislation may lay down what is good practice in any case. The outcrop of the seam must be sealed off with a thick bank of earth and rock. The pit floor may be bulldozed up against the seam but a more efficient method is to drill and blast the crest of the highwall so that it is thrown down into the pit. This both seals the seam outcrop and reduces the highwall slope to an angle at which it can be safely abandoned. Many underground fires (particularly in the United States) might never have started if this procedure had been observed. Once a fire starts to creep along underground it can be expensive and almost impossible to extinguish.

LEGISLATION AND RESTORATION OF PITS

The following paragraph is taken from the Eighth Schedule of the Coal Mines Regulation Act, No. 37, 1912, of New South Wales. It is typical of most acts.

> 'The owner, agent, contractor or manager of any open-cut working shall, unless otherwise directed by the Minister, cause to be removed separately for replacement as far as may be practicable, the top soil on such part of the area as may be disturbed, and such owner, agent, contractor or manager shall, as work progresses, cause the strata removed for the purpose of extracting a seam of coal or shale and all other residues, to be returned to the excavations made, or deposited on such sites, in such manner as the Minister may in writing direct. Such work shall be completed within six months of the giving of such direction. In all worked areas, the exposed coal seam shall be effectually covered with inert material to prevent fire

hazards and the sides of the cut shall be battered to a safe low angle and graded to slopes and contours consistent with the surrounding land, the top soil previously removed shall be replaced and all depressions effectually drained unless such depressions are, with the consent of the Minister, to be used for the purpose of ponding water or depositing other materials; such fillings, battering, grading and drainage shall be carried out to the satisfaction of the Minister.'

The Mining Act, No. 49, 1906 of New South Wales, which governs metalliferous mines, makes a similar statement and in addition may require the area to be replanted with shrubs and trees. It also requires proper drainage to prevent erosion and treatment of drainage water to prevent pollution.

This brings out a basic difference between the two types of deposits. With a bedded deposit the excavation can be filled in and contoured as the open cut progresses. With direct back casting the operation is simple. When the last cut is taken the material from the box cut is restored in its proper order and the surface made good. The swell of the broken ground over solid (bank) volume will usually compensate for the coal that has been removed.

A mined area can be used as pasture or farmland for a few years before it is used for building or other operations requiring stabilised ground. With fertilization and contour drainage the restored surface can be used to grow crops immediately and full fertility is restored within 2 or 3 years. Depressions may be left deliberately for reservoirs, stock ponds or recreational lakes. Shrub or tree planting may be adopted for reserves and amenity areas.

A large opencut mine may have to remain as a hole in the ground, but some mines have produced ambitious restoration plans. A natural terraced amphitheatre can be easily constructed by restoring the subsoil and top soil which should have been saved. The bottom of the pit can be a lake with picnic areas on the banks. The minimum requirement of safe slopes and fencing, if adopted close to an urban area, might well produce refusals for further mining leases.

DRAINAGE OF ABANDONED MINES

Most metallic ores have sulphides in association with them. Coal frequently has iron pyrites, FeS_2, in association with it. These sulphides will oxidise,

$$FeS_2 + 3O_2 \rightarrow FeSO_4 + SO_2$$

or, in the presence of water as in most mining environments,

$$2FeS_2 + 2H_2O + 7O_2 \rightarrow 2FeSO_4 + 2H_2SO_4$$

This reaction proceeds to

$$4FeSO_4 + 2H_2SO_4 + O_2 \rightarrow 2Fe_2(SO_4)_3 + 2H_2O$$

In an acid environment the ferric iron can be precipitated as the insoluble hydroxide. In any case one is frequently left with acid mine waters and unsightly yellow and brown iron staining.

If the natural drainage of a mined area runs into streams it can poison plant and animal life and discolour the stream and associated water for kilometres downstream. The same is true for water that runs through and leaches mine waste dump and tailings dam areas. These can produce high quantities of poisonous compounds of such elements as lead and arsenic.

Where possible the mine should be sealed against drainage. Any aquifers used for town water supplies must be carefully protected. If a mine has had shallow underground workings associated with it, then it may be impossible to seal. The water must then be carefully ponded and purified before it is realeased into local watercourses. Limestone treatment can reduce the acid content but chemical treatment is needed for other compounds. It may be necessary to permanently impound water from tailings dumps containing cyanide compounds, e.g. from gold extraction, and to specially divert surface drainage water away from these dumps.

Non-entry Mining

AUGER MINING

The main operation in auger mining is to bore large diameter holes into a coal seam from the limits of an open pit. This enables coal to be recovered from under the highwall without stripping further overburden. The method has been tried in Australia but has never been adopted as a routine operation. Mines in the United States have used it as a routine operation on a small scale, and the augers available are all American built.

Coal augers are similar to wood augers, but are larger. They have a cutting head and scroll rods to clear the coal. Diameters range from about 0.4 to 2.1 m and the flight lengths from which a string of rods is built up may be from 1.8 m to 12 m in length. Penetration can be up to 60 m if the seam is sufficiently flat and uninterrupted.

When a highwall is finished, and before it can weather and spall off, a contractor may move in and bore as many holes as possible with his own equipment. Hole spacing is such as to leave ribs between holes to prevent the highwall collapsing. Ribs will be 10 to 15 cm wide with larger pillars at intervals as necessary. Holes may be banked vertically in thick seams. In operation a path up to 30 m wide for the equipment to work on is cleared along the highwall or an outcrop. The auger may have one to three cutting heads; the coal is augered out and is loaded by a built-in conveyor, or else may be dumped for later loading. Outputs are relatively low and large operators are content to leave these methods to small contractors.

In the late 1940s some large machines were built to mine under remote control into the outcrops, towing a train of linked conveyors behind themselves. Experiments continued to the early 1960s but no information has been published recently on large machines. An article in Colliery Engineering, p. 87, February 1962 indicated an output of $6\frac{1}{2}$ tonnes/minute was possible. An underground continuous miner by the same firm (Joy) can produce 10 tonnes/minute from a less complicated and cheaper machine. It is now more efficient to construct a portal in the highwall and convert to underground mining.

An attempt at underground augering was made in Britain in the early 1960s with the Collins miner, but after extended trials it did not go into full production.

UNDERGROUND GASIFICATION

Extensive trials have been made in Britain, U.S.A., France, Belgium, Poland, Italy, and Russia to gasify poor quality or thin coal seams in situ. The Russians appear to have persisted the longest with trials but in no instance in any country has a clear economic case been made for gasification.

The principle is simple but applications are difficult. An inlet is needed to inject air, oxygen/air mix, steam, steam/air mix or steam/oxygen mix, all of which have been tried. There must be a gasification area and an outlet is necessary to draw off the gases. In some methods gas flow reversal was practised. In the gasification area there are three zones, the oxidation (and ash) zone, the reduction zone and the distillation zone where volatiles are driven off by heat in advance of the fire.

The main problem is to pick up the useful gases from the reduction and distillation zone without the gas flow short-circuiting through burnt out areas so that the fire extinguishes itself. Early experiments involved underground drivages which were expensive. Later experiments concentrated on boreholes in close proximity which were linked up by high or low pressure water flow to fracture or wash a path through the coal. Alternatively a path was burnt through more permeable coal. A series of boreholes was needed to continually extend the fire zone once it was started.

A comprehensive survey of the U.K. operations and of fundamental principles can be found in the articles[16, 17, 18] listed at the end of the chapter. The U.K. experiments were eventually abandoned.

IN-SITU LEACHING

In-situ leaching may be for metallic deposits what underground gasification is for coal; an elegant method with tremendous technical problems. The main problem is the same for both; it is not possible to process the mineral before it is attacked. Leaching should not be confused with solution mining which is described below.

Leaching involves the selective dissolution of a mineral from its gangue without disturbing the mineral deposit to any marked extent. The metallurgical processes (hydrometallurgy) for leaching ores after they have been processed on the surface are well developed. Technical information can be obtained from the extractive metallurgy books mentioned at the end of Chapter 3.

The surface operations known as *heap leaching*, in which large quantities of ore are treated by trickling leach solutions over them, are analogous to the desired underground operations. They are characterised by a low annual throughput and recoveries as low as 20 per cent. At the Rio Tinto mine in Spain it takes two years to exhaust the 1.25 per cent of copper from a 100 000 tonne heap of run-of-mine pyritic ore. If metallic ores are crushed to minus 6 mm then 10 000 tonnes may be tank leached in about 5 days. However very few underground deposits would be as broken or as porous as the crushed ore.

Two recent articles[19, 20] in Mining Magazine give the current approach to in-situ leaching. Both are extracts from papers presented at the A.I.M.E. Centennial meeting in New York in 1971.

The advantages claimed for in-situ leaching in favourable conditions are

1. minimum capital investment by mining standards
2. very short pre-production period (say 3 months)
3. minimum pollution of land, water and air
4. small labour force required
5. low safety hazard
6. suitability for inaccessible positions or very low grade ores
7. negligible waste disposal costs.

All the disadvantages are of course embodied in the unfavourable conditions. The two main ones are low porosity mineral ores and high porosity host rocks. The former will not allow the leach solutions to penetrate the ore and the latter allow *pregnant solutions* (i.e. mineral-containing solutions) to escape before they can be collected. There may also be a danger of polluting potable aquifers used for community water supplies.

The method depends on flow through rock and is governed by Darcy's Law which can be expressed as

$$\text{quantity flowing} = q = \frac{KA\,(P_1 - P_2)}{\mu L}$$

where K = a constant for the rock type which defines the permeability of the porous medium,

A = cross-sectional area through which the fluid is flowing,

μ = viscosity of the fluid, and

L = length over which the pressure drop $(P_1 - P_2)$ is measured.

The permeability, k, is normally expressed in millidarcys, where one darcy is equal to a flow of 1 cm^3/sec cm^2 for a pressure gradient of 1 bar/cm. Permeability and porosity are to some extent related, although a porous rock may not have continuous pores, only a high void content, and therefore be impermeable.

Limestones, shales and siltstones may have permeabilities from one millidarcy to less than 0.05 millidarcys. Sandstones are more likely to have high permeabilities which may exceed 200 millidarcys. The permeability needed for leaching is probably in excess of 5 millidarcys. It would be preferable to have a finely disseminated ore in a highly porous orebody to aid the leaching rate. Finally the host rock should either be a natural water trap to avoid solution loss, or the formation needs to be naturally flooded so that draw-down points can be established by pumping to establish controlled circulation. However groundwater flow rates through the area must be low. This combination of circumstances is likely to be rare. The geophysical and rock property surveys needed to establish a scheme may be expensive. The methods needed for surveys can be found in standard textbooks on oil-well technology.

A caved deposit of low grade copper ore in Ohio, U.S.A. has been leached in situ, but the rock was broken by previous mining processes. In situ leaching of an undisturbed uranium orebody in a sandstone bed is being tried by the Anaconda Copper Company[19] in U.S.A. and the Kennecott Copper Company are experimenting with a small porphyry copper orebody[20].

Apart from the straightforward operation of recovering naturally leached minerals from acid mine drainage water it is likely that full scale leaching will be confined to broken ore on the periphery of abandoned mines. Techniques are available for increasing the permeability of rock by hydraulic fracturing or with conventional explosives, but this is expensive. Studies on the peaceful use of nuclear explosives reported in World Mining[21] and by the U.S. Bureau of Mines[22] have suggested that an underground nuclear explosion could be used to fracture a large volume of low grade ores for subsequent leaching. Unless a government is prepared to underwrite the uncertain cost, and the uncertain environmental risks, and the almost certain opposition from sections of the community, it is unlikely that a mining company would be found to risk the technology involved.

SOLUTION MINING

Despite its name this operation does not really fall within the province of a mining engineer. Both rock salt and sulphur are recovered in solution from underground deposits. The total recovery of thick deposits leaves large caverns underground which can collapse and cause surface subsidence with consequent damage. This is dealt with in Chapter 11 and can be the concern of a mining or civil engineer.

America developed the *Frasch process* in which superheated water or steam is pumped down into the sulphur-bearing strata to melt the sulphur. An air lift is then used to raise the molten sulphur and excess water to the surface as a foam. The air lift is produced by forcing compressed air down the central pipe of an annular column. Expansion of the air, and the air content of the sulphur foam, make a low density vertical column which is displaced to the surface by the superheated water column. This solution method has the advantage of producing a relatively pure product. Modern methods of recovering sulphur from sulphur dioxide flue gases in metallurgical sulphide roasting is gradually making the Frasch process uneconomic.

Rock salt can be solution mined by dissolving the salt beds (a mixture of the various halides or potassium and sodium compounds) and pumping the resultant brines to the surface. Under many conditions selective mining of underground deposits (see Chapter 7 for methods) is more economic.

References

1. Griffith, S. V. *Alluvial Prospecting and Mining* (2nd ed.) Pergamon Press, London, 1960.
2. Anon. *Ocean Mining Symposium, 1967.* M. J. Richardson, California, 1968.
3. Macdonald, E. H. *Manual of beach mining practice: exploration and evaluation.* Dept. of External Affairs, Canberra, 1968.
4. Rodgers, H. C. G. and Bunting, H. R. *Recent developments and applications of bucket wheel excavators in Australia.* Proc. Australasian Institute of Mining and Metallurgy, No. 238, p. 85, June 1971.
5. Brealey, S. C. and Atkinson, T. *Opencast Mining.* The Mining Engineer, No. 99, p. 147, December 1968.
6. Atkinson, T. *Selection of open-pit excavating and loading equipment.* Institute of Mining and Metallurgy, Trans. A. **80**, No. 776, pp. A101-29, July 1971.
7. Atkinson, T. *Ground preparation by ripping in open pit mining.* Mining Magazine, **122**, p. 458, June 1970.
8. Peters, T. N. *Road haulage of ore by Peko Mines N.L.* Trans. Aust. I.M.M., No. 219, p. 11, September 1966.
9. Kress, R. H. *What big trucks need to grow on.* Mining Engineering, **23**, p. 40, May 1971.
10. Morgan, B. V. *Complex factors affect haulage costs* ('Off-Highway Haulage', Part 3). Rock Products, p. 59, August 1970.
11. Antill, J. M. *Civil Engineering Management.* Angus and Robertson, Sydney, 1970.
12. Price, L. and Hortie, R. *Truck selection and application at Hilton.* Canadian Mining and Metallurgy Bulletin, p. 65, July 1971.

13. Standards Association of Australia *A.S. CA 23-1967, S.A.A. Explosives Code*. Sydney, 1967.

14. Simchuk, G. *Selection procedures at Goonyella*. Coal Age, p. 88, June 1972.

15. Anon. *Mt Newman*. Mining Magazine, **121**, pp. 4-19, July 1969.

16. Lloyd Jones, C. R. *Underground gasification*. Trans. Inst. Mining Engineers, **109**, p. 1025, 1949-50.

17. Plane, A. S. *Engineering aspects of underground gasification*. Trans. Inst. Mining Engineers, **111**, p. 138, 1951-52.

18. Masterman, C. A. *Recent developments in underground gasification*. Trans. Inst. Mining Engineers, **117**, p. 742, 1957-58.

19. Anon. *In-situ leaching: a review*. Mining Magazine, **125**, p. 213, September 1971.

20. Anon. *In-situ leaching: pilot plant tests on copper extraction*. Mining Magazine, **125**, p. 27, July 1971.

21. Hansen, S. M. and Rabb, D. D. *Seek profitability answer to nuclear in-situ copper leaching at Safford*. World Mining, p. 48, January 1968.

22. Hardwick, W. R. *Fracturing a deposit with nuclear explosives and recovering copper by the in-situ leaching method*. U.S. Bureau of Mines, Report of Investigations R.I. 6996, 1967.

Chapter 6. Underground Metalliferous Mining

This chapter will deal with orebodies that have significant vertical dimensions. It is essentially three-dimensional mining in which lateral development roadways are driven at successive depths (see Figures 1.3 and 1.5) and connected vertically. Mining then proceeds upwards in *overhand stoping*, or downwards in *underhand stoping*. Steeply inclined thin orebodies (veins) or massive bodies are worked on these general methods.

A change in some thin bodies from working along the strike to working down dip is made when the footwall dips at less than approximately 45°. At about this angle the ore will no longer flow under gravity and has to be moved mechanically. As an orebody becomes substantially flat and is less than a few metres thick then it may be worked as a bedded deposit as described in Chapter 7. *Breast stoping* is a term sometimes used for the room and pillar methods of mining in Chapter 7, or for slot mining.

Only general methods of mining will be described. Current technical journals publish many interesting variations that are devised to meet local conditions. Techniques have been basically similar for many years but improvements in machinery, and a chronic shortage of mining personnel, have made intensive mechanisation a pre-requisite of most mining operations. This has changed the emphasis and application of techniques, and has altered many traditionally held viewpoints of the relative advantages and disadvantages of methods.

As the older generation of skilled craftsmen retire, and the richer remnants of ore disappear, it is likely that some methods such as square set stoping and top slicing may have to disappear with them. Small mines with a handful of men and low overheads may still find them profitable.

Underground Machinery

LOAD-HAUL-DUMP MACHINES (LHD)

The load-haul-dump machine, as its name implies, is capable of picking up ore and carrying it for some distance, and then dumping it. The machine is often abbreviated to LHD, or called a *scooptram* (which is in fact a trade name for one variety). The LHD is to modern underground mining what the front-end loader or electric shovel is to open pit mining. It is virtually an essential piece of equipment for large scale production.

There are three varieties of LHD. One is diesel operated and is similar to the surface front-end loader, but it has a lower profile so that it can operate in a 3 m high roadway (Figure 6.1). It is rated on bucket size which can range from 0.38 m^3 to 6.1 m^3. The 3.8 m^3 size is popular in cut and fill stopes. It can travel freely in a 4 m wide roadway with about 5 m width on sharp bends, but it needs a working width in stopes of about 6 m to facilitate manoeuvring for clean-up duties. When it is not loading rock or ore it can serve as a transport vehicle, or a working platform for roof bolting, or it can be used for erecting service pipes or ventilation duct. Its mobility enables it to travel rapidly between several working places in one shift.

In addition to the simple bucket machines, which may now be fitted with a turn-over bucket, a bottom-emptying bucket, or a blade-ejector bucket, there is the diesel powered transloader type of LHD. In this the bucket is used to fill a built-in hopper to increase the capacity of the machine. Whereas the single bucket LHD has an economic one-way haul distance of 150 to 200 m the combination loader-hopper type may have an economic haul of 300 m or more. LHD machines are manufactured in most countries, or are readily imported. The two makes most frequently used in Australia appear to be the Wagner and Eimco with the Wagner ST5 and ST5A and the Eimco 915 and 916 in the forefront. All four have 3.8 m^3 buckets.

Diesel powered machines which operate in a confined environment underground must have their exhaust gases purified. The usual method in Australia (at present) is to pass the gases through a water scrubber to remove carbon monoxide, oxides of nitrogens, and aldehydes, all of which are harmful to health. An alternative method is to use a combination silencer/purifier (muffler) which comprises a large tank filled with ceramic pellets coated with a platinum or platinum-palladium catalyst to chemically de-activate the exhaust gases. In any case a large volume of air is also necessary to dilute the exhaust gases. Government legislation lays down minimum quantities varying from 0.034-0.044 m^3/sec per kilowatt of engine power at the machine or about 0.095 m^3/sec per kW at the surface to allow for air losses in the mine.

Fig. 6.1 Load-haul-dump machine

Maximum concentrations of toxic exhaust impurities are also laid down. In practice a 3.8 m^3 LHD machine has an engine approaching 150 kW so each machine in a stope or roadway needs 6 m^3/sec of air. Rough design figures of 25 m^3/sec can be adopted for a stope and an auxiliary fan can be used to give a local boost past the machine operating area. Where LHD machines operate in dusty areas it is possible to mount a cylinder of compressed air on them and let the driver breathe from it through a face mask. Research has also been carried out into air vortex cooling jackets.

For older metal mines the use of diesel engines could mean a radical increase in total quantities of air to be forced through the mine, so the third variety of LHD machine is compressed air operated, such as the Atlas Copco Autoloader range. These are captive on a 75-100 mm bore air hose but can still tram for about 150-200 m. The hoses are not as vulnerable to crushing as are electric cables and are left lying on the road surface in the tramming area. Autoloaders are of the bucket-loading hopper type and discharge by rear dump. The machines are rugged and the air motors are simpler to maintain than are diesel engines although more expensive in 'fuel'. Their disadvantage is that of limited operating range on a hose length. Auxiliary transport or hose changing is necessary to move between operating areas.

UNDERGROUND DUMP TRUCKS

An LHD machine has a limited economic haul and its operating range is extended with the use of haulage trucks. These are diesel operated with scrubbed exhausts, and like the LHD machines are built with a low profile. They are not intended to work in the mud and slush of an open pit and can have a low ground clearance.

An LHD loading bucket has a normal loading height of 2-3 m so the dump truck side must be around 2-3 m in height. Dump trucks can be articulated for operation in stopes and levels or may be rigid frame if they are used for haulage from a draw-point up an incline to the surface. There may be special applications; Isa Mine filled deep level development rock by LHD machine into a dump truck. The dump truck was driven to the mine cage and wound to the surface, and was then driven to the rock dump for emptying. The cage and truck size were carefully matched. A more conventional approach would have used mine cars.

Underground dump trucks at present range up to about 21 m^3 capacity, 45 tonnes load. They may be bottom dump, end discharge or end tippers. For some special applications surface dump trucks may be used with de-rated engines and scrubbed exhausts. Alternatively underground dump trucks can be used for surface operations where their lower loading height, and better payload/dead weight ratio may be advantageous.

OTHER LOADING METHODS

LHD machines, especially of the diesel variety, are a relatively recent innovation. Ore can also be loaded by rocker shovel, gathering-arm loader, or scrapers. The hand shovel tends to be a last resort in modern mining. A *rocker shovel* is shown in Figure 6.2. It is crowded into the broken ore on its wheels for filling, and for emptying the bucket is raised rapidly to eject the material over the top of the loader; the bucket and body will slew on the chassis and are self-centring as the bucket rises. The rocker shovel can fill into a separate mine car or into a hopper. This is also the loading principle of the compressed air LHD machines.

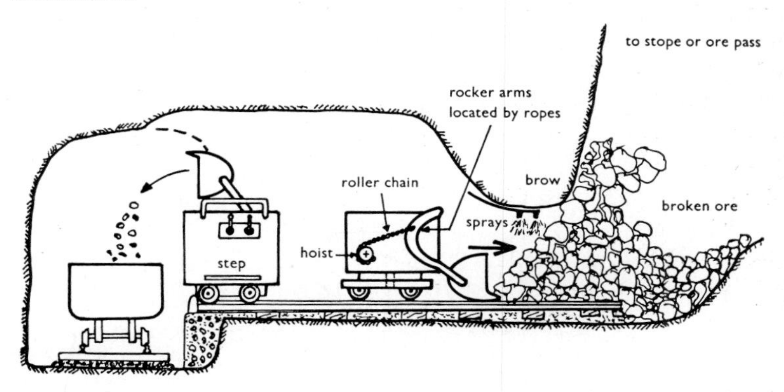

Fig. 6.2 Rocker shovel draw point

The *gathering-arm loader* (Figure 6.3) is a machine that has spread from coal mining to metal mining. Its gathering arms and flight conveyor provide a more continuous loading action than a rocker shovel.

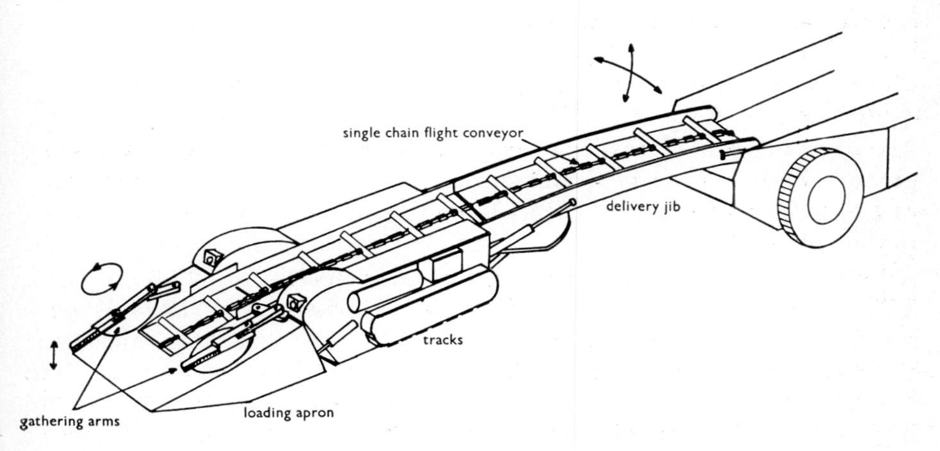

Fig. 6.3 Gathering-arm loader

The delivery of the conveyor can swing from side to side and distribute the load more evenly. They have been used in particular with underground dump trucks. (See also the description of continuous miners in Chapter 7.)

The *scraper (or slusher)* has been used for many years in both coal and metal mines. It used compressed air drive at first, but now both electric and hydraulic motors are also available. A scraper effectively provides a loop of rope with the loose ends wound on contra-rotating drums as in Figure 6.4. A hoe is pulled backwards and forwards on the loop to gather ore or rock. The hoe shape enables it to ride over the ore in reverse and dig in on the forward pull. The rope arrangement can be 3-drum as in Figure 6.4, or 2-drum. For a 2-drum scraper the return sheave is anchored by plug and feather in a series of holes across a stope or drive to give full width clearance. A scraper can load mine cars by pulling loose ore up a ramp, or load ore to an ore pass by scraping across an opening in the floor of the stope or drive.

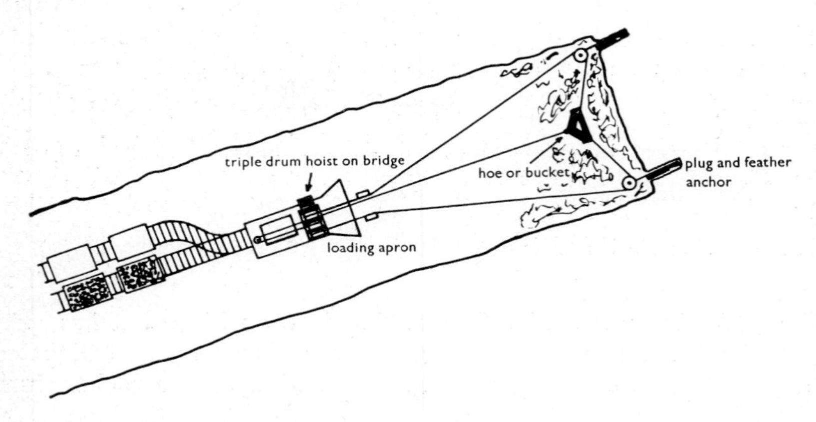

Fig. 6.4 Triple-drum scraper loader (slusher)

RAIL TRANSPORT

Metalliferous mines usually operate with levels driven on an arbitrary grade. The levels can be driven at about 1 in 200 to 1 in 100 in favour of the load, and whether they follow the sinuous curves of a narrow orebody, or are driven in a straight line, then locomotives can be used.

Locomotives can be battery electric, mains electric, or diesel powered. The battery electric are generally used for lower ore outputs, for stop-start duties, and as general runabouts with supplies. Their capacity is rated in ampere-hours and if they are used on heavy duty main line hauls they are not as cheap or convenient to operate because of the high rate of battery discharge. They may need to change batteries during a shift.

In contrast trolley or pantograph locomotives (thyristor d.c. control or a.c. for modern installations) are limited to main line haulage because they have to follow the overhead wire. However they are the most economic on long heavy duty hauls because they are rated for continuous working, and accept the temporary overloads of acceleration. Isa Mine uses two 20 ton pantograph locomotives (Figure 6.5) to

Fig. 6.5 Electric locomotive (*courtesy Mt Isa Mines*)

haul a train (rake) of sixteen 11.3 m^3 drop bottom mine cars of 25-30 tonne load capacity each. The track is 3 ft 6 in. gauge and uses 90 lb/yd rail to Queensland State Railways main line standards. (N.B. these figures are not yet officially changed: the S.I. equivalents are 1.067 m and 44.7 kg/m.)

Diesel locomotives are the most flexible but have to be rated on starting and acceleration loads. They have to be carefully maintained to avoid noxious exhaust emission, their maintenance costs are often higher, and availability (i.e. potential working time less maintenance time and breakdowns) is often lower than for trolley locomotives. For metal mines the choice is more usually of electric locomotives. In coal mines the extra safety of diesel locomotives which do not need a potentially dangerous overhead wire makes a diesel loco the more usual choice.

The locomotives can haul a train made up of any one of a variety of mine cars (or trucks). The simplest and cheapest car is virtually a large box on wheels, although it is heavy duty steel, must not leak dust, and has roller bearing axles. They have to be emptied by rotary dumping and are more favoured in coal mines.

The *rocker dump* car (Figure 6.6) is rather old fashioned and was of more use for hand tramming. They can dump on either side of the track into an open ore pass. The *gable bottom* car is high and narrow with a low capacity for a given volume, but they remain clean even when used for sticky ores. They empty at both sides of the track at once, so that a large mouthed ore pass is necessary. Dumping is a manual operation.

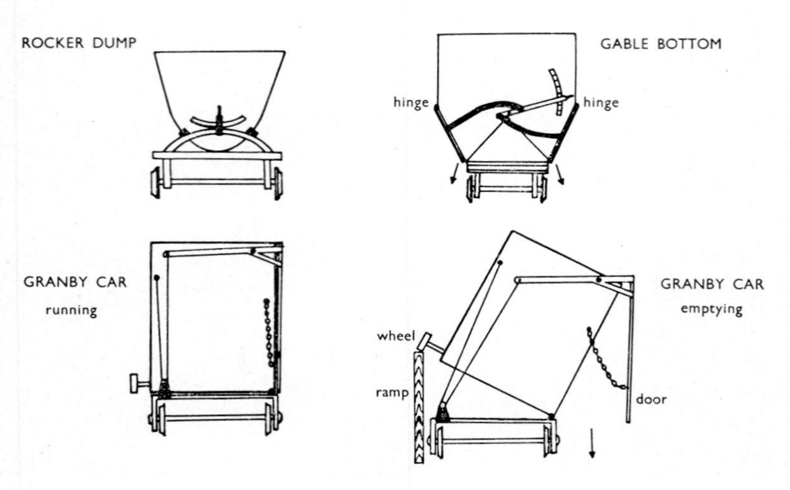

Fig. 6.6 Mine cars: rocker, gable bottom and granby

The *granby car* is the 'Ford/General Motors' of the underground mine car world, and ranges in value and finish from a model T to a luxury limousine. The basic principle is simple but effective. When the body of the car is tipped whilst the wheels remain on the rails a leverage system moves one whole side of the car up and away to allow the ore or rock to slide out. Dumping can be automatic by running a side tipping wheel over a ramp at low speeds to lift the body of the truck. The ramp is slid aside to allow free passage on the return journey. An alternative arrangement is to spot consecutive cars of a train over a lifting ram and tip the cars individually. This is a convenient arrangement where the other cars of the train are simultaneously stopped over a weighbridge for load measurement. Sticky fines can build up in granby cars and they have to be run out of service for cleaning.

The last major class are the *drop-bottom cars*. The type with multiple doors that drop between the tracks are not favoured in metal mines because of a generally slower unloading rate, a possibility of large lumps bridging over the axles, and the danger of doors not latching properly. Modern drop bottom cars, such as the system shown in Figure 6.7, drop the whole bottom of the car, wheels and all, whilst the body of the car remains suspended on side rollers.

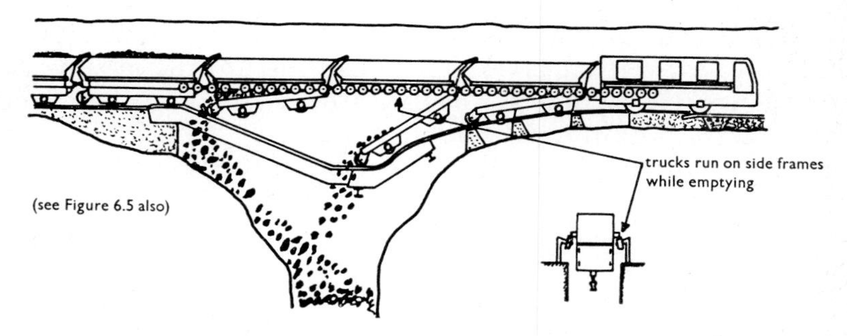

Fig. 6.7 Mine cars: drop bottom

Cars such as the Swedish OK system (used at Mount Lyell) drop the bottom sideways. The Hudson Roc-Flo cars used at Isa Mine on 17 and 19 levels drop the bottom of the car from one end. The locomotives at Mount Isa are fitted with side rails and run over the ore pass side rollers on these, because the normal ground rails have to be broken. There is in fact a 20 tonne a.c. locomotive at each end of the train so that one is always available to drive the train through. The train driver can remain in the loco or use remote radio control while unloading.

It should be noted in passing that before large equipment can be used (for economy of scale) in a mine, there has to be a large shaft to get the equipment down. Any machinery is limited in size by the largest single component that can be slung in the shaft. Locomotives are built as a basic body with the ballast needed for traction added on as side bars and end bumpers. The ballast is removed and large locomotives are slung end-on down the shaft, using the winding rope after the cage or skip has been removed.

CONVEYORS

Conveyors will be described in Chapter 7. The general application of conveyors favours relatively small material which is not sticky. Run-of-mine metal ores are not usually in this class until they have been crushed so that conveyors are generally the last link in a metal mine transport system.

DRILLING MACHINES

Although rotary machines are available the majority of production holes underground are drilled with compressed air operated percussive drills. The small ones for casual jobs such as secondary blasting are light and are hand held. Holes for small production rounds are drilled with heavier machines mounted on an air leg which provides both support and thrust for drilling from horizontal to vertically upwards, although the drill and airleg are generally in one in-line unit for vertical and near-vertical holes. These one piece drills may be called *stopers*. Downward drilling is done with heavy *sinkers,* whose weight, plus that of the operator, provides the thrust.

The heavier production drills for development ends and for cut and fill stopes are now usually mounted on drilling jumbos. These are rubber-tyred and diesel-driven between jobs. On arrival they couple on to compressed air and water mains for drilling. From one to four drilling machines with automatic feed may be mounted on a jumbo. Some drills are mounted for either horizontal or vertical drilling of relatively short holes (say 3-5 m). Other more powerful drills are mounted one or two to a jumbo for *ring drilling*. These will drill a fan of holes in a vertical plane (or near vertical) through a 360° range to depths of up to 30 m or more. They enable one operator to drill about 120 m total length of hole of about 50 mm diameter in a 5 hour working shift, which is all that may remain of 7 hours nominal shift after travelling and a meal break.

Rubber-tyred four boom development jumbo
(*Courtesy Mt Isa Mines*)

It should be realised that holes underground may have to be drilled upwards as well as downwards when working in large stopes. It may be recollected from Chapter 3 that mention was made of underground grade control and exploration. The diamond drills mentioned earlier are used to drill fans of holes to the stope limits for routine determination of ore grade cut-off values. This will be done at spacing of 20-50 m on each level as necessary. About every third ring of diamond drill holes may be extended to 50 m or so to look for blind ore. A full time team may also be employed under a geologist's direction to carry out a planned search for deeper orebodies.

UTILITY VEHICLES

A variety of auxiliary vehicles are available and needed for underground use. They are usually diesel-engined and rubber tyred for mobility. There are general runabouts fitted with flat-tops or personnel carrier bodies. Some may be four wheel drive and even four wheel steer. The more expensive ones may be fitted with hydraulic winches, or small hydraulic cranes for self-loading. A particularly useful item is the scaling platform. This is a self-erecting work platform mounted on the back of a flat-top vehicle (Figure 6.8). It will lift to 8-10 metres to enable workmen to inspect the high backs of a stope and to bar down loose rocks, or insert extra rock bolts for roof support.

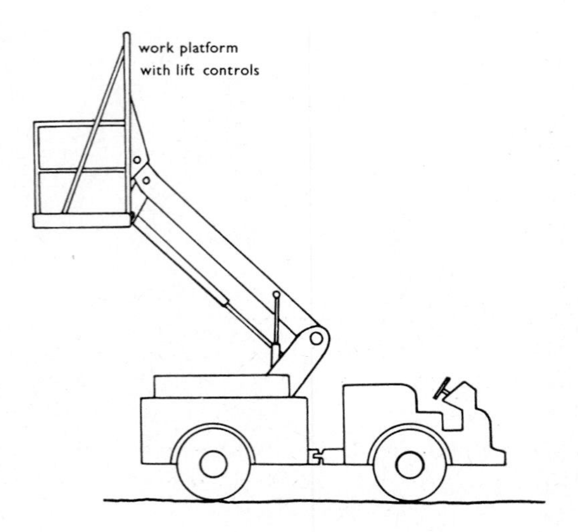

Fig. 6.8 Utility vehicle fitted with work platform

Classification of Stoping Methods

Ores are won by stoping, which is a general term used to cover any mining method. The basic classification of methods devised by the U.S. Bureau of Mines in 1936 is still valid: the names are the same, but techniques and relative importance have altered, and some additions must be made. This classification is based on a transition from strong rocks and ore to weak rocks and ore. These are the primary factors in a choice of mining method. In general the larger the opening that can be left unsupported, then the greater the degree of mechanisation that can be adopted, and consequently the greater the safety and economy of the operation.

CLASSIFICATION OF STOPING METHODS BY ROCK STRENGTH

A. Stopes Naturally Supported:
   1. Open Stoping:
      (a) open stopes in small ore bodies
      (b) sublevel stoping
      (c) longhole stoping
   2. Open Stopes with Pillar Supports.
      (a) casual pillars
      (b) room (or stope) and pillar (regular arrangement)

B. Stopes Artificially Supported:
   3. Shrinkage Stoping:
      (a) with pillars
      (b) without pillars
      (c) with subsequent waste filling
   4. Cut and Fill Stoping.
   5. Stulled Stopes in Narrow Veins.
   6. Square-set Stoping.

C. Caved Stopes:
   7. Caving (ore broken by induced collapse):
      (a) block caving; including caving to main levels and caving to chutes or branched raises.
      (b) sublevel caving.
   8. Top slicing (working under a mat, which, together with caved capping follows the mining downward in successive stages).

D. Combination of supported and cave stopes, (as shrinkage stoping with pillar caving, cut and fill stoping and top slicing of pillars, etc.).

TABLE XI.

APPLICATION OF UNDERGROUND METAL-MINING METHODS

Type of Orebody	Dip	Strength of ore	Strength of walls	Possible Method of mining	Class No.
Thin bodies	Flat	Strong	Strong	Room and pillar Casual pillar Open stopes	2b 2a 1a
		Weak or Strong	Weak	Top slicing Longwall	8 See Chapter 7
Thick bodies	Flat	Strong	Strong	Sub-level stoping Room and pillar Cut and fill	1b 2b 4
		Weak or Strong	Weak	Top slicing Sub-level caving	8 7b
		Weak	Strong	Square set Cut and fill Sub-level stoping	6 4 1b
Narrow veins	Steep	Weak or Strong	Weak or Strong	Resuing in (a) Open stopes or (b) Stulled stopes	 1 5
Thick veins	Steep	Strong	Strong	Open stopes Sub-level stoping Shrinkage stope Cut and fill method	1a 1b 3 4
	Steep	Strong	Weak	Cut and fill stopes Square-set stope Top slicing Sub-level caving	4 6 8 7b
	Steep	Weak	Strong	Open casual pillar Square-set stope Top slicing Block caving Sub-level caving	2a 6 8 7a 7b
	Steep	Weak	Weak	Square-set stopes Top-slicing Sub-level caving	6 8 7b
Massive		Strong	Strong	Shrinkage stope Sub-level stoping Cut and fill stope	3 1b 4
		Weak	Weak or Strong	Square-set stope Top-slicing Sub-level caving Block-caving	6 8 7b 7a

Table XI gives the application of underground mining methods to different shapes of orebodies. It is by no means an exhaustive list but does serve to show that no method is unique for a given set of circumstances. Several factors must be assessed before a choice is made. The physical factors are dealt with in more detail in Section C of this book.

Underground Development

SHAFTS AND INCLINED DRIFTS

One of the first problems that requires a decision is whether to develop the orebody through shafts or through inclines, or a combination of these. Four factors are involved; depth of ore deposit, time for development, cost, and choice of haulage methods out of the mine.

Conveyor transport drifts are inclined at not steeper than about 1 in 3, truck haulage drifts are graded at about 1 in 7 to 1 in 9, and shafts are vertical or near vertical. Obviously the deeper the initial depth of working, then the less attractive an inclined drift becomes because it is three to nine times longer than a vertical shaft. The handicap is not so much in cost of construction as in operational costs.

Dependent on cross-sectional area, sinking methods, and rock type, a shaft may cost in the order of $600 to $800 per lineal metre sunk, plus lining and furnishing costs, and with grouting or freezing as extras. Development drifts shallower than about 1 in 3, when driven by themselves, can cost around $150 to $200 per metre, plus extra for special linings or treatment. Consequently, for inclined drift grades less than 1 in 4 a shaft may have a cost advantage for entry purposes. However if a drift is driven in conjunction with ore production from stopes, so that the machinery can be used continuously in more than one working place, the cost of drivage may fall to $70 to $100 per metre. If the ore body is close to the surface, and production can start soon after an entry is made, it is as economical to drive a shallow angle drift as it is to sink a shaft.

Advance rates for a drift may be around 25-30 metres/week with conventional methods (tunnelling machines are not at present used for this sort of work), although 8 m per day can be achieved if desired. For shallow shafts with no special equipment the sinking may only be 5 to 10 m per week. For shafts of 500 m or so, where mechanisation is worthwhile, then sinking rates of over 30 m per week can be attained. These high rates are obtained from trained crews with special equipment so that there is a start-up delay while tenders are called, shaft winders bought, etc. In contrast a declined drift can be started almost immediately with normally skilled miners using ordinary production equipment.

A major decision is what type of ore haulage will be used out of the mine. Fully loaded ore trucks travel up a 1 in 9 incline at about 8-10 km/hr but can be loaded in the stope or from a chute and can then travel straight to the ore dump. If a shaft is used then mine cars have to be loaded on a level via an ore pass and chute and hauled to a shaft for winding. This system is not as flexible as are trucks. However the speed of conveyance in the shaft is about 45-50 km/hr for the major part of the wind, with a cycle time of 40-80 secs to complete a wind with 10-20 tonnes of ore. As the ore body is worked to greater depths then a shaft becomes a necessity to move large quantities of ore economically.

When a complete cost study is made the use of trucks on an inclined haulage is currently uneconomical below about 180 to 240 m vertical depth. However an access decline is still useful for rubber tyred loading and service equipment. Consequently current practice in mining massive deposits is to develop an orebody as in Figure 1.6. For the first few years ore haulage will probably be up the incline. This gives time to plan a shaft properly. Then as the truck fleet reaches its 4 to 5 year replacement life a change can be made to shaft winding, if economics do not necessitate an earlier change. Two declines may be driven at first in the orebodies, but one decline with raise bored upcast shafts may be used.

In narrow veins, where barren rock would have to be taken to widen roadways for haulage trucks, then mine cars may be used with a shaft almost immediately. Mine car haulage up drifts to 1 in 2 is possible but speeds are only 16-25 km/h and the drift has to be straight, not a spiral. Because drifts have to be kept in the footwall to avoid sterilising part of the orebody to provide a protective pillar, then spiral drifts are preferable to straight drifts which would travel too far away from the orebody. For mine cars the choice must usually be of a shaft and not an incline if the deposit extends below a few tens of metres.

Strength of rock, or the presence of excessive water, or of running sand, can over-ride an economic choice for technical reasons. A vertical circular shaft is both stronger and easier to sink in difficult conditions than is an inclined drift.

OREBODY DEVELOPMENT

An orebody is entered by shaft or drift as appropriate and levels are driven off at pre-determined intervals. The level interval is primarily a function of the raising techniques necessary to provide vertical connections in the stopes, with the mining method and angle of dip of the orebody also having an influence. In fairly steep orebodies current level intervals are from 50 to 90 m, but not all levels are equally developed for haulage. Figure 6.9 shows a simplified arrangement for an ore pass system at the Isa Mine, where several levels are connected into a crusher system. Comparison should be made with Figures 1.3 and 1.4.

Levels are connected vertically for access and ventilation as necessary. Raising in the solid can be by cyclic drill, blast and load methods; or by raise borer. Short raises to 8 m can be driven by hand from simple staging. Higher raises must be mechanised or else driven as compartmented raises; a small cribbed portion provides access and air way, the broken rock is 'shrinkage stoped' in the major part of the raise. Most raises over 15 m are driven by cage raising, Alimak raising (using a raise-climbing trolley) or possibly longhole raising. Details can be found in a more technical book.

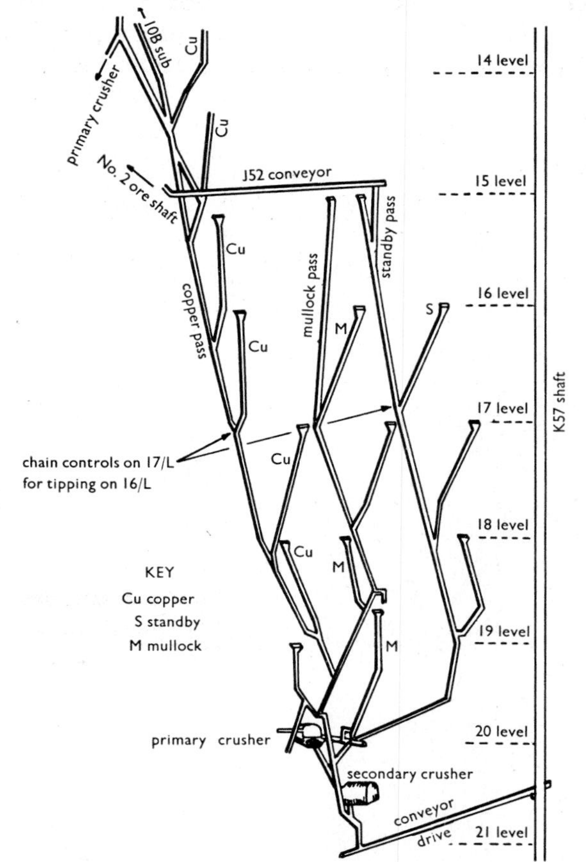

Fig. 6.9 Ore pass system at Isa mine

Within a stope, raises can be left through the fill by erecting rectangular wooden or steel cribs before filling operations. A *crib* is made by laying lengths of timber or steel alternately crosswise, like a hollow tower built with matchsticks. The outside is boarded across if necessary to prevent fine material from falling through. Alternatively circular steel ducts in prefabricated sections can be used. Internal diameters are about 1 to 1.2 m to allow ladders to be fixed. Service raises and hand-driven raises may be about 2.5 x 2.5 m with hoisting equipment installed. If diesel equipment is to be lowered into stopes then the raise must be virtually a small inclined shaft and may be around 4 m x 3 m cross section running down the footwall (Figure 6.23). Most raises are in the orebody and follow the footwall in grade in narrower stopes.

DRAW POINTS

These are also known as *mill holes.* They may be developed at the bottom of open stopes as an inverted cone by drilling and blasting, or may be developed at the bottom of an ore pass system. Their form is determined by the way in which the ore is to be loaded.

A large chute can be used to load ore from a main ore pass into a dump truck, or smaller chutes may be installed on each of several ore passes along a level to load directly into mine cars. These chutes are of the form shown in Figure 6.10. Ore can gravitate down a natural rock pass and into a chute as in A. An alternative arrangement is shown in B where the chute and an ore pass have been built up on the stope floor and protected by square-set timber levels.

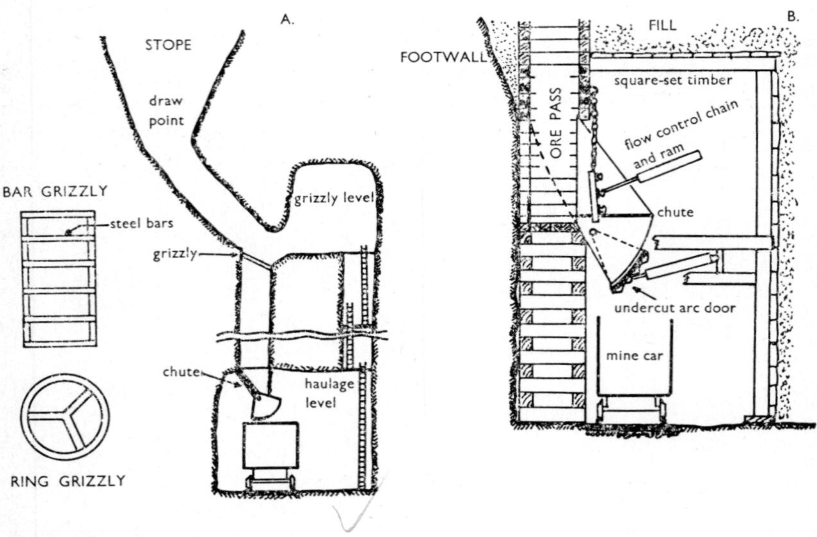

Fig. 6.10 Ore loading chutes

Chutes cause production hold-ups if they become blocked so large lumps are excluded by feeding ore through a *grizzly,* which is a grating made up of steel bars. The spacing is such as to exclude really large pieces of rock or ore whilst allowing smaller lumps to flow freely through. Dependent on equipment size the bars may be from 0.3 to 0.6 m apart. Where an ore pass is raised in a filled stope a prefabricated circular grizzly may be used on top to make it easier to move up for each lift. Travelling and vent raises of 0.6 to 0.9 m diameters are also guarded with these.

Lumps which do not fall through grizzlies are broken with a hammer or an air pick, or occasionally by blasting. *Hang-ups* in the draw point are broken by 'bombing': an explosive charge is poked up on a long pole, and then detonated. Workmen must not enter the draw point to break lumps loose for safety reasons.

Where mine cars or dump trucks are loaded with a rocker shovel the draw point may be of the form shown in Figure 6.2. The low brow gives good ore flow control and the higher portion is necessary to allow the bucket to discharge.

Figure 6.11 shows a *scram drive* draw point layout. Ore is broken in the stope by blasting and gravitates down into the drive. A scraper bucket travels backwards and forwards in the drive to scrape ore along the floor and drop it through a grizzly down a millhole into mine cars. Scraper sizes can range from 22 to 110 kw with wire ropes from 12 to 40 mm diameter. Soft floors may be lined with concrete with inset steel rails to give better wearing properties. Draw points may be strengthened with concrete or lined with full-column grouted rock bolts to prevent premature collapse of the brow. It is difficult to water spray this type of draw point and ventilation is arranged to take dust away from the workmen and straight into the return airway. This airflow is adopted at other drawpoints wherever possible in addition to spraying.

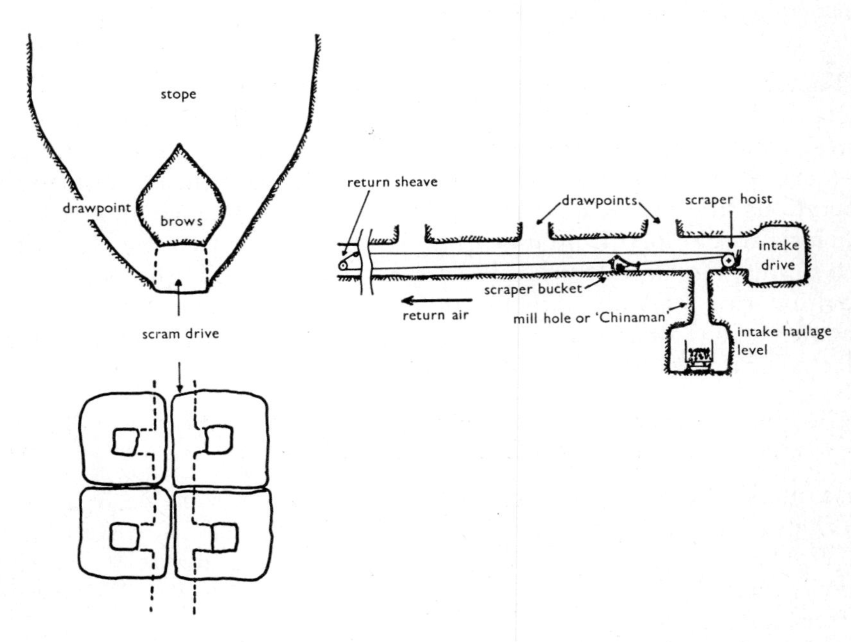

Fig. 6.11 Scram drive draw points and mill hole

Scraper loader bucket in a scram drive.
Mill hole in foreground, drawpoint on right.
(*Courtesy Joy Manufacturing Co. Pty Ltd*)

Figure 6.12 shows a layout that can be adopted where a sudden increase in production is needed. A diesel LHD machine can load direct from a temporary draw point in an open stope into mine cars in a cross-drift off the main haulage level. It is more usual to keep the operations of LHD machines and locomotives independent. An LHD machine would load from a draw point into an ore pass, this functions as a storage bin from which mine cars can be loaded. Each 3.8 m^3 LHD machine can load about 500-700 tonnes a shift dependent on rock fragmentation and haul distance, and its production rate can be affected if it has to wait for other transport.

STOPE DEVELOPMENT

In a massive orebody a network of drives is made on each level to block-out and undercut the orebody with draw points, for example as in Figure 6.13. Stopes are opened out above the drives by vertical development (or inclined development) of the draw points. This creates a free space into which rock can be blasted to enlarge the stope to working dimensions.

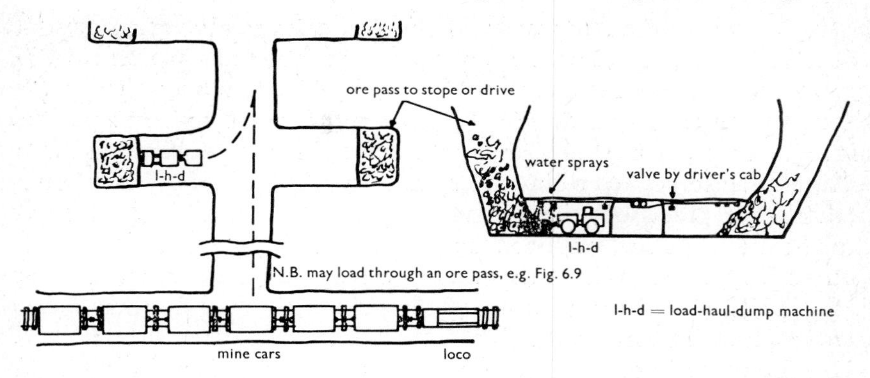

Fig. 6.12 LHD draw points

In some mines construction of individual draw points for open stopes is not carried out. The stope bottom is percussive drilled from the draw point level and blasted into a continuous V shape as in Figure 6.18. Broken ore is loaded out from the bottom drive as it rills down. However it is still necessary to bore or drive a raise to form an initial cut-off slot.

The mining methods will be one or more of those described below as appropriate for the rock and ore strengths. These strengths can vary within an orebody. Vertical pillars (rib pillars) may be left between stopes to protect access drives and raises. A *sill pillar* is left horizontally around and above the level drives to protect them and provide height to develop draw points. This sill pillar is commonly about 15 m thick.

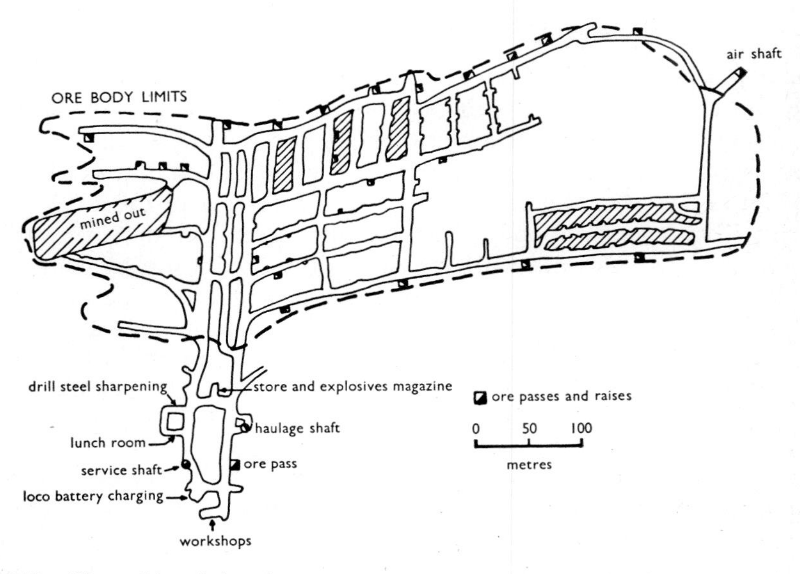

Fig. 6.13 Plan of level development

As stopes are worked upwards to meet the level above a horizontal *crown pillar* is left below the superjacent drives to stop them collapsing. In most orebodies means will be devised to recover these pillars before the level is abandoned. Pillars can be recovered by square set mining, or by cut and fill for *competent* (strong) ores capable of supporting themselves over limited spans. In weaker ores, where caving methods are practised, the pillars may be mass blasted or caved out into the stope below, or into a salvage drift.

In vein type orebodies only one drive may be necessary at each level and this will follow the contour of the vein with a single row of drawpoints at required intervals.

Development in any type of underground mining is usually planned ahead and is completed before production starts. However a need for a sudden increase in production may arise and this can be met with the use of *leading stopes*. In narrow orebodies the backs (roof) of a drive can be blasted down to a height of several metres. The drive is then square set and ore raises and chutes are developed in fill instead of rock. This can be adopted as a normal working method. Development of conventional draw points, drives and cut-off raises can take as long as 1½ to 2 years for a large stope, but a leading stope can be in production in 4-6 months.

In massive orebodies the equivalent of a sub-level open stope or a sub-level cave can be developed quickly in the upper part of a major stope to provide ore while the proper draw points are completed on the main level. Ore can be dropped down a raise which will later be used to create a cut-off slot.

ORE PASS SYSTEMS

In many mining methods, particularly cut and fill, it is necessary to drop the ore down to a haulage level. It then quickly becomes convenient to drop ore down a network of passes which join up at the lowest working level, or at a major interval of a few hundred metres (see Figure 6.9). On this level a major haulage system can be installed which will give economy of operation over several smaller installations. In a large mine the ore will gravitate into major underground ore pass storage bin systems; one for each major grade or type of ore, e.g. lead/zinc, and copper, plus another bin for waste rock. Each bin will be drawn off in turn through a device such as a Ross chain feeder into the primary crusher. This will reduce all ore to less than (for instance) 0.3 m for easy handling on a belt conveyor or into mine cars. The ore will be taken to the shaft and loaded into skips.

Most large tonnage ore passes are mined in rock alongside the stope. Steel millholes in a filled stope have a maximum life of 100 000 to 150 000 tonnes of ore. In the case of a 100 m long stope, 12 m wide, this would limit the vertical lift with a single ore pass to about 30 m.

For maximum economy the vertical interval between levels should be as large as possible. With steel millholes it would then be necessary to install more than one pass. At a cost of about $300 per metre for the steel prefabrication plus installation charges this would be very expensive. It is possible to increase the life of a steel ore pass by fitting extra wear plates, or else to put wear plates and RSJs into a timber ore pass at the bends, but this increases the initial capital cost and has to be offset against the advantages of cribbed ore passes in a filled stope. The main advantage is that a raise does not have to be driven in advance in the rock, but can be built easily in the stope.

Mining Methods

Mechanisation is used wherever possible to give the maximum output per manshift employed. If necessary small ore bodies may be aggregated and will be mined as one unit along with the waste rock between them. This is subsequently removed in the mill (beneficiation plant) and the higher mill costs are offset by greatly reduced mining costs. In any case compromises are made in most stopes because the mean profile of a stope wall must be fairly even and in some cases waste is taken, and in others small amounts of ore must be left.

1. OPEN STOPES

Open stoping can only be used in hard rocks and strong ores. The ore is mined and no support is provided for the wall rocks or for the backs. Stope dimensions may be the whole of the orebody if small, or an arbitrary division of a much larger block. A hole in the ground of 30 m x 120 m plan and 60 m high could be considered typical. It would be unwise to open a larger plan area than this in any rock as there is a limit to the span that a brittle rock can bridge over. In softer rocks it may be necessary to keep open stopes down to about 20 m width.

Essentially an open stope has no support inside it. In practice casual pillars of low grade ore may be left, or roof bolting and timber cribs may be used to support occasional patches of weak roof, particularly in orebodies with low dip angles.

Class 1a—Small Open Stopes

A small orebody is likely to be worked by hand drilling and blasting. The ore will be loaded with either a small bogger into a mine car or by a scraper clearing the ore into a mill hole. A *bogger* is a general term for any form of mechanical loader; hence 'bogged out' means loaded out. These methods are improvised as necessary.

Class 1b—Sublevel Open Stopes

Sublevel stoping is essentially a method for steep orebodies because it relies on gravity to bring ore down to the loading level. It can be applied in flatter measures but more development work is needed. The older methods of sublevel stoping relied on workmen building scaffolding on *stulls,* i.e. wooden posts pocketed horizontally into the hanging wall and footwall, or mounted in the eyes of timber pins driven into boreholes in the walls. From the scaffolding the workmen would drill holes and charge and fire them. Broken ore dropped down through the stope, sometimes through wooden chutes as described in stull stoping.

Modern methods of sub-level open stoping can best be described by use of two case studies, one from Australia and one from Africa. The former is in an area where labour is difficult to recruit and is very expensive. The latter has plentiful labour and wage rates are lower. The first case study is given in some detail to illustrate the general methods used.

Cleveland Tin N.L., Tasmania. Carter[1] has described the operations at this mine near Luina in Tasmania, and has also given the reasons for their choice. It is worth adding some general facts first. The company spent about $1 million on exploration and then nearly $9 million to start the mine in 1968, including $1½ million for the town and $1½ million for mine development work. Underground production costs are reported to be around $2 per ton of ore. When evaluating potential filling costs these were variously derived to be between $1 and $1.50 per ton of fill placed underground. If cement were added then the cost could be $6 per ton of fill. In the comments on use of mill tailings (i.e. the final reject project from the mill) for use as underground fill it should be remembered that there is a recovery problem for tin ores. Cassiterite (tin oxide), alone or with sulphide ores, is usually in very fine grains thinly dispersed through a rock matrix. The rock is pulverised to recover the ore but only around 60 per cent recovery is possible by current methods. Consequently the tailings contain 40 per cent of the tin mined. They are carefully stockpiled in anticipation of a future process for recovering this ore.

The orebodies mined at Cleveland Tin consist of a series of parallel lenses of ore which are nearly vertical and plunge *en echelon* to the South, so that they must be worked as a series of stopes at increasing depth. The lenses vary in width from 1.5 m to 18 m with an average width of 7.6 m, and an average length of 365 m. Average grade is 0.92 per cent metallic tin and 0.40 per cent metallic copper in the ores.

Selection of the stoping system was controlled by the following factors:

1. the ore and wall rocks as exposed in old open cuts, exploratory drives and diamond drill cores was strong and competent

2. the ore body is steeply dipping
3. the strike is relatively uniform
4. the ore has a high sulphide content (N.B. If a sulphide ore is left in a broken state in a stope it may heat spontaneously and catch fire: in any case it oxidises and can then present beneficiation problems)
5. the surface topography is steep-sided heavily wooded valleys, and the orebodies are discontinuous. Neither of these factors favour open cut operations
6. the remote locality and high labour costs favour a mechanised method
7. the ore width was sufficient to permit use of recently developed diesel-powered equipment
8. there was no economic source of surface fill and the mill tailings could not be used because of the tin content.

The company investigated mining methods in many countries, and carried out feasibility studies and eventually adopted a mechanised sub-level open stoping method, working underhand. This enables diesel engined drilling jumbos and LHD units to be used.

The generalised layout of Cleveland Tin is given in Figure 6.14 and a more detailed arrangement in Figure 6.15. Each stope is developed by cross-cuts and then sublevels along the strike of the stope, although only one stope would be fitted with main fans. Sublevels are driven at 18 m vertical intervals and are 3.6 m high, leaving a 14.4 m block of ore to be blasted. The drives are also 3.6 m wide and in regular sections of ore can be silled out to the full width simultaneously to give a better output per firing and more economic working. In irregular orebodies lateral test drilling is carried out from the drive before silling out. This avoids accidental drilling and blasting of country rock instead of ore.

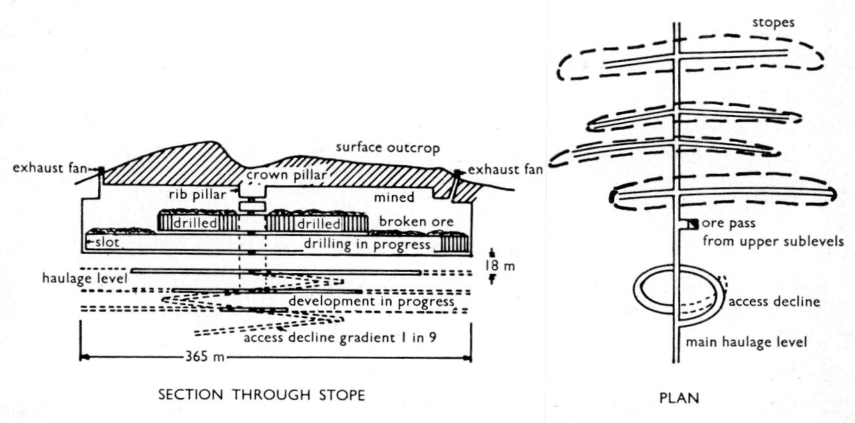

Fig. 6.14 Generalised layout of mechanised sublevel open stoping

At the end of the sublevels, i.e. at the end of the orebodies, a raise is developed to the sublevel above and then slotted out to give a free face for blasting. The survey department has the geological plan for each level and draws up the drillhole pattern for the sublevel. Holes are drilled vertical to the strike but angled sideways to match the slope of the walls. Holes have a 1.2 m spacing and a 1.5 m burden; diameters are 50 mm and firing is with ANFO primed with AN60 and No. 6 detonators. Average total hole length drilled per shift is 90 m and 4 to 5 tonnes of ore are broken per metre of hole dependent on stope width. Details of machine operation are given in Table XII. The ore is loaded by scooptram into the main ore pass and taken out of the mine by haultruck.

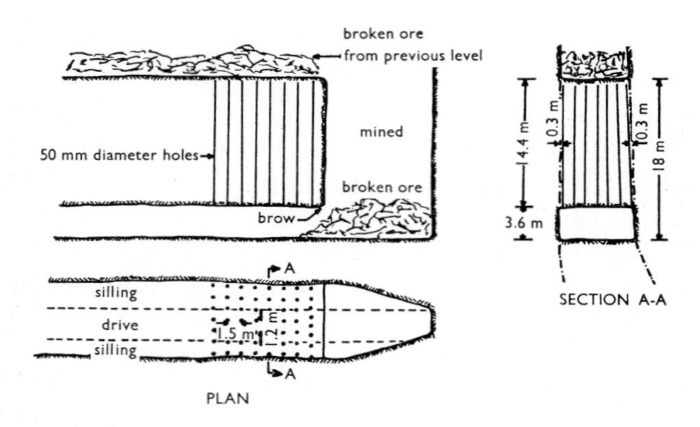

Fig. 6.15 Detail of sublevel open stope

Ground support by artificial means is negligible. Weak patches such as faults or major cracks may be supported by leaving small pillars of ore. About 10 per cent dilution of ore by slabbing of rock from the walls is allowed for in calculation of ore grades. (N.B. *Dilution* = waste/ore + waste.) It should be noted that the broken ore and slabbed rock left on each stope sublevel floor is recovered from the sublevel below; machine operators must not drive out beyond the brow of solid rock. The open stope is stabilised by a 25 m thick crown pillar at surface level if the orebody outcrops. This pillar also controls ventilation and keeps groundwater out. In addition the stope has a central rib pillar as a spine about 25 m wide along the strike length and solid through the mine. Both pillars will be extracted at a later date.

The advantages of this system of sublevel stoping can be summarised as:

1. continuous operation with no breaks for filling
2. low cost per ton and small workforce
3. early production from top levels as no drawpoints are developed

4. low backs in drives give safe working and men do not have to enter the stopes. (N.B. Men should keep back from the stope brow in case falling rocks from above ricochet outwards. Loader operators are protected by the machine.)
5. there is good ventilation through the sublevels
6. retreat mining results in minimum maintenance of roadways
7. all machinery is easily brought out for servicing.

TABLE XII.

OPERATING DETAILS OF CLEVELAND TIN N.L., TASMANIA

MINE STATISTICS		
Ore produced per 28-day period	22400	tonnes
Average broken ore grade	0.92%	Sn
	0.40%	Cu
Average development distance per period	168	m
Average underground workforce including Staff	45	
Tons per manshift including Staff	24.9	tonnes
Tons per manshift excluding Staff	31.0	tonnes
Average strike length of lodes	366	m
Average width of lodes	7.6	m
Average tramming distance ST–5A (one way)	130	m
Average haulage distance MTT (one way)	1143	m
Average bucket load. ST–5A Scoop Tram	6	tonnes
Average load MTT Ore Carrier	18	tonnes
Average load Euclid Dump Truck	15	tonnes

EQUIPMENT AT MINE	
Wagner ST–5A Scoop Trams	3
Wagner MTT423 38 Ore Carriers	2
Euclid B7 FD Dump Truck	1
Gardner-Denver 3-Boom Development Jumbo	1
Atlas Copco 2-Boom Development Jumbo	1
Gardner-Denver 2-Boom Stoping Jumbo	1
Wagner PT10 Personnel Vehicle. Flat Top	1
Wagner PT10 Personnel Vehicle. Basket	1
Ingersoll Rand XLE 1620c.f.m. (7.65 m^3/sec.) Compressors	2

MANNING AT MINE	
Underground Manager	1
Mine Foreman	1
Shift Bosses	3
Mining Engineer	1
Surveyor	1
Assistant Surveyor	1
Chainman	1
Jumbo Operators	11
Hand held Machine Miners	5
ST–5A Operators	6
Haulage Drivers	6
Diamond Drillers	3
General	5

There are some disadvantages:

1. selective mining is difficult; all ore in the vertical plane must be taken. Headgrade is maintained constant by working several faces concurrently.
2. maintenance costs on equipment can be high. The underground mechanical engineer spends more money than the underground mining engineer.

Mufulira Copper Mines Ltd, Zambia. This mine is not described in detail but is intended to provide an example of overhand sublevel stoping in steeply-dipping and flat-dipping areas. The full text can be found in an article by Airey[2]. For the steeply dipping seams (Figure 6.16) the ore is again taken on the retreat but in this case the lowest sublevels are taken first because the ore is to be dropped into draw points for gravity loading. By keeping a lead on the lower drives workmen receive better protection from falling ore or rock. The stope is about 160 m long by the width of the orebody and about 120 m high. The rib pillars are about 6 m across.

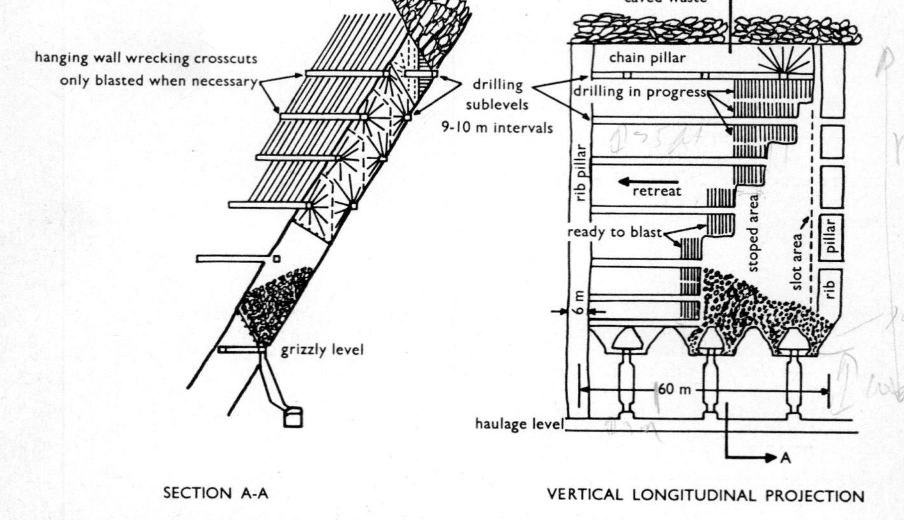

Fig. 6.16 Open-stoping layout in steeply-dipping areas

It can be seen in this case that the hanging wall is deliberately caved and the crown pillar is wrecked as soon as possible. The stress on underground openings is proportional both to the depth of working and to the area of ground left open. A wide span between two abutments can be unstable both across the span and at the abutments. If the span is deliberately broken to leave a cantilever with an arch of broken rock above it, then the abutment stresses are spread over a wider area and production takes place in a safer de-stressed area.

The caved waste is prevented from *rilling* (falling) prematurely down into the stope below by erecting a barricade of timber stulls and boards.

The same principle of caving the waste is also shown in Figure 6.17 which depicts the method of sublevel stoping in flatter areas. Ore will not gravitate to the draw points so it is scraped down dip to a series of mill holes and ore passes. These drop the ore to loading points on a main haulage level.

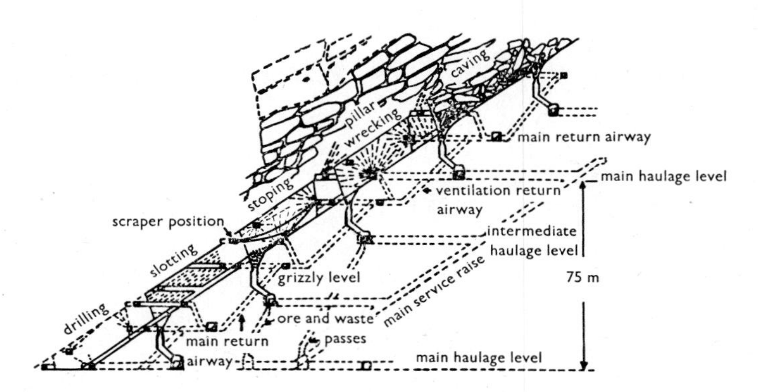

Fig. 6.17 Open-stoping layout in flat-dipping areas

Class 1c—Longhole Open Stopes

Improvements in percussive drilling machines particularly from the introduction of hydraulic booms, and independent rotation with stronger thrusts, have enabled longer holes to be drilled at higher penetration rates. It is now expedient to drill holes of up to 30 m or more in softer ores. Open stopes of over 50 m in height can be drilled and worked with only a drill drive at the top and bottom main levels. The ore is blasted a slice at a time and loaded from draw points at the bottom of the stope.

Figure 6.18 shows a form of longhole open stoping used at Isa Mine[3] in narrow and medium width orebodies. The draw points at the bottom of the stope are developed for diesel LHD machines. Drilling is carried out from the bottom drill drive and a number of sublevels. An initial cut-off slot is blasted from a raise-bored hole and then the stope is worked along its length. Ring holes are drilled at 1.8 m intervals well ahead of the stope face with a spacing of 2.4 m between the toes of adjacent holes in each ring. The holes fan out and the collars (open ends) of every second and third hole will not be charged with explosive so as to compensate for their closeness and to avoid pre-splitting the ring. The burden for each ring is 1.8 m and two to four rings may be blasted at a time with delay detonators to give good fragmentation. About 2-5000 tonnes of ore per ring is broken per blast to provide several shifts loading output. Figure 6.19 shows the method adapted for extraction in massive orebodies, where a network of rectangular stopes is taken.

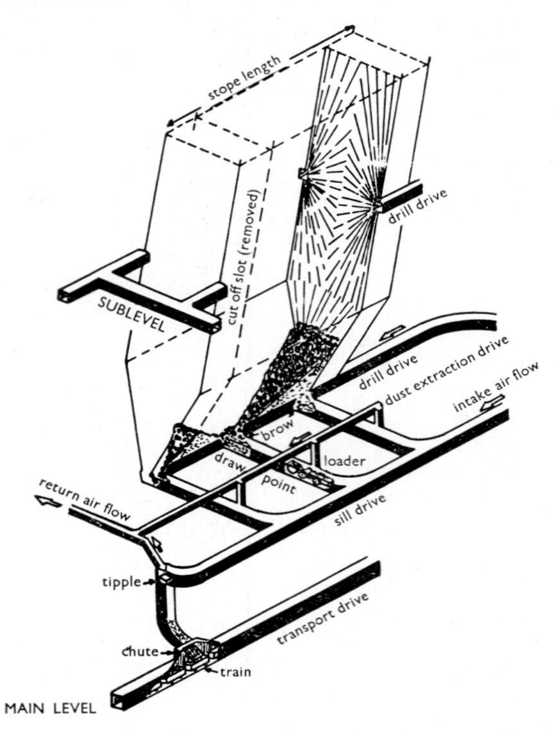

Fig. 6.18 Longhole open stoping in narrow and medium width orebodies (*original courtesy of Mt Isa Mines*)

The New Broken Hill Consolidated mine at Broken Hill has started to use longhole stopes in its lead/zinc orebody. A typical stope is about 73 m long by 10 m wide and 49 m high. It is silled out at the base (as in Figure 6.15) and drilled vertically upwards for 25 m. The crown is also silled out and holes are drilled downwards to overlap the bottom holes by about 2 m. A cut-off slot is developed by cage raising and the ore is then blasted about three rings of holes at a time and loaded out by LHD machines from under the intact brow.

The stopes are worked in groups of three separated by two 6.5 m pillars, and with each group of three separated by a 14 m pillar. It is intended to fill the empty stopes subsequently with a 10/1 sand/cement fill, developing about 0.8 MPa strength, and then the pillars will be extracted by ring blasting.

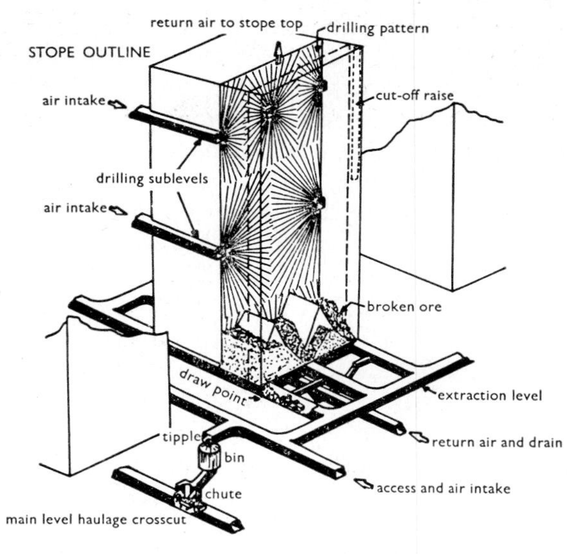

Fig. 6.19 Longhole open stoping in a massive orebody

2. OPEN STOPES WITH PILLAR SUPPORTS

With slower handworking methods it was possible to work open stopes in slightly weaker rocks than Class I by leaving in pillars of low grade ore or waste. If the arrangement was random then that was the casual pillar method. In mechanised mining it is more convenient to drill and blast everything within the stope to keep mining costs to a minimum. In any case, it is bad mining practice to leave casual pillars in deep mines because this causes irregular breaks to develop in the walls and will cause areas of high stress, rock falls, or even rockbursts. Consequently Class 2a is obsolescent.

When orebodies become flatter, then the stope with rib pillar arrangement is converted into a room and pillar method, unless the ore body is thick enough to block out into vertical stopes. Room and pillar methods in metalliferous mines may also be called *slot mining*, or *breast stoping*.

The Western Mining Corporation at their Kambalda nickel mines in Western Australia use two methods of slot mining, both with essentially the same dimensions. The nickel is in a vein-type orebody which is fairly continuous and consists of disseminated sulphides overlying a thin band of massive sulphides. The dip is relatively flat near the top of the dome-like deposit.

Where the ore thickened to over 3 m, and was less than 1 in 5 grade, then mechanised slots about 15 m wide were driven to the dip between 8 m wide pillars. The slots were drilled using three-boom diesel jumbos and bogged out with Wagner ST5A LHD machines. Loading was to the dip to enable the loader bucket to be crowded into the ore successfully. There would be too much wheel spin and tyre wear if the machine tried to force its bucket into the ore up-grade. After the slots had been mined between sublevels 30 m apart, then the slots were filled with sand brought in from the surface, and the pillars were mined out by the same method. Because of the low room height the sand fill does not have to be stabilised. Loading by LHD was direct into haul trucks and up an inclined drift to the surface.

Where the ore is less than about 3 m thick, or is steeper than 1 in 5, then slots are driven 8-10 m wide with airleg-mounted hand held drills and are scraped down dip with a slusher into a mill hole. (It would of course be possible to use a small bulldozer in deposits to about 20°.) The broken ore was picked up by LHD machine and either trammed direct to trucks on the incline, or dropped down an ore pass, dependent on the proximity of the stope to the incline. Figure 6.20 shows a typical arrangement. The stopes are separated by 8 m pillars. These stopes are close to the surface and are unsupported except for occasional rockbolts in weak patches of the roof. The slots are filled with sand after mining and then the pillars are extracted. It should be noted that this is not 'cut and fill' which is a cyclic process, but filling after extraction so that the pillars can be mined. Waste rock from development can also be dumped in the stopes. The steeper parts of the Kambalda ore bodies are worked by cut and fill methods.

The function of the pillar in room and pillar mining is to support the roof, which is now a more convenient term than backs or hanging wall. The maximum roof area for a given strength rock is the same for a vertical or horizontal stope. Hence room and pillar stopes contain

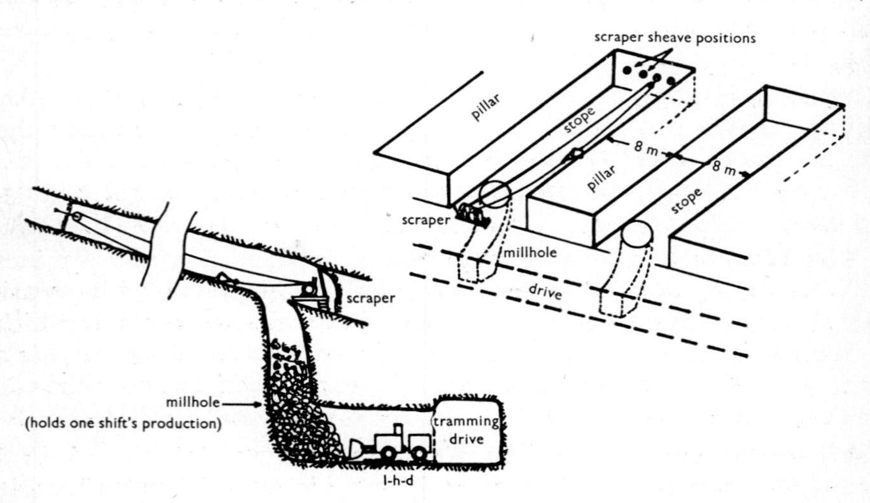

Fig. 6.20 Slot mining (room and pillar)

a smaller volume of ore than steeper stopes and a longwall method may be attempted to give greater outputs. Longwall methods described later are necessary in deeper deposits subject to rockbursts.

SUMMARY OF OPEN STOPING

Open stoping is only possible in strong rocks and ores. It can be extended to slightly weaker rocks by the use of sublevel or longhole stoping where men do not enter the stopes but work from small drives, protected by the unmined ore.

Small deposits can be usefully mined by open stoping as hand work is essentially selective, and waste and low grade ore can be separated out. However large open stopes with extensive mechanisation are not selective.

The advantages of mechanised open stoping have been listed in the case study for Cleveland Tin N.L. An extra advantage of direct loading from open stopes, as distinct from shrinkage stoping or caving, is that large lumps of ore can be separated out and broken before they can block an ore pass or a chute.

3. SHRINKAGE STOPES

These are classified as a method using artificial support but the support is in fact broken ore left in the stope. After a rock surface is exposed it gradually weathers and relaxes, and is subject to induced stresses from mining operations. If an open stope is left in a medium strength rock the sides will spall and slab off, but if the stope is filled with broken ore then the fractured rock is restrained from falling. However shrinkage stoping cannot be used in weak rock because the sides of the stope would squeeze together and trap the broken ore so that it cannot flow.

Shrinkage stoping relies on the fact that when a solid rock is broken by blasting then the broken fragments occupy a larger volume. This expansion is sometimes known as the *swell factor*. It can be around 1.3 to 1.5 (i.e. a volume increase of 30 to 50 per cent) dependent on the amount of fragmentation. If it were not for this swell there would be no output during the stoping period.

Figure 6.21 shows a typical stope layout. The stope is silled out and draw points are prepared. The draw points must be closely spaced because they will have to flow under choked conditions; about 8-10 m has been found to be a practical interval which gives free drawing characteristics and an even working floor. Similarly the draw point aperture is important. With ANFO blasting in the ore and hole spacings of 1.2 to 1.5 m and a burden of 1 m, then chute sizes need to be about 1.8 m x 1.2 m to avoid blockages.

A raise is driven at the centre or at one or both ends of the stope to give through ventilation and to act as a cut-off slot. Cribbed manways are raised on the footwall at about 50 m centres for access and for fresh air intakes. Auxiliary fans will be used to circulate air in the stope.

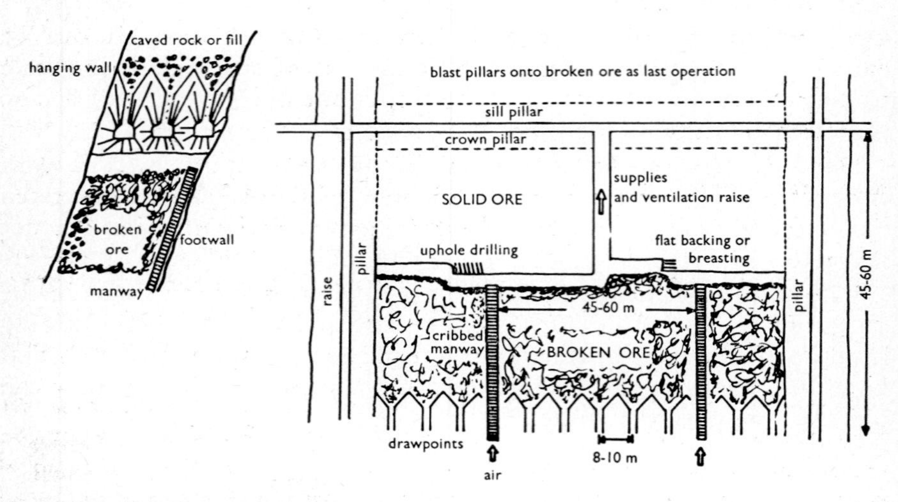

Fig. 6.21 Shrinkage stope layout

The ore is mined by successive overhand lifts of about 2 m which can be flat backed with horizontal holes or broken with inclined holes (uppers). If upholes are used along with delay blasting then more ground can be drilled and broken per blast, also drilling and loading are relatively independent operations. The men in the stope use the broken ore as a floor to work on.

After each blast the stope is practically blocked with broken ore over a short length. About 40 per cent of the broken ore is then removed through draw points below the blast until a working height of about 2 m for uphole mining is regained. If the ore is flat backed then broken ore is drawn before a blast. The amount of ore drawn must be measured: if excess ore is drawn then a working platform would have to be built in the stope or, even worse, the floor may appear to be at the right height, but could have a cavity under it. If this cavity collapsed any men working above it could be trapped.

For successful shrinkage stoping the ore should be free flowing. Too much fine or clayey material will cause 'hang-ups'. These may have to be 'bombed' from above to free them, but can be extremely difficult to clear. In addition the ore must not cement together from oxidation of pyrites, nor must it develop spontaneous heatings or deteriorate during storage in the stope.

Stopes may be excavated by taking slices along the vein (especially in narrow veins) from one end of the oreshoot to the other. Sometimes in wider bodies the ore is mined in a series of transverse stopes. Each stope will be separated from the next by a pillar of solid ore to reduce the unsupported span. The Zinc Corporation lead/zinc mine at Broken Hill uses both methods[4]: for transverse stopes both the pillar and stope are 10 m wide. The pillars will be recovered after filling the stopes. To facilitate this the fill must be supported with side mats or consolidated with cement.

In some instances of shrinkage stoping casual pillars may be left to support local areas where the walls are weak, and in others pillars of lean ore or waste may be left because it is uneconomic to mine them. These casual pillars also provide support. When pillars are left they must be carefully tapered at the top to slopes of about 50-60° to avoid ore hang-ups over them. Rock bolting may also be used to support slabby sections of the walls to avoid excessive dilution of ore with spalled rock. It is preferable to use rock bolting rather than to use casual pillars.

After a shrinkage stope has been mined out to the top then the broken ore is drawn off leaving the stope empty. The backs under the crown pillar may be rockbolted to increase their stability and avoid premature collapse when drawing off ore. The Zinc Corporation install water sprays on the backs before the stope is drawn off. The ore is saturated from above to help suppress dust at the draw points.

An alternative method of abandoning a stope is to blast the crown pillar and cave it along with the stope. If intervening pillars are ring blasted then the whole level can be *block caved*. This may be a way of starting a block cave, which is described later.

A list of the relative merits of shrinkage stoping is given below. Although the disadvantages apparently outnumber the advantages no economic assessment has been attempted. All the relative factors must be assessed when this method is considered.

Advantages of Shrinkage Stoping

1. Shrinkage stopes in good conditions can be cheaper than cut and fill operations.
2. Broken ore in the stope can act as a stockpile to even out production operations.
3. Uphole shrinkage stoping is basically more efficient than cut and fill because it is not a cyclic operation.
4. There are no ore passes in the stope to wear out or need maintenance or erection.
5. No ore has to be handled in the stopes, although some scraping may be necessary to level out a working platform for a mobile drill.

Disadvantages of Shrinkage Stoping

1. Wall and roof conditions in the stope must both be good. Ideally the orebody should be vertical but dips to 50° can be coped with.

2. Stripping a footwall is difficult because sharp ledges must not be formed. These would cause hang-ups: either ore must be left in walls or mullock taken to provide a smooth profile.

3. Spalling rock from the walls will cause dilution of the ore.

4. Broken ore may oxidise with possible attendant difficulties in drawing-off and in flotation.

5. There is also a possibility of spontaneous combustion in ores with a high sulphide content.

6. Closely spaced draw points are needed to give good ore flow. Draw control is difficult and working floor conditions are potentially dangerous. Travelling in the stope is rough.

7. Large blocks of ore that would normally require secondary blasting may be buried in the stope. These can block a draw point or cause hang-ups.

8. Shrinkage stopes are slow producers in the mining period. Once the stope is started it is difficult to change to another method.

9. Because approximately 60 per cent of the broken ore remains in the stope until it is completed the interest charges for the money spent on breaking this ore should be charged in the economic analysis.

10. Shrinkage stoping with transverse pillars needs more expensive development than does a cut and fill stope. About 50 per cent of the ore remains as pillars which can present a recovery problem.

11. Large amounts of sand and development rock are required to fill an empty stope. If the fill is urgently required for ground stability then there is a big problem.

12. Manpower requirements of a shrinkage stope can be high because it is difficult to mechanise the drilling operations, and drawing is not independent of the drilling operations while mining. Even at a later stage men may be needed to clear hang-ups in the stope during the emptying operations.

4. CUT AND FILL STOPING

In cut and fill stoping the ore is excavated by successive flat or inclined slices, working upwards from the level, as in shrinkage stoping. However, after each slice is blasted down, all broken ore is removed, and the stope is filled with waste up to within two to three metres of the back, or else completely filled, before the next slice is taken out. Just enough room is left between the top of the wastepile or sand fill and the back of the stope to provide working space. It is this repeated cycle of operations that is characteristic of the method.

Fill is introduced primarily to support the walls of the stope, and in small scale operations may consist of waste sorted from the ore in the stope (in which case (b) and (c) are done together), or of waste rock from development work or rock from waste stopes excavated to provide filling material. In larger operations the fill may be sand and gravel, or deslimed (sizes less than 10-20 microns removed) mill tailings. Stulls, props, cribs or rock-bolts may be employed to support local patches of insecure ground.

Prior to the introduction of reliable diesel powered drilling jumbos and loaders, cut and fill mining was just one of several mining methods that could be used. In modern mines, certainly in Australia, cut and fill tends to be an automatic first choice for the feasibility study unless the ore is known to be soft enough for caving. The same equipment can be used for either mechanised cut and fill or for open stoping. The mine can be designed for a cut and fill operation and then if conditions warrant it (i.e. the rock is proved strong in practice) a change can be made to an open stoping method.

Cut and fill is the most versatile of mining methods. Massive orebodies can be blocked out and stoped. The use of wall mats or cement stabilised fill will enable the pillars to be recovered. Where a mine consists of interbanded high and low grade deposits these can be grouped together for an economic series of cut and fill stopes as in Figure 6.22. Orebodies are stoped from one level to the next on a relatively horizontal plane and the fill prevents large scale collapse of the country rock. The arrangement of ore passes and ventilation raises is such that even when some are blocked by mining operations another will always be easily available.

In tabular steep-dipping deposits of narrow or moderate width the ore is usually mined by longitudinal stopes in which the ore is stoped out from wall to wall. Stopes may be carried the full length of the oreshoots or may be limited in length by pillars, which will lessen the exposed span of the walls. Stopes of up to 20 m width may be taken under strong ore backs. If the orebody is wider then an E-shaped stope may be driven as described later.

Cut and fill stoping can be used in irregular orebodies where stringers of valuable ore run out into the country rock. These stringers are followed out and then the cavity will be filled so that another lift can be taken. In some cases narrow veins may be taken by *resuing*. This is used in veins which are narrower than a practicable stoping width so that one of the walls has to be stripped out. In resuing one wall may be shot down first and left in the stope for fill, after which the vein is stripped. Alternatively the vein is broken first and loaded out and then the opening is widened to provide working room and to fill the stope with the swell of the broken rock. Consequently resuing can be classed as either cut and fill, or as vein mining.

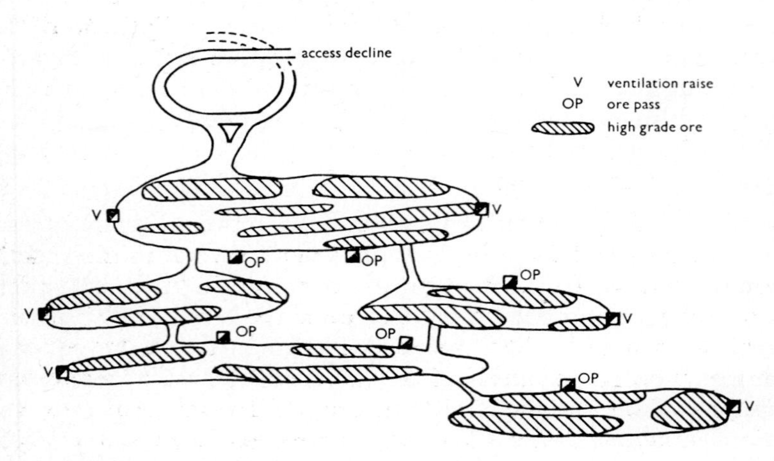

Fig. 6.22 Plan of interconnected cut and fill stopes

Mechanised Cut and Fill

The term 'mechanised' refers specifically to the use of diesel or air driven LHD machines to load the broken ore into trucks or ore passes.

The Henderson mine of Campbell Chibougamau Mines Ltd, in Quebec, was probably the first important metal mine to apply mechanised cut and fill methods about 1960. It was of course a development of earlier non-mechanised methods. Mechanised cut and fill methods rapidly spread, particularly in Australia, where the mines are probably in the forefront of practice and development.

Figure 6.23 shows the layout of a MICAF (Mount Isa Cut and Fill) stope as used at Mount Isa mine, Queensland. The basic principles of the system are clear.[3, 5] The stopes are about 180 to 700 m long, and from 4 to 11 m wide in the silver/lead/zinc orebodies. Dips are from 60° to 65° and the vertical level interval is 58 m, but stopes are mined to a double level interval.

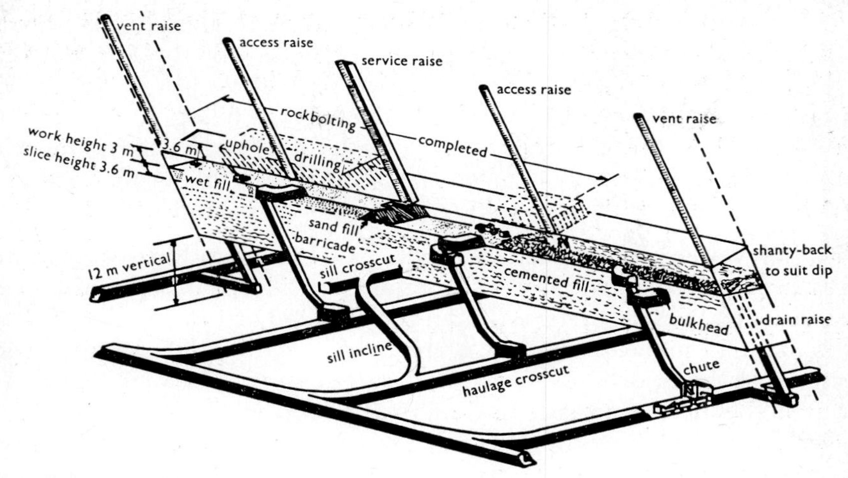

Fig. 6.23 Diagrammatic layout of a MICAF stope (*original courtesy of Mt Isa Mines*)

Stope Access and Development

A 1 in 8 incline is driven from the bottom haulage level to establish a sill level 12-15 m above the haulage level. The sills for the different orebody stopes are connected by a crosscut. A 2.7 m by 3.3 m service raise on the footwall of the orebody at the centre of the stope connects the sill with the top haulage level. After the sill access incline is sealed and the first fill placed, this raise is the only main access into the stope. It has a ladderway for personnel and a hoist for handling equipment. The raise serves as the stope intake airway, exhaust is through 2.4 m by 2.4 m ventilation raises, one at each end of the stope. Initially there was no ore pass development, passes were cribbed through the fill. Now 2.4 m by 2.4 m passes are developed in the footwall country rock, about 8 to 10 m from the ore contact for maximum stability.

The major advantage of access through raises is that development is cheap and rapid. Stopes can be quickly brought into production. The method is probably best suited to small or medium tonnage operations per stope. A disadvantage is that the major units of equipment, loaders and jumbos, are captive in the stope. Servicing and maintenance must be carried out in the working stope where conditions are not ideal. In the event of a major breakdown the units must be dismantled and hoisted out through the service raise to the workshops.

Units cannot be moved easily from place to place to deal with fluctuating work demands, or as replacements in an emergency or breakdown. They have to be dismantled before removal. Thus equipment capacity must be closely matched to the stope production requirements.

Cobar Mines Proprietary Limited (Figure 1.4) already uses access inclines in the copper and copper/zinc orebodies of the C.S.A. Mine at Cobar in New South Wales.[6, 7] These orebodies dip at 70° to 80° and widths range from 4.5 m to 21 m the average being 10.7 m. Stopes are opened up for the full strike length of the orebodies, ranging from 150 m to 365 m. Cobar had the advantage of being planned as a new mine (1965) which would use mechanised cut and fill.

Main levels are established at intervals of 180 m, and major servicing areas such as underground workshops are provided on each main level. There are two incline systems driven at a gradient of 1 in 7 in the country rock about 15 m away from the ore contact. Each incline system is connected by a transport cross-cut at intervals of 45 m, which also provides access to the main ore passes feeding the underground crusher. Access to each stope lift of 4.5 m is provided by 4 m x 3 m cross-cuts driven from the incline system. Ore from the stope is trammed directly to the main ore passes when the stope access is close to a transport cross-cut, otherwise the ore is trammed to a subsidiary stope pass off the incline system with a drawpoint located off a lower transport cross-cut to minimise tramming distances. Crushed ore is conveyed to underground storage bins for delivery to the automatic shaft hoisting system.

Renison Ltd operate a shallow tin mine at Renison Bell in Tasmania. In the lode orebodies LHD machines fill broken ore directly into 18 tonne capacity diesel trucks which carry the ore up a 1 in 9 incline to the surface.

New Broken Hill Consolidated[8] uses a different method of cut and fill. The stopes in the zinc/lead orebody are silled out on the haulage level and a double gangway of square set timber is erected. Chutes are installed and the timber is barricaded off and fill is poured around it. The stope then works by raising steel and timber cribbed raises in the stope as in Figure 6.24.

The orebody is divided into a series of panels. Each panel is in an E-shape with three 10 m wide stopes between hanging and footwalls and a 10 m wide strike stope along the footwall. This leaves two buttress pillars 6.4 m wide to support the hanging wall. The panels are separated by 14 m pillars which are left intact.

In the south stope of each panel a 2.4 m by 3 m raise in ore against the footwall connects the working level with the upper level. This raise is used for transport of heavy equipment to and from the stope, it also serves as the exhaust air-way and the route for fuel and sandfill lines. All other connections to the stope are from below through cribbed passages in the fill. Each stope in the panel has a cribbed 1.8 by 1.8 m internal dimension ore pass on the footwall side, a cribbed manway and timber hoist in the centre, and a cribbed intake vent pass on the hanging wall side. More recent panels have a combined intake ladder/timberway. All raises, etc. are situated on the pillars to facilitate their final recovery, which will probably be by cut and fill, or by square set stoping.

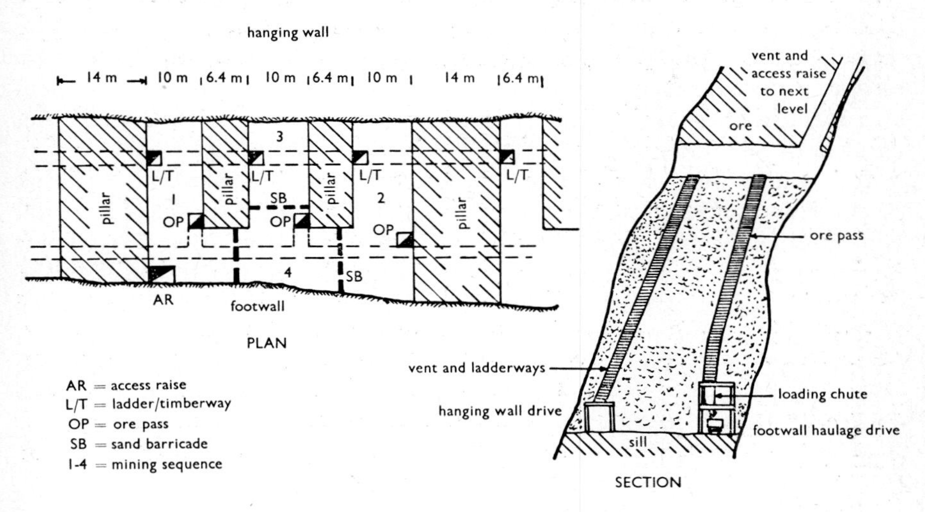

Fig. 6.24 Cut and fill stopes at New Broken Hill Consolidated

Mining Practice

It would be tedious to describe the individual practices for each of these mines; they can be obtained from the listed references.[3,5,6,7,8] Instead a composite practice is given which compounds the operating experiences of these and other mines over the last few years.

All the cases listed use upholes bored in the backs. Holes are drilled about 50 mm diameter with a one, two, or three boom jumbo. Spacing and burden are 1.2 to 1.8 m. For maximum efficiency and grade control the drillhole positions are marked out by the surveyor and numbered. A reference datum line is marked on the stope wall. The driller is given a chart of hole angle and depth and will only be paid contract rate for efficient drilling. Blasting is with ANFO unless the rock is very wet when a semi-gelatinous explosive (e.g. A.N.60) will be used.

Experience has showed that the holes are best bored at about 10° to 30° off vertical, leaning away from the loading direction. This gives a looser rock pile which is easier for the loader to tackle. The broken ore rills more easily down to the loader. It should be realised that an average slice height is 3 to 4 m and the broken rock pile may be 6 to 7 m high. The LHD machine crowds in at the base of the pile and sweeps the bucket upwards to fill. The exceptions are when a large lump must be worked out and pushed aside for secondary blasting.

A large stope is divided into working sections, or several stopes are interconnected. In this way a few thousand tonnes of ore can be broken in one blast. There will be several shifts loading in each position for an LHD machine. Concurrently there will be a cleared section which will be prepared for filling, and a filled section that will have a drilling team at work. In this way the cyclic operations can be performed continuously. Production scheduling is the responsibility of a senior mine engineer to ensure smooth operation, and to avoid expensive machinery standing idle: three-boom jumbos and LHD machines cost about $60 000 each.

It has also been found from experience that it is better to retreat stopes from the ends to a centre access; if the roof has a cleavage that is only suitable for angled drill holes in one direction, then the stope is retreated from one end to the other. Retreat working has the jumbo drilling holes in advance, followed behind by the loader. The jumbo is not then locked in by a pile of broken ore and can keep well in advance of the LHD machine. Also filling can be carried out in small increments behind the machine, instead of waiting until a stoping slice has been completed. This avoids intermittent runs of large quantities of fill which are inconvenient to stockpile.

Ventilation

Ventilation air travels better if the used air, which warms up, is taken in on lower levels, brought up into the stope, and then taken out to the rise. Consequently early practice was to bring air up into cut and fill stopes through cribbed raises in the fill. This presented two problems. Slimes from wet fill percolated into the raises and made a mess in the airways. Also, when a large blast was fired, the airway would be covered over, so that the airways were usually twinned at the end of the stope.

The introduction of raise borers and access declines has brought a change in practice. Raise bored holes are put into the backs through the ore at the end of the stopes. The air comes in the access crosscut and leaves up the stope end raises. No slimes can creep into the ventilation raises, there are no airway blockages, and in addition the vent raises acts as a cut-off for blasting.

Stope Filling

The fill for a mechanised stope is likely to be sand or finely ground mill tailings: development rock may also be added. Where there is no mill at the mine, and a shortage of sand, it is possible to mine broken rock or to use waste from old mine dumps and drop it through fill passes from the surface into a stope. This is obviously easier in shallow stopes, but has been applied in some deeper mines. If rock fill is used it has to be scraped out along the stope or spread with an LHD.

Sand fill is popular because it can be reticulated hydraulically. It can be dropped from the surface through a 150 mm borehole above the stope and then distributed through 100 mm pipes along the stope. Barricades are erected where necessary and the mixture of sand and water is run into the section to be filled. Where mill tailings are used there is also a cash benefit in that arrangements do not have to be made to dispose of them on the surface, nor to reclaim the site on which they were dumped.

Sand fill is cycloned if necessary to keep the minus 10 micron fraction down to less than 10 per cent to aid drainage in the stope. The sand or tailings will be stored in a surface bin then, as required, will be slurried to about 65 per cent solids and dropped by gravity down a borehole or a rubber-lined shaft pipe column. When it is run into the stope it must be dewatered to give a firm surface. Dewatered sand fill (without cement) will support men walking on it almost immediately and will take machinery within a few hours.

In a small mine without much machinery it is possible to line the bottom of a stope with plastic sheeting to stop seepage of slimes, which blocks gullies and erodes pump impellers. Water drainage must then all be by decantation.

More normal drainage is by a mixture of percolation and decantation, although decantation removes most of the water. Three or four decant pipelines, or a hessian wrapped cribbed tower, may be provided for each filling block. These are connected into a drainage sump. (N.B. Pipelines are capped off and covered over during drilling and loading operations.) Slotted pipes wrapped in hessian sacking are laid along the sill floor or through access bulkheads as necessary to allow percolation of water down through the sand.

Each access to a stope must be sealed as the sand rises past it. The Zinc Corporation at Broken Hill measured hydraulic heads on the barricades and found them to be low in well drained sand. Only a light timber barricade is used but soil drains of coarse rock and matting are placed over drainage pipes which pass through the base of the seal to prevent any possible increase in hydrostatic head at a later date.

There has been discussion on the amount of regional support that sand fill provides. The sceptics do not appear to have read coal mining literature which shows that sand fill run in horizontally on bedded deposits can limit surface subsidence to 10 per cent of extracted height, whereas a caved deposit shows 90 per cent subsidence. India and some central European coal mines have been using sand fill for over seventy years.

The U.S. Bureau of Mines has measured convergence in narrow sand filled stopes in the Coeur d'Alene district of Idaho and shown it to be between 10 to 20 per cent laterally, even when crown pillars were removed. Because of the massive deadweight involved this figure should still apply in wider stopes. In any case the sand certainly prevents discontinuous slabbing of walls and general deterioration of working conditions.

Cemented Cut and Fill

Sand fill can be stabilised with the addition of 6 to 10 per cent cement and can develop compressive strengths of 0.7-1.0 MPa. Crown pillars can be recovered by taking regular spaced open stopes, filling them with cemented fill, and then mining the pillars between stopes in a small scale version of the method shown in Figure 6.19. Isa Mine opens a MICAF stope to full width (the orebody width or about 10.7 m, whichever is least) over about three lifts. The sand fill in each of these lifts contains 10 per cent Portland cement to solidify it. This permits recovery of the sill pillar with minimum dilution from the fill above.

Dilution From Fill

Several sophisticated methods have been developed to prevent dilution of ore with fill. Guidelines have been painted on stope walls. Cement dust has been mixed into the top few centimetres of sand fill with a rotary cultivator to give a hard crust. In general however it is accepted that some of the broken ore will be ground into the new surface of the fill and will act as a floor for loading. The final operation in loading is for the loader operator to recover the broken ore from the surface of the fill. About 5 per cent dilution of ore with sand would be accepted, but experienced operators should reduce this amount considerably.

Stope Support

The advantages of mechanised cut and fill are such that its use is extended into as weak a range of rocks as possible. Uphole stoping requires a maximum height of about 8 m and a minimum height of about 3 to 4 m of hanging wall exposed. If the hanging wall, or the backs or footwall, are slabby then they can be rockbolted, possibly with the addition of mesh or guniting. Further details are given in Chapter 8. Occasional cribs may be installed to hold bad patches of backs, provided that the loaders can manoeuvre around them. The cribs are withdrawn before blasting.

When a crown pillar is approached the last few lifts of backs may be bolted systematically with a set pattern of rock bolts as an extra security measure. The bolts are blasted along with the rock and can be a nuisance in loading and crushing operations. Some mines have adopted systematic bolting of the backs as a routine safety measure at all levels in the stope.

Flat-backed Stopes

To extend the range even further a mine can use the original mechanised cut and fill method of flat-backing, as illustrated in Figure 6.25.

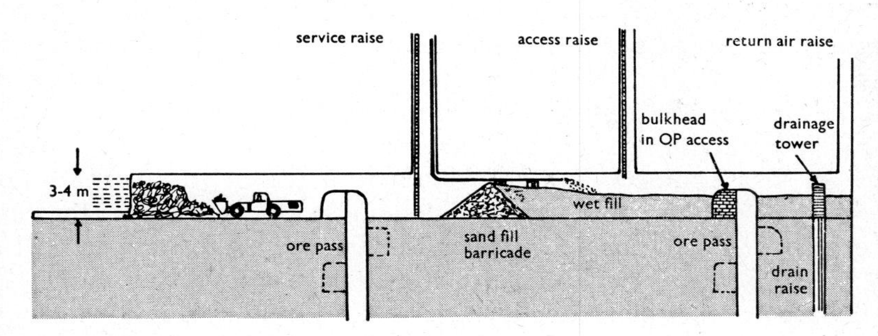

Fig. 6.25 Flat-backed cut and fill

This method is also called horizontal cut and fill, and the working face advances in the same way as a development end. With scooptram type LHD machines a minimum working width of about 5.5 m is needed, and a height of about 3 m. With a rocker-shovel such as the Cavo a height of 2.5 m and a width of about 2 m is sufficient. This enables a stope to be worked with only 2.5 m of wall exposed.

Flat backing may be used to follow odd pockets of high grade ore outside the normal outline of a stope. The tonnage per blast in a flat backed heading is obviously restricted by the effective depth of the round, say 3 m, so that the output per manshift is likely to be reduced.

Filling in a flat-backed stope cannot be carried out until the stope has reached its limit, when the machinery is withdrawn and the stope is filled to the roof. Settlement of the fill and roof irregularities, will provide an airway through the stope, or else forced ventilation through ducts is necessary.

Other Uses of Cut and Fill Stoping

Cut and fill can be used without LHD machines. The ore broken in a stope of short length can easily be scraped back to an ore pass with a slusher. The slusher can be rotated around the ore pass and may have a run of over 50 m in either direction. In this case cribbed ore passes in the fill must be used. Fairly wide stopes can be worked with this method but ore may have to be cleaned up by hand from the edges of the stope into the slusher hoe run.

Where slusher loading is used in a stope which is filled with rock waste, then the floor of the stope can be inclined downwards to a central ore pass. Fill is dropped down two end raises and will rill down towards barricades at the stope centre: this makes the filling operation easier. In both rock and sand filled stopes a temporary board floor may be laid for the scraper hoe to run on to avoid excessive dilution of the ore with fill.

Narrow veins can be worked by cut and fill. Western Mining Corporation at Kambalda mines the steeper pitches of its nickel vein by this method. Stopes are developed above a drive and a one-man party (by tradition in this area) flat backs the stope with an airleg-mounted drill. After the ore is blasted he scrapes it back to an ore pass with a slusher. Only two raises are carried—one for the ore pass and one as a manway. The compressed air exhaust from the drill and the scraper provide part of the ventilation. When the stope has reached its limits a recess is made in the backs over the manway, the scraper is lifted into it and sand is run down into the stope through a borehole from the surface or from an upper level. The fill borehole is inclined to pass down through the ore.

Cut and fill is also a good method to recover pillars. It can be used in conjunction with underhand or overhand square set stoping, described later.

Experiments are being made with underhand stoping by cut and fill. A top slice is taken and then a laminated wooden mat is laid and the stope is filled with cemented mill tailings. Mining of the next slice below the mat can then take place. Laboratory experiments indicate that in a relatively narrow stope uncemented sand fill will bridge over at a height of three to five D, where D is the width of the stope. Experience in underhand square set pillar recovery at Broken Hill South indicates that loads are not high under sand-filled sections. It may thus be possible to mine underhand without using cement or square sets, but by setting the mat on stulls. Sideways convergence of the stope would progressively consolidate the sand fill.

Summary of Cut and Fill Stoping

Applicability

Cut and fill is applicable under a wide range of conditions; from small to large deposits of irregular outline, and from flat deposits to dips of 90°, although filling operations are easier in steep deposits. Its greatest application is in stoping ores of reasonable strength between weak walls which require permanent support. The back must be self-supporting over the width of the stope, although occasional roof bolts may be used.

Advantages of Cut and Fill

1. It is estimated that the preparation and development costs for mechanised cut and fill stoping are only one-half those for longhole, sublevel or shrinkage stoping (no draw points or sublevels needed) and one-third less than the older style of cut and fill with slusher loading.

2. Stopes can be brought into production more quickly because of the reduced development needs.
3. Ore is removed as fast as it is broken so that capital is not tied up. Fire hazard and oxidation problems are also removed.
4. Manpower requirements are low.
5. The manpower works in a concentrated area and supervision is easier.
6. The method has a good safety record: men only work in newly exposed areas which do not have time to deteriorate.
7. Stopes are easily ventilated, the small space open enables a good current of air to be maintained.
8. There is less dilution of ore from slabbing of walls than in other methods.
9. Secondary breaking can be done in the stopes. Provided that ore pass and chute sizes match bucket sizes hangups are minimised.
10. The fill supports weak walls and also maintains the general stability of the mine. There are no old open stopes to collapse suddenly. Reduced subsidence can enable mining to take place below bodies of water.
11. Returning the bulk of mill tailings underground avoids surface disposal problems.

Disadvantages of Cut and Fill

1. Output from small stopes is irregular unless several stopes can be programmed concurrently.
2. Filling must be readily available when needed.
3. The cost of fill can be high: it may be about 50 per cent of the total mining cost.
4. The disposal of tailings slimes without coarse sand can give problems of tailings dam stability and tailings pond restoration.

5. STULLED STOPES IN NARROW VEINS

Two methods of vein working by cut and fill and resuing have already been mentioned. However one advantage of a narrow vein is that it is easily possible to brace the hanging and footwalls apart by using stull timbers directly between them, and these stulls may provide the only artificial support during excavation of the stopes. Stulls may be placed at irregular intervals to support local patches of insecure ground, in which case the stopes are virtually open stopes. Filling may be introduced where necessary to provide better and more lasting support for heavy ground. If close filling is required, and the regular cycle of operations of cut and fill is followed, then the method may be classified as cut and fill. The method is specifically stull stoping where stulls are placed at regular intervals throughout the stope as in Figure 6.26 and serve as the principal or only support of the walls or back, and incidentally as a working platform, with no systematic filling. The system is used in working narrow veins of fairly strong ore with moderately strong walls, and is used principally in small mines, preferably with rich ore to pay for the high labour costs. Stoping can be either overhand or underhand.

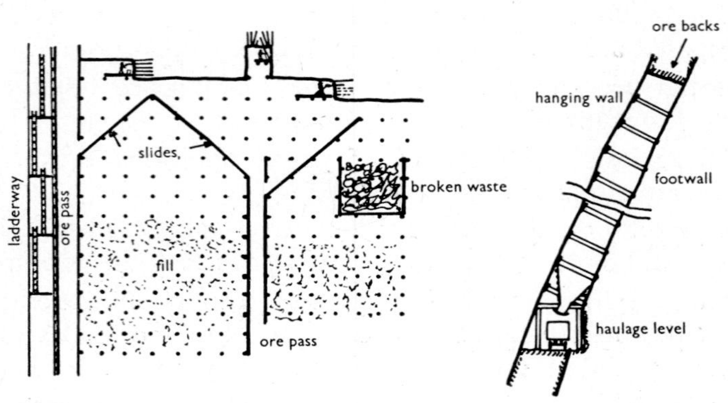

Fig. 6.26 Stull stoping in a narrow vein

Stulled stopes are better where timber is relatively cheap and the stope width is equal to the vein width—i.e. the walls of the vein form the walls of the stope. It is specially applicable to thin, tabular deposits having an immediate hanging wall which is weak, and which will slab off for a few metres back unless supported in some manner. Where the weak hanging wall is soft and incompetent stulls cannot be depended on to hold it up and the method is not applicable.

As ore deposits run deeper, say below 2-300 m, it becomes essential to fill these stull stopes as soon as possible to give permanent support to take over from the stulls. If the stopes are not filled then large areas of the mine can collapse, although a weak hanging wall would progressively cave, as for a bedded deposit. The filling operation need not be cyclic as for cut and fill.

6. SQUARE SET STOPING

This is a labour intensive high cost method because of the massive timber sets which have to be erected. It also has a high material cost for the supports. As a result it is now unlikely to be used for large scale mining. It was originally developed in America and was introduced to Broken Hill in 1888 at the BHP mine by W. H. Patton, an American mining engineer who was general manager of the mine. It is still used at Broken Hill[9,10] for the recovery of high grade silver/lead/zinc deposits.

The term 'square set stoping' is applied to that method of mining in which the walls and back of the excavations are supported by regular framed timbers forming a skeleton enclosing a series of connected hollow, rectangular prisms in the space formerly occupied by the ore, and providing continuous lines of support in three directions at right angles to each other. The ore is excavated in small rectangular blocks just large enough to provide room for standing a set of timber. The essential timbers comprising a standard square set are respectively termed 'posts' or 'legs'; 'caps'; and 'girts' or 'ties' or 'struts'. The legs are the upright members, the caps and struts being horizontal (Figures 6.27 and 6.28).

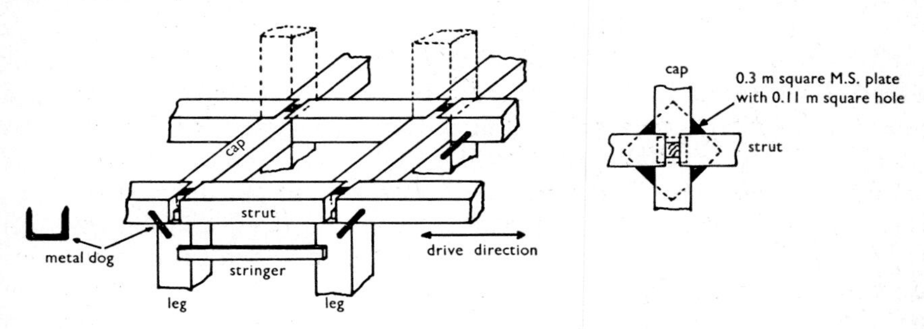

Fig. 6.27 Square set timbers

In the original square sets the legs, caps and struts were all 0.23-0.25 m square but now sizes are reduced wherever possible. Square set stopes are usually waste-filled or sand filled so that leg sizes can be reduced to about 0.20 m square. A mild steel cap plate is used to provide sufficient bearing surface for the top ring timber which has to remain about 0.23 m square (Figure 6.27).

The ends of the members are framed to give each a bearing against the other two at the corners of the sets. Stopes can be mined out in floors, or horizontal panels, with the sets of each successive floor framed into the sets of the preceding floor, or they may be mined in a series of vertical or inclined panels. Usually square set stoping is accompanied by filling of the stoped ground: in heavy ground the sets are filled with waste shortly after they are put in, leaving only a small volume of unfilled stope at any one time.

The main application of square set timbering is in the recovery of badly fractured ore remnants and pillars where the ore backs and walls are too weak to stand open for even one to two metres for more than a short time.

Stopes can be mined overhand. They are silled out at the base and operated by cut and fill for a few lifts until conditions deteriorate whereupon the square sets are introduced throughout the stope. Boards are laid on the last layer of fill for scraping ore to passes. The next lift of ore is broken down by flatbacking along above the first layer of square sets. As each round is blasted a second layer of square set timber is superimposed on the first. The stope will be worked in a series of longitudinal or transverse slices until the second layer is extracted. The first layer can be filled section by section keeping ore passes and ladderways open, and then a third layer is taken, dropping the ore into the second section. In really bad conditions only one layer may be kept open.

Where a pillar is designed to be removed from within and below fill a bottom layer of boards (sollars) is left in the filled stope, and side mats are put in. The pillar stope is opened below the bottom boards and benched underhand. In this way the condition of the backs is known. Permanent angle braces are used to deflect stress sideways onto consolidated fill (Figure 6.28). The pillar is started from an ore pass left alongside the pillar (see Figure 6.24) and is benched downwards for about 3 sets and then along as shown so that the broken ore can rill downwards to a scraper gangway. As soon as a block of a few sets are mined they are filled onto a board platform and mining continues beneath filled sets. It is obvious that this is a slow and expensive method and demands skilled workers. It is almost a last resort to recover valuable ore. About one-third of the labour force is employed to bring in timber, and the large amount of timber can be a fire hazard when it dries out.

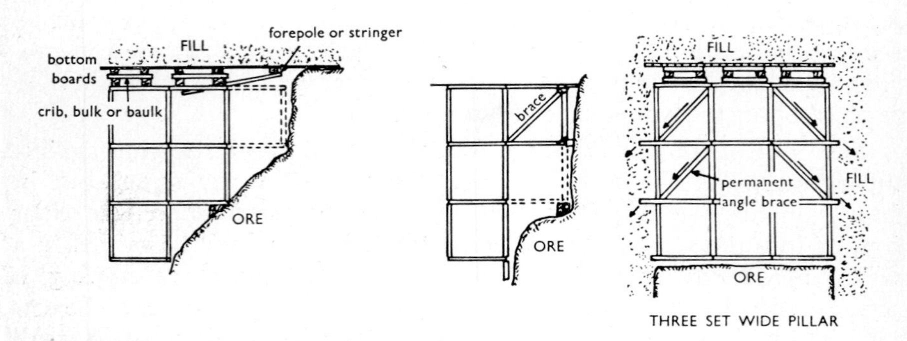

Fig. 6.28 Underhand square set stoping

It should be remembered that although square set stoping is expensive, it is often necessary to square set a drive as a single or double gangway in weak ground. This provides haulage levels and loading chutes. Otherwise the only advantage of square set stoping is that it will succeed in recovering weak ore where all else may fail.

7a. BLOCK CAVING

When the wall rocks, or the rocks capping an orebody, are weak then caving methods can be considered. If the ore is weak enough to cave under its own weight, then it can be block caved. If the ore is relatively strong, then it is sub-level caved so that it can be blasted.

An outline method of block caving is shown in Figure 6.29. The ore is undercut with a series of drawpoints and grizzly levels as described earlier. Below the grizzly level a series of transfer raises are used to drop the ore down into haulage drives. To start the ore caving the drawpoint raises are drilled and blasted into expanded inverted cones until they meet. Temporary supports are used when necessary and these are withdrawn from a distance.

The ore is also weakened at the edges of the block by cut-off raises which define the block and give a firm pillar. The cut-off raises may have to be interconnected by cut-off drives or even by a shrinkage slot stope to break the ore at the edges. Some longhole drilling may be necessary to fragment the ore sufficiently to start a cave. The amount of development work needed to break the rock obviously depends on its strength. This strength cannot be defined in terms of the compressive strength, or more usefully the tensile strength of a small specimen. For a caving operation a parameter such as the *rock quality designation*, described in Chapter 9, is more useful. It is more important to know how easy it is for the ore to break into small pieces than it is to know how strong the small pieces are. The former is a function of bedding planes and joints, and induced cleavage. The latter on the usual small laboratory specimens tells one how strong an unbroken rock is.

Once the ore is undercut and starts to run it breaks away in tension, under gravity loading, and breaks should develop and run up to the surface. The large blocks of ore and rock capping, as they move downwards, mutually crush each other to grind the ore down to a size which will pass through the draw points. Hence a soft and friable rock is more easily caved.

Drawing ore is an important process and must be watched carefully for two aspects, safety and dilution. For safe operation the ore must not be allowed to hang-up in large chambers. If one of these chambers collapses then it can produce an *air blast* by sudden compression of the air inside the cavity. The air blast can have the same effect as an explosion, but since it may occur unexpectedly it can cause injuries and damage to drawpoints and equipment. Many block-cave operations are monitored through boreholes from the surface. If the ore hangs-up then it must be broken by extra blasting.

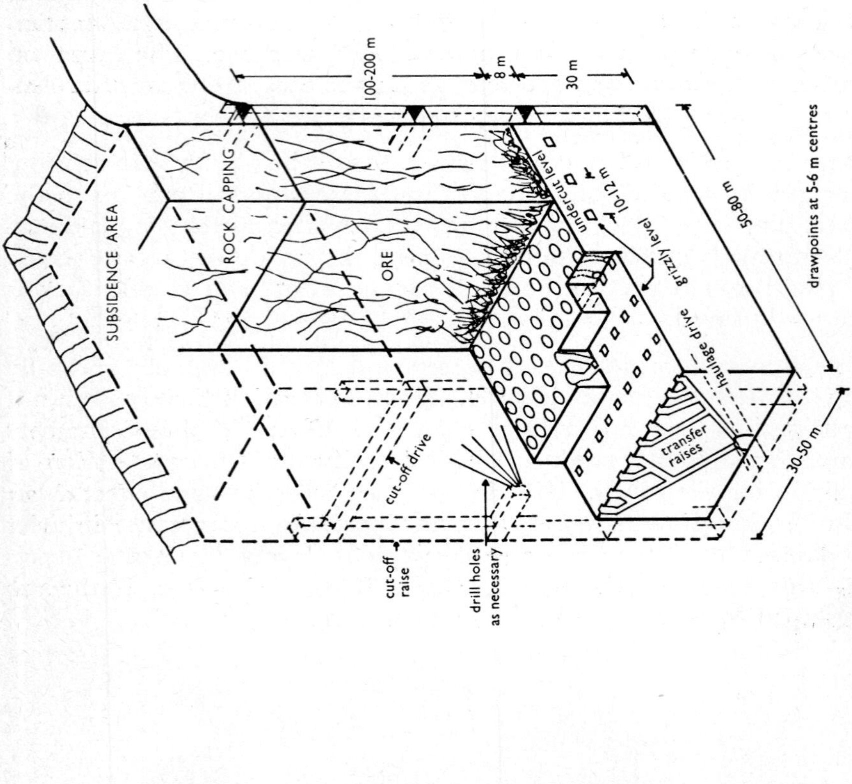

SUBSIDENCE AREA
ROCK CAPPING
ORE
100-200 m
8 m
30 m
undercut level
10-12 m
grizzly level
haulage drive
transfer raises
50-80 m
30-50 m
drawpoints at 5-6 m centres
cut-off drive
cut-off raise
drill holes as necessary

Fig. 6.29 Typical block caving layout

broken capping and fill
broken ore
abandon drive and cave to lower level
SECTION
25% drawn
50% drawn
75% drawn
completed
caving completed on this level
PLAN

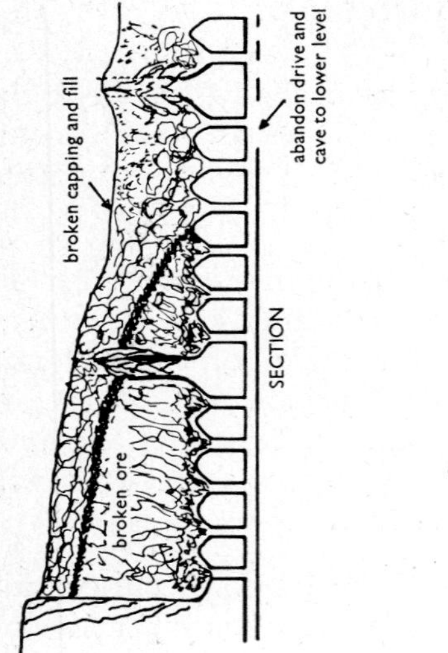

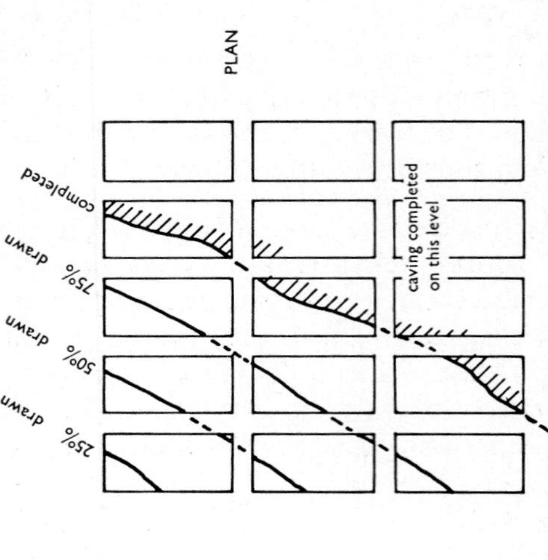

Fig. 6.30 Draw control for block caving

In scientific caving the volume of ore from each drawpoint must be carefully measured (usually by a conversion of mine car loads to weight or volume of ore in situ) so that excessive dilution of ore can be avoided. When broken ore is drawn from a single point an ellipse of ore flows from above it; this is dealt with in more detail in sub-level caving. If one point were drawn to excess then a cylinder of rock would travel down through the ore instead of remaining with the capping. When an adjacent point is drawn then the cylinder of broken rock would mingle with the adjacent ore stream and dilute its edges. Consequently each draw point must be drawn in regular small quantities to lower the capping evenly. Where adjacent blocks are to be worked then the new blocks must be developed and partially drawn off before the old ones are abandoned. Figure 6.30 illustrates the principle. Detailed plans showing this information would be kept up to date and draw point attendants would be issued with instructions on which points to draw and how much ore to release from each. An article by Tobie and Staley[11] gives details of the practice at San Manuel Mine, Arizona; a chalcopyrite operation producing 50 000 tonnes per day.

Final clearing of draw points will be monitored by increasing the frequency of analysis of ore samples to, say, one per 50 to 100 tonnes drawn. When three consecutive samples fall below the ore cut-off grade the draw point is abandoned. In cases where there is a clear visual difference between ore and rock then the supervisor may make his own decision.

Block caving suffers from the same problems as shrinkage stoping: long development times and a limitation to non-sticky low-sulphide ores to give free flowing and non-oxidising conditions. In addition blocky ores can cause frequent hang-ups in the draw points.

The effects of block caving, which is essentially a large scale operation, extend up to the surface where marked subsidence can be caused. To minimise the effects of subsidence, and to help stabilise the hanging and footwalls, rock fill may be tipped into the surface opening. Block caving often follows on below an old open pit operation and the hanging wall side of an open pit can become unstable. Suitable precautions must be taken to fence off or to stabilise the affected areas.

Because of the broken rock above a cave it is easy for surface water or ground water to enter and to flood workings. While a cave is close to the surface pumping capacity must be provided to meet the maximum expected stormwater flow. As a cave progresses deeper, and sidewall collapse or fill provides more cover, then this rock acts as a sponge and rainwater flows are smoothed out. If there is a large flow of groundwater then heavy pumping is necessary to lower the water table below the caved areas.

Boldt and Queneau[12] describe the history of the Creighton Mine in Canada. This nickel operation progressively worked a high grade orebody down from a surface open pit through a series of stopes. In the process a large orebody of low grade nickel in the hanging wall was extensively undermined and fractured. With an increase in demand for nickel a feasibility study showed that this low grade ore could be developed as a block cave. The previous mining operations proved an asset in helping to produce good cave conditions.

More recently methods of breaking ground with a nuclear explosion to stimulate a block cave have been suggested. A U.S. Bureau of Mines report[13] suggests that although the technique is feasible it would not be economical in hard ores, but nuclear blasting might be a better technique for leaching.

Advantages of Block Caving

1. Mining cost is low; it is nearly as good as open-pit mining.
2. A high rate of production is obtained once caving has been started.
3. Conditions can be standardised, making for safe and efficient operation.
4. The accident rate is fairly low.

Disadvantages of Block Caving

1. The capital expense is large and the development time is long.
2. The ore is diluted with waste and there is some loss of ore.
3. There must be careful supervision of ore drawing.
4. Low grade ore in the capping and at the margin of the orebody, or odd stringers, is lost *OR* there is excessive dilution if caving is uncontrolled.
5. There is no chance of selective mining of high and low grade ore. It must be taken as it comes.

7b. SUBLEVEL CAVING

The main advantages of sublevel caving are that the ore can be mined selectively and only small openings are needed in the rock and ore. The sublevels are orientated to gather the maximum ore from between hanging wall and footwall. The drives can be either longitudinal or transverse dependent on the width of the orebody. Figure 6.31 shows the layout of a typical sublevel cave operation. Drives are made on successively lower sublevels and the ore between the drives is blasted down and loaded out. In this respect it is similar to sublevel open stoping, but in caving the walls and capping rock are weaker. They break up and follow the operation down (Figure 6.32).

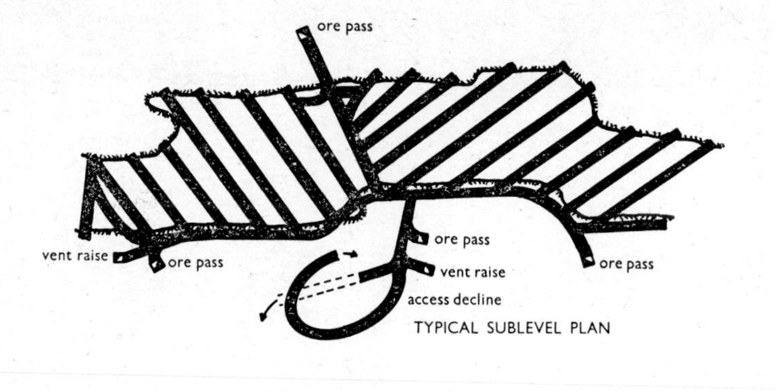

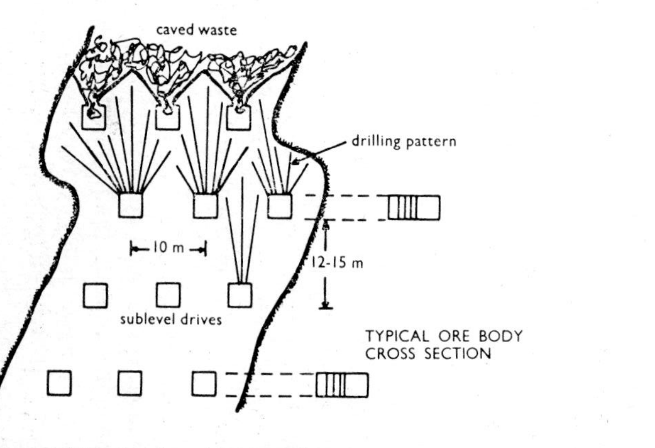

Fig. 6.31 Sublevel caving layout

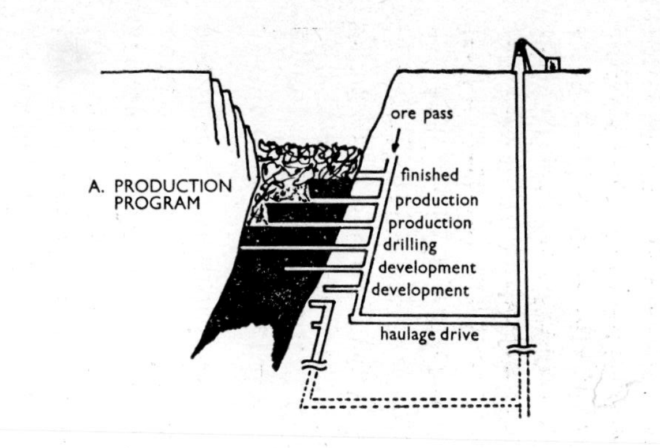

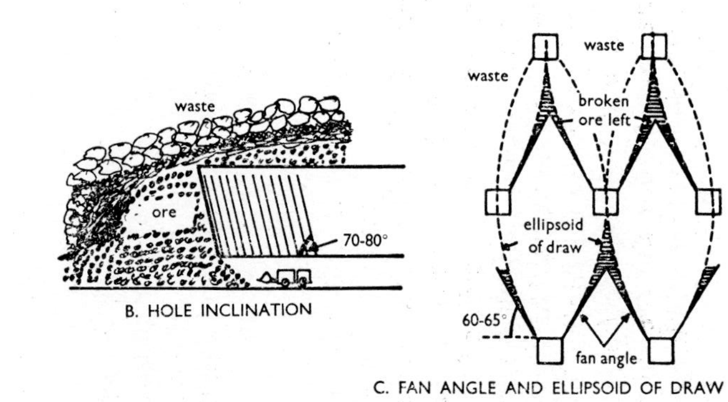

Fig. 6.32 Sublevel caving details

Where an orebody contains small and irregular ore shoots surrounded by weak rock then block caving would cause excessive dilution. If in addition the ore is hard, or contains hard pockets, then block caving will not operate properly, and sublevel caving is chosen. A large orebody with variable grade ore can also be selectively worked by sublevel caving.

The openings required for production are quite small; it is only necessary to have a drive large enough to admit the LHD machine chosen. Hence a drive 2.5 m high by 2 m wide can suffice if a rocker shovel is used. Better production is achieved from a diesel LHD and the drive then has to be about 4 m wide by 3 m high although smaller machines are available. To enable a larger drive to stand in incompetent rock Craigmont Mine[14] shotcreted weak portions in addition to using rockbolts and occasional posts and caps.

The plan layout of drives is adjusted to traverse the orebody in a manner to suit the loading equipment. If diesel LHD equipment is used then angled drives give easier access. Mineral grades are usually zoned across an orebody and drives angled at 30-40° from the footwall enable ore to be mined selectively from each drive. Feasibility studies at Kiruna and practice[15] showed this to be effective. If air-powered LHD machines are used then drives at right angles to the footwall may be necessary to suit the short haul characteristic.

Although orientation of the drives is fairly free, the horizontal and vertical spacing is critical. Drive sizes are kept small because they are usually in weak rock or ore. Conversely the pillars must be as wide as possible for stability and to minimise the number of drives, because the drive is more expensive than pillar extraction. However the controlling factor is the geometry of the draw properties of broken ore. Janelid and Kvapil[16] have dealt with the theory of this and Cox[17] has dealt with the practical aspects of ore flow at Mufulira Copper Mines and several Swedish iron ore mines.

When ore is drawn down through a broken mass it tends to flow from within an ellipsoid (Figure 6.32c). Thus the amount of ore that will flow through an individual opening is limited at the base to an angle of 65-70° from the horizontal. This rapidly changes to a vertical cut-off and in a coarse ore there is a tendency to bridge over and hang-up. Movement at adjacent draw points will usually destroy the bridge if the draw points are close. To allow for unevenness of the blast trimmer holes in the fan cross-section and to reduce the sliding friction the outer holes of a fan are drilled at 60-65° from the horizontal. Occasionally the angle may be flattened to 50° but the ore does not slide. A wedge of broken ore is usually left on top of each intermediate pillar and is recovered from below.

Given the included angle of a fan of holes it is then necessary to determine the vertical height between sublevels. The geometry of the layout indicates that alternate sublevels should be above each other and intermediate levels should be offset by half a pillar width. The height between alternate sublevels is then dictated by two factors: the accuracy with which the longest fan holes can be drilled, and the most efficient flow pattern. A 15 m floor to floor sublevel interval is indicated as a maximum on present drilling practice because the longest holes in the fans are then around 20-25 m. Theoretical and model studies for flow properties show an ideal spacing of 18-20 m, although practice indicates that ore fragmentation is not good at these heights. Sublevel intervals tend to be around 10 m both for grade control and good flow properties.

Once the height interval is established the width interval follows. In practice it is usually close to 10 m between road centres. Even where the height interval is reduced the 10 m width is both practical and is an economic minimum. Closer spacing would be virtually the top slicing method.

In addition to fan angles the hole inclination in relation to the pillar must be considered as in Figure 6.32b. Unlike open stoping these holes are fired against brocken rock and ore which have a confining effect. The best fragmentation and brow configuration is obtained by sloping the holes forward. A typical angle is about 20-30°, i.e. the holes have an angle of 70-80° from the horizontal. Hole burden is about 1.5-1.8 m and two to three rings of holes are fired by delay detonators at each blast. Charges are reduced at the brow level to avoid excessive shattering because the machine loads out from under the brow.

Grade Control

Sublevel caving is perhaps the most sensitive method of mining with regard to dilution. A capping of waste rock is being lowered over individually blasted layers of ore. Each time ore is loaded out from a drive then the rock vertically above drops the fastest. The volume of ore drawn must be pre-calculated and the quality drawn must be carefully sampled. The average grade per blast is calculated from samples in the four adjacent drifts. About 500-1000 tonnes may be blasted per round. If a rapid analysis instrument (such as an x-ray fluorescence meter) is available about 10-20 samples may be taken from the broken ore, both for quality control and grade monitoring[17]. When three successive samples from the end of the run fall below the cut-off grade then loading is stopped. Kiruna mine[15] has a computer program method of scheduling optimum ore grade levels from about 45 loading machines and up to 100 loading points.

Dilution rises rapidly towards the end of a run. At 70 per cent recovery there can be about 15 per cent dilution and at 90 per cent recovery there may be about 25-30 per cent dilution. Careful records are necessary to calculate an economic recovery rate. (In fact 90 per cent recovery is very good for any mining method: in many stoping methods small pillars and pillar remnants can easily leave 10 per cent of the ore unmined.) Once the flow characteristics of a sublevel cave are ascertained the volume of ore drawn from each blast is adjusted to compromise between dilution and recovery.

Older methods of sublevel caving used a mat of crushed timber between the ore and the rock capping (as in top slicing). Timber is expensive, so is the labour to install it. Also a timber mat can jam and cause hang-ups which are undesirable in mechanised mining. None of the modern mines appear to use timber mats.

Advantages of Sublevel Caving

1. It allows selective mining of weak or medium strength orebodies in weak rocks.
2. Small orebodies can be mined and the method is flexible.
3. The operations can be highly mechanised.
4. If orebodies are weak near the surface and stronger at depth the same equipment can be used for a change to mechanised sublevel open stoping.
5. Development is less than for block caving and very few openings have to be maintained for long intervals of time. Production is obtained quickly.
6. The ore is mined rapidly after being broken so that it does not deteriorate or catch fire.
7. It can be used in wet and sticky ores which will not block cave or shrinkage stope.
8. It is much more economical than other weak ground methods of mining, especially square set and top slicing.
9. No pillars of ore are lost.
10. It can be used to recover large pillars of ore by caving between fill.

Disadvantages of Sublevel Caving

1. Either high dilution (20-30 per cent) must be tolerated or a poor recovery results.
2. Ventilation of stopes is difficult; each production drive usually requires duct ventilation if diesel equipment is used.
3. The method causes surface subsidence.

8. TOP SLICING

Top slicing is more suitable for large horizontal deposits. It belongs to the same vintage as square set mining and needs cheap labour and a cheap and plentiful timber supply; this does not have to be high grade hardwood because it is not used for support but to provide a flexible mat as a roof between the ore and the rock above it. Under modern conditions either open pit, block caving or sublevel caving is more likely to be used. Small pillars that could be top sliced are more likely to be recovered by cut and fill methods or to be abandoned.

To develop an ore body for top slicing the top of the deposit may be levelled off and the ground broken up by square set mining. A generous mat of timber will then be laid on the floor of the drives and the standing timber drawn off or broken by firing charges in the posts. Mining then proceeds by taking parallel slices of ore from under the mat, allowing the capping to subside as each slice is worked out as in Figure 6.33. If necessary extra timber is added on the floor of each slice to provide a flexible roof of broken timber which will separate the roof waste rock from the ore.

Advantages of Top Slicing

1. Safe method to use where overhand stoping cannot be employed and timber is plentiful and cheap.
2. High recovery is possible with very little dilution.
3. Method is suitable for intermittent operations.
4. Once initial development is done the method can be reasonably cheap provided that labour is cheap.
5. Can be employed under sand or other loose material.

Disadvantages of Top Slicing

1. Method causes surface subsidence.
2. Ventilation can be difficult and the timber mat in conjunction with sulphide ores can cause a fire hazard.
3. A considerable number of working ends is needed for a large output and the rate is not flexible.
4. Period of development to production is fairly long.
5. Handling of timber and laying of mats is expensive in labour and time consuming: as for square set mining about one third of the labour force is used for timber handling.
6. Sudden collapses of broken roof, which has hung-up in blocks, can be dangerous in the slice below.
7. Waste or low grade ore cannot be easily left in place.

9. LONGWALL STOPING

Longwall mining for soft minerals is described in detail in Chapter 7. The technique is modified in hard rocks and for hard ores. South African gold mines work thin reef deposits at relatively low angles of inclination. The host rock is a hard quartzite with a uniaxial compressive strength of around 200 MPa. When worked at a depth of 1500 m or more these hard rocks are liable to rockbursts if pillars are left.

The reefs are mined by longwall methods shown in Figure 6.34. A face is opened out down dip to be worked advancing along the strike, although retreat mining is possible. Mining is carried out by drilling short holes angled to the line of the face. The holes are fired in relays from the bottom of the face to the top or an intermediate position. As far as possible the line of the face is kept straight so that there are no stress-raising corners, although the face is angled with the dip end lagging so that the scraper loader keeps into the face. Blast mats are erected to stop the broken ore from flying back into the waste.

The ore is fired primarily because of its hardness, but the heavy shotfiring also helps to release stress or to induce potential rockbursts while the men are withdrawn. Experiments are in hand to cut the reef with diamond saws to help mining. In this case the cutting machines could well be mounted on a conveyor in due course as in coal mining.

It is essential to break off the roof and cave the waste in deep mining. A strong row of breaking-off supports are used and gravity loading snaps the unsupported beds, or else they sag and lower in large blocks. The breaker props can be bundles of timber stulls chained or fastened with wire bands. A more modern method is to use hydraulic chocks which are recycled and moved forward after each face strip (see Chapter 7).

COSTS

It is neither possible nor advisable to try and give costs of all types of mining. For all mines it is the total cost of mining, milling, development, exploration and administration which matter. If a method of mining such as sublevel caving is used which gives high dilution then mining costs can be low but the mill costs go up because extra waste ore has to be removed and disposed of.

It is true too of mining costs that each method has low cost and high cost components and a surprising number of mines arrive at mining costs of about $2 per tonne for underground methods. However at, say, one million tonnes a year, a saving of one cent per tonne can help dividends enormously. The awful truth of mining is that costs cannot be accurately predicted to a few cents, although Chapter 4 shows a need for this.

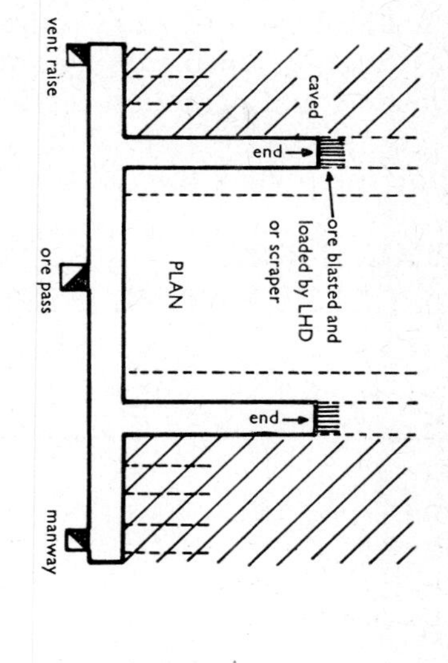

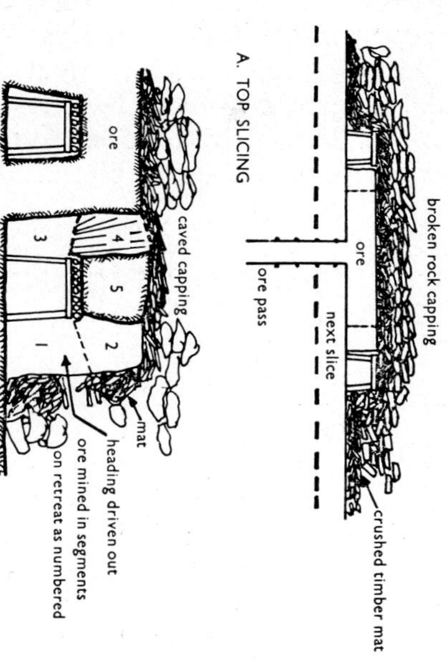

Fig. 6.33 Top slicing

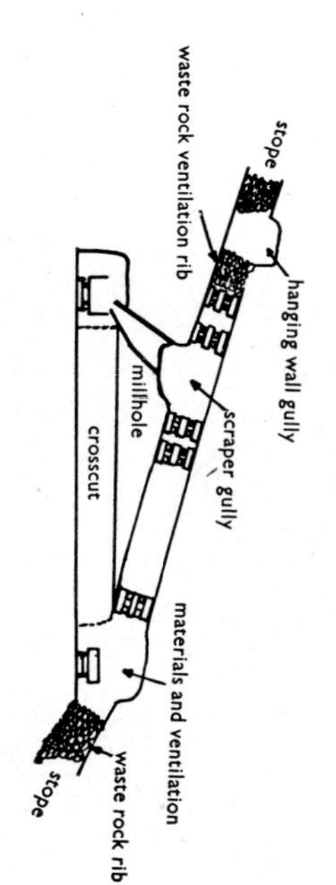

Fig. 6.34 Longwall reef mining

The U.S. Bureau of Mines report quoted earlier[13] gives a comparison of costs for open stoping and for block caving. These were deliberately made equal, although the actual range was from $U.S. 1.15 to $U.S. 2.50. The intention is to show the way in which sectional costs vary with the method. Table XIII shows the comparison of costs.

TABLE XIII.

COMPARATIVE COSTS OF BLOCK CAVING AND OPEN STOPING

Activity	Open Stope	Block Caving
	$	$
Development	0.384	0.296
Drilling and Blasting	0.335	0.004
Drawing Ore	0.275	0.338
Drive repairs	0.007	0.465
Haulage	0.323	0.308
Hoisting	0.116	0.083
Ventilation and Safety	0.074	0.119
Supervision and engineering	0.200	0.135
Workshops and Stores, etc.	0.116	0.082
TOTAL	$1.83	$1.83

The experienced mining engineer keeps his pocket book of costs for all operations and would synthesise a complete operation along the lines of Table XIII.

References

1. CARTER, K. J. *Mining practice at Cleveland Tin N.L., Tasmania.* Proc. Australasian Institute of Mining and Metallurgy, No. 237, p. 1, March 1971.
2. AIREY, L. D. *Introduction of the cascade method of continuous-retreat open stoping at Mufulira Copper Mines Ltd, Zambia.* Trans. Inst. Mining and Metallurgy, 75A, p. A137, 1966.
3. DAVIES, E. *Mining practice at Mount Isa Mines Limited, Australia.* Trans. Inst. Mining and Metallurgy, **76A**, p. A14, 1967.
4. HART, E. K. *Shrinkage stoping at the Zinc Corporation Ltd.* Aust. I.M.M. Monograph No. 3, 'Broken Hill Mines 1968', p. 259. Aust. Inst. of Mining and Metallurgy, Melbourne, 1968.
5. ISOKANGAS, T. A., PURDIE, T. F. and BOYD, R. J. *Mount Isa cut and fill (MICAF) stoping.* Proc. Aust. Inst. of Mining and Metallurgy, No. 226, p. 185, June 1968.
6. BRADY, J. T., OWERS, N. F. and ANNEAR, C. H. *Ore breaking and handling at the C.S.A. Mine, Cobar.* Proc. Aust. I.M.M., No. 229, p. 7, March 1969.
7. MULLER, H. B., RICH, H. J. and GRANT, A. B. *The preparation, storage and placing of hydraulic fill, C.S.A. Mine, Cobar.* Proc. Aust. I.M.M., No. 229, p. 27, March 1969.
8. RUTTER, W. T. *Mechanised cut and fill at New Broken Hill Consolidated Ltd—'B' Lode.* Aust. I.M.M. Monograph No. 3, 'Broken Hill Mines 1968', p. 269.
9. TRELOAR, J. H. *Square set overhand stoping.* Aust. I.M.M. Monograph No. 3, 'Broken Hill Mines 1968', p. 243.
10. ROBJOHNS, M. J. *Underhand stoping at North Broken Hill Ltd.* Aust. I.M.M. Monograph No. 3, 'Broken Hill Mines 1968', p. 251. Aust.
11. TOBIE, R. L. and STALEY, E. K. *Draw control at San Manuel.* Mining Engineering, p. 66, June 1967.
12. BOLDT, J. R. and QUENEAU, P. *The Winning of Nickel.* pp. 91-148. Methuen, London, 1967.
13. HARDWICK, W. R. *Fracturing hard rock with nuclear explosives and extraction of ore by a modified block-caving method.* U.S. Bureau of Mines Report of Investigations, R.I. 7391, 1970.
14. PILLAR, C. L., PETRINA, A. J. and COKAYNE, E. W. *How Craigmont chose sub-level caving and why it proved successful.* World Mining, p. 26, June 1971.
15. ANON. *Production scheduling at Kiruna.* Mining Magazine, **124**, No. 2, p. 107, February 1971.
16. JANELID, I. and KVAPIL, R. *Sub-level caving.* Int. J. Rock Mechanics and Mining Science, **3**, p. 129, 1966.
17. COX, J. A. *Latest developments and draw control in sub-level caving.* Trans. Inst. Mining and Metallurgy, **76A**, p. A149, 1967.

Chapter 7. Underground Mining of Bedded Deposits

Introduction

A mining engineer tends to think in terms of either metal mining, or coal mining. Coal mining implies working bedded deposits of varied thickness but of wide extent. Mine sizes are thought of in terms of several square kilometres. The deposits can spread out under water involving special working restrictions. They can also lie under urban areas where precautions have to be taken against subsidence.

However coal mining methods are also applied to halite and sylvite deposits and other soft ores which can be cut by machines. Pillar and stall methods are applied in limestone and bedded copper ores. Generally this chapter will refer to coal mining for simplicity.

Classification and Properties of Coal

Marketing and utilisation of coal requires a knowledge of some of the terms used to describe it. Details of the properties of coal and other fuels can be found in a book by Francis[1]. The tests for these properties are laid down in the standards of the appropriate countries. Australian Standard A.S. K181-1968 gives the terms used for coal preparation (coal beneficiation); A.S. K152 (in sixteen parts) gives the methods of analysis and testing of coal and coke. This is an Australian endorsement of British Standard B.S. 1016. Sampling of coal is dealt with by A.S. K153-1967, or B.S. 1017. Part 1 of these standards contains details of sampling from skips, conveyors, waggons, etc. and these methods can be applied to any material.

The terms used in the analysis of coal are

Proximate analysis—the analysis in terms of percentages of moisture, volatile matter, ash, fixed carbon and sulphur, plus the calorific value. This is not approximate, but 'Proximate' and very precise.

Ultimate analysis—the analysis in terms of the percentages by weight of the elements present; usually carbon, hydrogen, oxygen, nitrogen and sulphur.

Moisture Content—given as two values,

(a) Free—the percentage lost when the naturally moist finely ground coal is allowed to reach equilibrium with the atmosphere at 15.5°C

(b) Fixed—the percentage of moisture present in the air-dried coal

Ash Content—the percentage of residue obtained when coal is burned in air at 800°C in a muffle furnace under standard conditions.

Volatile Matter Content—the percentage of products evolved when coal is heated in a covered crucible to a temperature of 925°C under standard conditions.

Fixed Carbon—one hundred percent minus the sum of the percentages of ash, volatile matter and moisture.

Calorific Value—the gross calorific value of the coal as determined in a bomb calorimeter.

The bases generally used for reporting analyses are

1. as received
2. dry
3. moisture and ash free (daf—dry, ash free)
4. mineral matter free (mmf)
5. moisture and mineral matter free (dmmf—dry, mineral matter free)
6. the calorific value (CV)

Mineral matter is usually calculated from an empirical formula e.g.

The Parr Formulae (using percentage values)

Total Inorganic Matter = Moisture + 1.08 Ash + 0.55 Sulphur, or

Mineral Matter Content = 1.08 Ash + 0.55 Sulphur + 0.8 Carbon Dioxide.

The general intention is that after removing the mineral matter and the moisture only pure organic coal should be left, which can then be compared (dmmf) with other varieties on a CV basis.

SEYLER'S CLASSIFICATION

This is generally used as a pictorial representation of coal types (Seyler's chart). It has several sets of axes to show different parameters which are summarised in the table given below.

TABLE XIV.
SEYLER'S CLASSIFICATION SIMPLIFIED (DMMF BASIS)
(based on British coals)

Coal Rank	Fixed Carbon Range %	Hydrogen %	Volatile Matter %	Calorific Value Joules per g.	Calorific Value Btu/lb	British Standard Swelling Index
Anthracite	>93	3.0–3.8	5–10	35 800	15 400	1
Semi-Anthracite	93–91	3.8–4.4	10–14	40 700	17 500	1
Semi-Bituminous	93–91	4.4–5.0	14–20	36 800	15 800	$3\frac{1}{2}$
Bituminous	91–84	4.4–5.7	20–36	34 900–36 500	15 000–15 700	6–9
Lignitous	84–75	5.0–5.7	36–49	30 900–33 300	13 300–14 300	2–1
Lignite	<75	5.0–5.7	49–59	27 200	11 700	1
Brown Coal				20 900–29 300	9 000–12 600 daf	
Peat				16 300–20 900	7 000–9 000 daf	

Yallourn brown coal has a moisture content of about 65 per cent and a wet CV 'as received' of 6750 J/g.

COKING PROPERTIES

Before a coal can be carbonised into coke it has to possess two desirable properties. It must 'cake' together and it must swell. Not all caking coals will produce coke. Roughly the two properties combined determine the porosity and hardness of the resultant coke.

The range of coals which will produce coke have a volatile matter range of about 20 to 38 per cent. Maximum swelling cokes of maximum strength have a VM 28 to 32 per cent. Factors such as temperature of carbonisation and the size to which the coal is ground before coking also affect the quality of coke. The best metallurgical coke will be found from Gray-King types G4 to G9. They should also be low phosphorous (less than 0.01%) and low sulphur (less than 1.0%). Coals of G1 to G9 can be used with swelling indices of 4 to 9. Coals outside this range can be blended with high grade coking coal or used to make lower grade cokes.

CAKING INDICES

(a) Crucible Swelling Number

Heat 1 g of coal at 825°C for $2\frac{1}{2}$ min. in a standard closed crucible. Compare the profile of the coke button with those outlined in a series of standards of swelling indices which range from 1 to 9 by $\frac{1}{2}$ units.

(b) Carbonisation Assay: Gray-King Method

Heat 20 g of coal in a standard horizontal retort at 600°C for $1\frac{1}{4}$ hr and compare the profile of the carbonized residue with those formed from a series of standard coals producing coke types A to G3 and G4 to G10.

NUMBER CODING OF COALS

There are three major classifications of coals all based on the proximate analysis. The N.C.B. Coal Rank Code is used in Britain and modified for use in Australia. America uses the A.S.T.M. classification which is based primarily on the proximate analysis and partly on the calorific value. Europe, other than Britain, uses the Economic Commission for Europe (ECE) classification which is a mixture of the N.C.B. and A.S.T.M. classifications.

The Australian system is given in Australian Standard K184-1969 'Classification system for Australian hard coal'. The coal is coded with a four digit number. The first digit gives the volatile matter and calorific value of the coal as shown in the table headed 'Coal Class'. This is followed by the coking Group and Sub-group, and the ash content derived from the appropriate tables.

COAL CLASS

Class Number	Volatile Matter (dmmf) %	Gross Calorific Value (daf)	
		J/g	BTU/lb
1	$\leqslant$ 10	—	
2	10–14	—	
3	14–20	—	
4 A	20–24	—	
4 B	24–28	—	
5	28–33	—	
6	>33	$>33\,700$	$>14\,540$
7	>33	32 000–33 800	13 770–14 540
8	>33	28 400–32 000	12 220–13 770
9	>33	29 400–28 400	11 650–12 220

The dmmf basis for Australian coals is not in accordance with international practice which uses the daf basis for volatile matter. However most Australian coals have a relatively high ash content and for these coals the dmmf basis was chosen because it gives a more accurate expression of the volatile matter. A relationship has been established between the daf and dmmf values for Australian coals.

The group of coal, which is shown by the second digit, is based on the crucible swelling number. This is an index of the caking power of coal on rapid heating whereas the sub-group, (third digit) is based on the Gray-King coke type, an index of caking properties at a slower heating rate, which is more characteristic of conditions during metallurgical coking.

COAL GROUP

Group Number	Crucible Swelling Number
0	0 –½
1	1 –2
2	2½–4
3	4½–6
4	6½–9

COAL SUB-GROUP

Sub-Group Number	Gray-King Coke Type
0	A
1	B–D
2	E–G
3	G1–G4
4	G5–G8
5	G9

The fourth digit, which is shown in parentheses, gives the ash content of the coal.

ASH NUMBER

Ash Number	Ash (Dry basis) %
0	$\leqslant 4$
1	4–8
2	8–12
3	12–16
4	16–20
5	20–24
6	24–28
7	28–32
8	>32

Some Queensland classifications based on the analysis of washed coal are (Queensland Govt. Mining Journal, May 1970, p. 205).

1. Blair Athol 501(2) Moisture 7.5%, Ash (air dried) 8.2%
   VM (dmmf) 32.0%, CV (daf) 13,970 Btu/lb
   C.S. No. ½, G-K Type B.
   (This is power-house and fuel coal)

2. Moura 543(2) Moisture 2.2%, Ash 8.6%
C.V. (daf) 15,330 Btu/lb V.M. (dmmf) 31.0%
C.S. No. 7½, G-K Type G3.
(Reasonable coking coal)

UTILISATION OF COAL

The use of coal has some bearing on its mining. The dominant market demands appear to be for a low cost fuel and for metallurgical coke. The older markets for large coal for hand firing in industry, in railway engines and in homes are disappearing. Consequently the size of the coal product does not matter. In modern use it must frequently be finely ground for coking, or pulverised for boiler use. It must also be as cheap as possible, particularly to compete with oil as a fuel.

Electric utilities take about one-third at least of coal production in Australia and in Britain. The trend is to build the generating station adjacent to the mine. The coal is won mechanically, conveyed to the surface, cleaned, and conveyed to the power station boilers untouched by human hand. The New South Wales Electricity Commission operates coal mines as well as power stations. Munmorah Colliery for instance produces and conveys 6000 tonnes of coal per day from the mine to the adjacent power station with only 250 men involved.

Coking coal commands a higher price than fuel coal so extra money can be spent to mine in more difficult conditions. In addition, or alternatively, it is economically feasible to spend money on separating coking coal from non-coking coal within one seam. Fortunately the coking portion often tends to be powdery and soft within the harder bright coal, and a simple size separation can often upgrade coking properties.

Restrictions on Coal Mining

GAS AND DUST

In the normal processes of coal mining a certain proportion appears as fine coal dust below about 200 microns; this is explosive when suspended as a dust cloud. The proportion below 5 microns is a danger to health. Coal seams also desorb *firedamp*, a mixture of hydrocarbon gases, but principally methane, CH_4. They can also give off carbon dioxide. As a result of various explosions of coal dust and gas in the seventeenth and eighteenth centuries all countries now have legislation which controls coal mining more closely than it does metal mining.

All mining engineers have to know their local mines acts. Typical legislation prescribes roof supports, transport regulations, dust suppression, and mine ventilation, plus all aspects of mining in a general sense and the conduct of mine officials. However mine legislation keeps up with modern practice and mines inspectors are willing to cooperate in safely planned experiments. The following methods of working are generally in compliance with mines acts but there are exceptions. For instance modern longwall mining contravenes the existing New South Wales Coal Mines Regulations Act, 1912, but is permitted by special exemption. This type of situation is common when new types of mining are introduced to a coalfield.

ELECTRICAL

All electrical equipment used in coal mining must be *flameproof*. This implies that the equipment is either incapable of producing incendive sparks capable of igniting gas or dust, or else is securely contained in an explosion-proof chamber. Low voltage (less than 32 volts) equipment may be *intrinsically safe* as an alternative. This means that it is designed so as not to produce incendive sparks. In any case electrical equipment must not be used in areas which contain more than $1\frac{1}{4}$ per cent methane in the general body of the mine air. Careful attention must be paid to ventilation standards to dilute any gas emitted. In some cases techniques such as methane drainage have to be adopted. Mines can be operated with all machinery driven by compressed air but this power is expensive, noisy, and inefficient for most purposes except drilling. Electrical equipment is used wherever possible.

The $1\frac{1}{4}$ per cent methane limit is monitored continuously. For instance a continuous miner may be fitted with a catalytic platinum filament 'pellistor' gas detector, adjusted to flash a warning at about 1 per cent methane and to cut the power supply off at $1\frac{1}{4}$ per cent methane. This can affect mining practice. In a three heading development the coal in each heading would be exposed at a greater rate than if a miner was driving a five heading development. In a gassy seam the number of entries per miner may have to be increased to allow more time for gas to drain off from coal in between working intervals.

STRATA CONTROL

Some further legal requirements are added to control undersea working and to provide for compensation for damage caused by subsidence. This is discussed in a later chapter. When bedded deposits are mined the workings spread over a large area and the surface will lower as a slice is taken out from below it.

The general field of rock mechanics, formerly known as strata control, is dealt with in Section C. A point must be made in advance about the strength of rock and the depth of working. An underground mine can only be worked if its roadways and working areas stay open. Rocks will either flow in a plastic manner or fail in tension if the pressure is high or the roof span is large. The overburden loading pressure is roughly 25 kPa/metre depth for rocks of a density of 2500 kg/m^3.

This vertical pressure due to superincumbent rock is called the *cover load.* At a depth of about 150 m the softest shales and mudstones will undergo pseudo-plastic flow. Harder shales will flow into roadways and roofs will crack up at 200-300 m. Large areas of soft sandstone will show breaks at about 300 m or so. Hard sandstones and conglomerates form good roofs to depths of 500 m or more but the coal ribs will fail under their loading and if the seam floor is a soft seat-earth then that will flow.

As a generality in the harder strata and shallow depths of the United States and Australian coal mining areas pillar mining is possible because the mine workings are in good condition. At depths of about 300 m the layout of pillar workings must be modified in hard rocks, or a change to shortwall or longwall mining is made in softer rocks. At greater depths, particularly in shales, longwall mining becomes essential because strata control in pillar working is too difficult. It is for this reason that longwall mining is almost universal in Europe.

CHOICE OF MINING METHOD

With present mining equipment, apart from a few exceptions, the highest output is obtained from pillar mining methods. The Joint Coal Board report for New South Wales in 1969-70 showed an average output per man shift at the coal face of underground coal mines for all Australia as 38.78 tonnes with 9.89 tonnes as the output per manshift for all men employed. Open-cut mines overall was 28.66 tonnes per manshift. This compares with a British underground O.M.S. (output per manshift) of about 7 tonnes at the coal face and 2.1 tonnes overall. Individual mines can perform much better: in the Burragorang Valley in New South Wales a team of 10 men on a unit can mine over 1000 tonnes per day. One mine with a total manpower of 101 men frequently produces over 10 000 tonnes per day from pillar mining in good conditions.

For this reason pillar mining is abandoned reluctantly as ground pressures become higher or roofs are weaker. Mechanised shortwall and longwall are then adopted.

Underground Machinery

Coal and some other ores are soft and can be won by excavating machines which will both cut and load coal simultaneously. For full efficiency their operation should be as continuous as possible. Coal is a low density high bulk material and belt conveying is particularly attractive.

To keep prices down production is generally mechanised. New South Wales practice is typical of experience in high labour cost countries. The N.S.W. Joint Coal Board Report for 1969-70 states that in June 1970 there were only two mines in N.S.W. using hand mining methods. One employed 15 men and the other 38 men. In total hand mining supplied 0.1 per cent of all coal won in New South Wales: 98.65 per cent came from fully mechanised mines.

Standard groupings of machines exist and the Report gives further statistics from which the following is abstracted:

Method of handling coal	Percentage of Total Tonnage (Total = 159 588 tonnes/day)
Machine loaded to shuttle cars, then coal transported by	
Conveyor belts	78
Conveyors and locos	8
Conveyors and rope	less than 1
Locomotives and rope	1.4
Rope	less than 1
Total via shuttle cars (including sundry methods)	95
Machine loaded (other than to shuttle car) then coal transported by	
Conveyor belts	3
Conveyors and locos	1.5
Rope (and horse)	less than 1
Hand loaded	— negligible.

Production in the generally shallower mines of Australia and the United States is principally by continuous miner, but as mines become deeper longwall mining is showing increasing acceptance. (This is dealt with later.) In Britain longwall mining is already the principal method. The trend in New South Wales is given in the Joint Coal Board Report.

In what is still occasionally called conventional mining coal is broken by a cycle of drilling, blasting and loading. However the table (p. 224) shows that the conventional mining practice in New South Wales is continuous mining. A similar situation exists in Queensland.

DAILY TONNAGE OF RAW COAL WON IN NEW SOUTH WALES

	Longwall Units	Continuous Miners	Coal Cutters	Mined by Hand or Grunched*	Total
June 1968	248	109 711	12 499	2178	124 636
June 1969	5580	127 703	10 984	1968	146 235
June 1970	7371	136 680	10 235	2781	157 067
Percentage at June, 1970	4.7	87.0	6.5	1.8	100.00

* Grunching is blasting coal off the solid; i.e. with only one free face to break to.

COAL CUTTERS

The cutting action of a pick is to rip a groove in the solid coal. To break the coal successfully a staggered sequence of picks is used so that each pick rips out part of the ridge between grooves from earlier picks, as shown diagramatically in Figure 7.1. The actual lacing sequence is not as simple as is shown; it may have up to 9 lines in staggered order. The outside row of picks must cut clearance for the jib and the resultant *kerf* for a single jib may be about 15-20 cm wide.

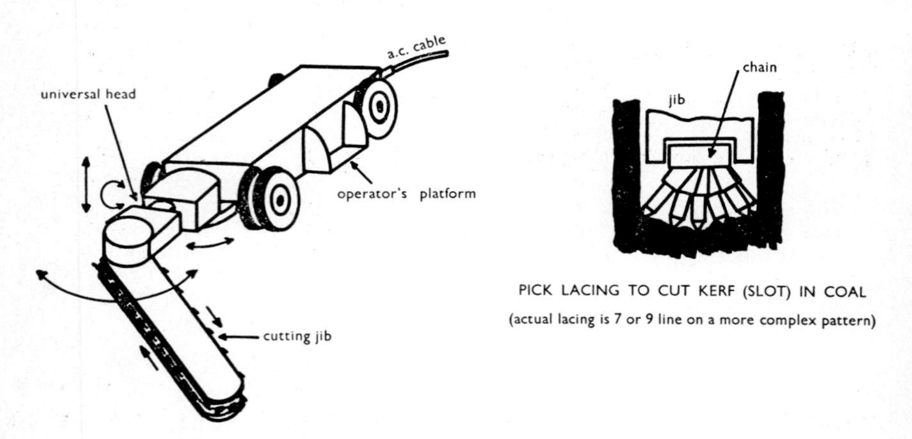

Fig. 7.1 Shortwall coal cutter

All the picks are replaceable and are located in boxes on a flexible chain. Several designs are available, mostly tipped with tungsten carbide to give a strong and re-sharpenable cutting edge. The chain runs around a jib which can be as long as 4 m on a modern shortwall coal cutter as sketched. The drive is by a sprocket and gear train in the universal head which enables the jib to swing through at least 180° laterally and rotate through 360° in a vertical plane. The shortwall cutter shown can cut horizontal, vertical, or diagonal slots in the seam to give a second free face for coal to be blasted into. Other versions are available which will only cut in a lateral plane. The cutters can be mounted on tracks or tyres, the latter giving greater mobility. They can have two or four-wheel drive. Shortwall cutting machines are still widely used, especially in the rubber-tyred versions. Earlier models slid along the floor, pulling themselves by rope haulage, and were *flitted* from one face to another on buggies.

In addition to the shortwall cutter shown there are also longwall versions, which tend to be long and thin, rather than relatively short and fat. Their function was to fit between a conveyor and the face as in Figure 7.2 Later they were modified to run on top of an armoured conveyor. The longwall coal cutter cuts horizontal slots but can have a turret mounted jib to give a range of cutting heights. Modern longwall techniques have caused them to become almost obsolescent.

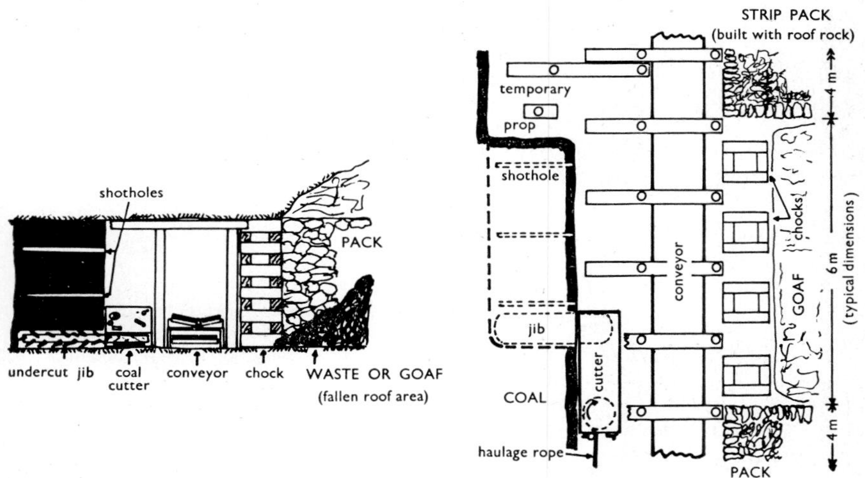

Fig. 7.2 Longwall coal cutting

A specialised application remains. The cutter can be used to drive headings in coal. It cuts an arc under the coal which is then blasted down. Loading flights are fitted to the cutter chain which then is used to scoop the broken coal onto an armoured conveyor. This is known as *flight-loading*, or *waffling*.

The function of both types of machine is to cut a stress-relieving slot in the coal and give a second free face for blasting. Shot-holes are drilled on one or both sides of a cut as necessary and the holes are loaded with explosives and blasted to break the coal or other mineral in which the cutter may be used. A tungsten carbide tipped cutter pick can cope with minerals or rock up to a uniaxial compressive strength of around 80-100 MPa without excessive wear.

CONTINUOUS MINERS

Coal cutters could only be used in cyclic mining with shotfiring, and in a successful attempt to virtually eliminate shotfiring in coal the continuous miner was developed.

The cutting head of the first continuous miners consisted virtually of several coal cutter jibs mounted in a vertical plane side by side above a gathering-arm loader apron. Because of its chain type construction it was known as a ripper-type miner. The general configuration of the machine was otherwise as in Figure 7.3 which shows the latest generation of continuous miner. To operate the ripper miner it is manoeuvred into the centre of a heading on its tracks and remains stationary while the cutting head is lowered and sumped (pushed forward on slides) with the pick chains running into the bottom of the seam about 0.6 m and then lifted up. Most of the broken coal is carried back on top of the chains to a central conveyor and loaded out over the conveyor jib. The cutting head is slewed from one side of the heading to the other, repeating the sump and vertical cut process, until a cut about 5-6 m wide is completed. The resultant roadside has a saw-tooth form which is susceptible to spalling in weak coal.

The individual chains were rapidly developed into a chain mat and these miners, such as the Joy 1CM and Joy 6CM, are still available.

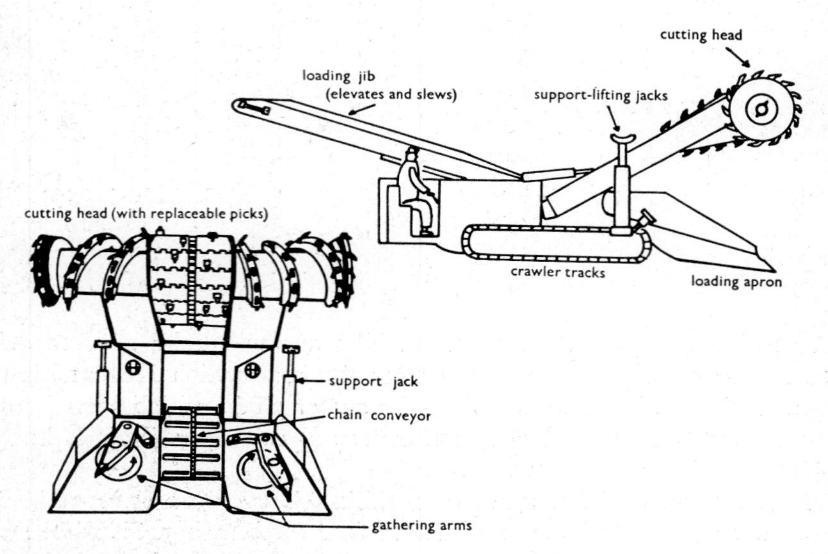

Fig. 7.3 Continuous miner

Joy 6CM continuous miner
(*Courtesy Joy Manufacturing Co. Pty Ltd*)

JOY

The Joy 6CM has a rating of 240 kW and a coal winning and loading capacity of about 3-5 tonnes per minute. The width is about 1.8 m and maximum working height about 3 m.

Of the same development period are the continuous miners with oscillating heads. The Lee Norse continuous miner in its early models had two pairs of pick-carrying disks instead of a chain mat. The disks were mounted with their axes of rotation normal to the coal face and to be able to break out all the coal the disks had to oscillate from side to side. The head did not slew and the width of cut obtained by sumping and cutting from one track setting was about 2.6 m. Wide headings were driven by taking successive parallel cuts. These machines were sumped into the coal at roof level by using the tracks, and mined at 5-10 tonnes per minute by cutting downwards through the coal.

An intermediate Joy continuous miner (the 8CM and 9CM series) was fitted with oscillating heads either side of a central chain mat, and then machines of the type shown in Figure 7.3 were produced. The disks have been replaced by a spiral cutting head mounted either side of a central ripper mat. The head is raised to roof level and the tracks are used to sump into the coal seam about 0.6 m. It is then forced down to floor level with the replaceable picks on its rotary head tearing the coal off: broken coal falls onto the loading apron where two gathering arms pull it onto the central conveyor. The loading apron can be raised and lowered about 0.5 m. The central single strand chain conveyor carries coal out over the loading jib which can be raised to over 2 metres and slewed about 45° to either side. It can load into a shuttle car or on to another conveyor.

These machines are typified by the Joy 10CM or the Jeffrey Heliminer. They are about 3 m wide and will cut a 6 m wide heading about 1.5 to 4 m high in two passes. This is done by manoeuvring the machine in and out on its tracks. Another common technique is to take a central cut followed by a sideway swivel in each direction to give a 'sawtooth' heading. This is an easy method but can result in an excessively wide road which may deteriorate if it is kept as a permanent roadway. It may also exceed the legal limit of 5.5 m laid down by the N.S.W. Coal Mines Regulation Act.

With a power of about 360 kW these machines can load from 8-12 tonnes of coal per minute. When this is compared to a maximum output of around 1000 tonnes per shift it can be seen that the miner only loads for intermittent periods to a total of about 2 hrs per shift for various reasons, such as travelling from one heading to another, waiting for supports to be erected (aided by built-in jacks), coal clearance problems, etc.

With the lower capacity machines it was necessary to keep them operating for longer periods. It was common to pass the coal over the machine and drop it on the floor of the heading, which acted as a surge bin. A *gathering arm loader* was used to lift the coal onto a conveyor or into a shuttle car. The new bigger machines load so fast that they can fill a large shuttle car in 1 to 1½ minutes and they would fill a heading in 2 to 3 minutes. Consequently they are not used for floor-dumping.

There is another major type of continuous miner which bores out a full-face heading in one operation. Instead of relying on ripper-type picks attacking the face end-on the borer-miner is fitted with milling-type cutter heads. The heads rotate parallel to the face being cut. Individual picks are mounted on the heads and may either provide a trepanning action or full-face boring (Figure 7.4). The Goodman and the Marietta miners are of this type. As with the cutter miners the unit is crawler mounted and has a central loading conveyor which loads out over a flexible rear jib. Three or four cutting heads may be fitted. Adjacent heads are contra-rotating, both for stability and to sweep the broken mineral in towards the loading conveyor. Loose mineral on the floor is ploughed up by an apron.

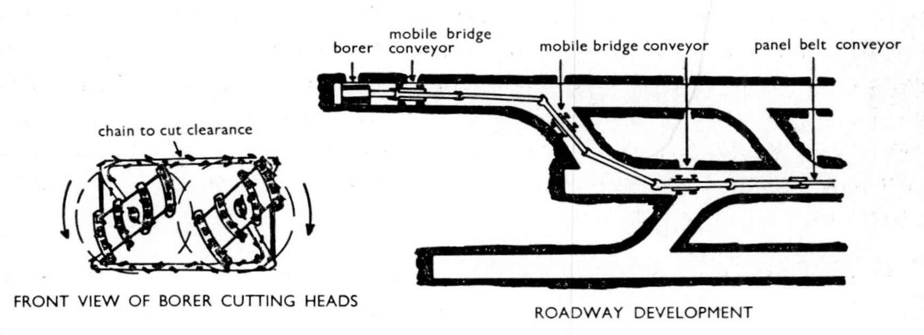

Fig. 7.4 Borer continuous miner and bridge conveyors

When roof conditions in some Australian coal mines have deteriorated these borer-type miners have been introduced for heading drivage. Roof conditions have been improved for two reasons. The rounded fillets of coal left at the corners of a roadway do have a strengthening effect in woody coals, but in brittle coals the fillets break away. The primary increase in strength is because the roadway is of a uniform narrow width and is taken in one pass. A nominal 5.5 m roadway is 5.5 m across the mid-height of the roadway, not 6.5 to 7 m with crumbling edges.

The borer miner in a one-pass roadway has a marked disadvantage. If a cutting head needs attention or pick changes in between cross-cuts the machine has to be backed up to a cross-cut and someone has to walk 'round the block'. If it breaks down then a small opening has to be mined around it by hand to obtain working room.

There is a secondary disadvantage that cross-cuts made with a borer miner cannot be set off at right angles but must be turned off in a tight curve to give angled crosscuts as in Figure 7.4. The resultant diamond-shaped pillars are not as strong as a rectangular pillar. The acute included angle is likely to deteriorate under load and will thus expose a wider span of roof. Borer miners in Australia have been used with shuttle cars but it is possible to use them with a series of mobile bridge conveyors as shown.

An application of borer type miners that has found favour is in mining bedded deposits or thick deposits where there is not a hard rock-to-seam boundary which causes high pick wear. Sylvite of Canada, in a potash mine in Saskatchewan, uses a Marietta miner 8.5 m wide and 2.5 m high for room and pillar mining. The rooms are driven 20 m wide and 1500 m long at a rate of about one per month. Each room contains 150 000 tonnes of ore.

The miner is used with an extensible conveyor which holds enough belting to permit an advance of 0.3 to 0.4 m per minute for 60-120 m before another length of belting is added as convenient. The miner tows the belt and a ventilation curtain wall, plus its own 1100 kW, 4160 volt cable. A single pass is taken to 1200-1500 m then the machine backs out, taking a second pass to widen the room to 14 m, then a third pass to reach 20 m. The mining system has a three man crew and can produce about 7000 tonnes a day. The high output is offset by the high capital cost of about $750 000 for each of the two machines in use at the mine.

GATHERING-ARM LOADERS

These preceded continuous miners and provided part of their development. They are virtually a miner as shown in Figure 7.3 with the cutting head removed (see also Figure 6.3). They can be used for loading rock or coal and were developed for electrical loading before diesel LHD machines were available. The shales and mudstones found in many coal mines are less abrasive than sandstone. In addition the floor of the coal seam or a natural bedding parting provides a smooth loading platform. Consequently the flat apron of a gathering arm loader can be pushed in easily to give a high rate of loading. Electrical power is more efficient than either diesel or compressed air and is preferred in the relatively thin workings of a coal mine where reticulation is easier, and output is more concentrated.

OTHER TYPES OF LOADER

Electrically powered and compressed air powered bucket loading machines, as described in Chapter 6, are used in coal mines. Rocker shovels, side-tipping and roll-over buckets are all available. Diesel units must be fitted with flame traps on air inlets and exhausts in coal mines; the water scrubber for exhaust gases can be the flame trap as well.

A *duckbill loader* which loaded by a reciprocating action may still be available but was only really efficient in a rise heading. It was an adjunct of shaking conveyors and both are obsolescent, if not obsolete. Scraper loaders (slushers) can be used in coal mines but are not popular for coal or rock loading. They have been used in specialised packing methods in advancing longwall in Britain.

SHUTTLE CARS

These are the workhorse of pillar mining and mine development in Australia and the United States. Their primary function is to carry coal over a short haul from the continuous miner to a conveyor belt. They have a shallow load pan with a central conveyor (Figure 7.5).

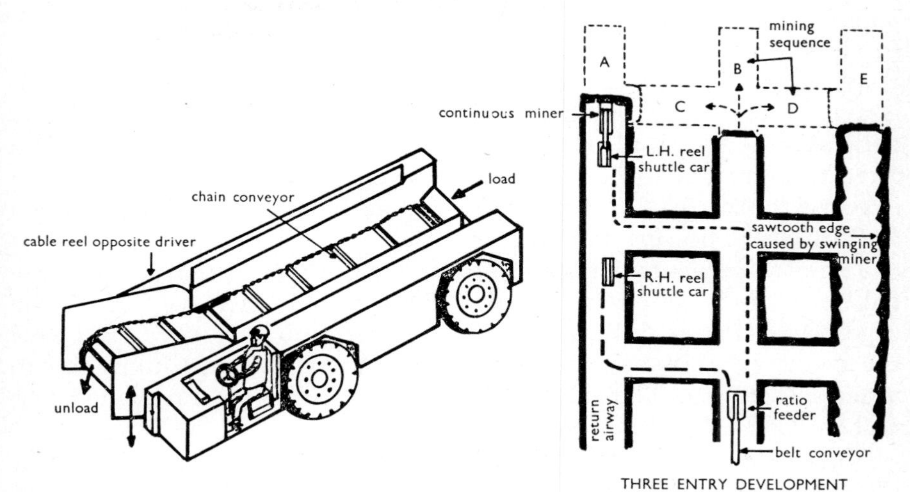

Fig. 7.5 Shuttle car and development

The coal is loaded at the rear end and the conveyor is run intermittently at slow speed until the pan is full. The car is then driven off at up to 8 km/hr to an unloading point where the conveyor is run at a controllable high speed to empty the coal over the elevatable front end. The driver reverses his seating position to return the car.

Cars are four wheel drive and four wheel steer for good traction and manoeuvrability. Capacities are from about 5 to 10 tonnes but can be increased in favourable conditions by fitting side-boards. Drives can be diesel, battery electric, or a.c. mains. The shuttle car operated from a.c. mains gives the best performance and is the most popular.

A powered reel is used to pay out and retract a cable as the car travels. The reel is fitted opposite the driver and cars are bought in pairs with a left and right hand reel. In multi-heading development (Figure 7.5) they 'wheel' around a block partly to avoid the electric cables becoming tangled together but mainly because they can then travel independently. They are too wide to pass each other in a normal roadway. In the central heading on a short haul they wait in an opening on their respective sides to pass each other. Alternatively they can be used in tandem on long hauls. One car is loaded and travels to meet the other. The loaded car fills the empty car at high speed then returns for more coal. The newly loaded car takes the load to the conveyor belt and returns.

Diesel and battery powered shuttle cars retired from active duty on coal loading make useful supplies cars, although special self-powered or tractor hauled supplies cars are available.

RATIO-FEEDERS AND FEEDER-BREAKERS

The feed from a shuttle car at high speed would overwhelm a normal belt conveyor and would cause excessive spillage. There is a choice of emptying the shuttle car more slowly or of using a ratio-feeder. This is a large receiving pan fitted with a heavy duty chain conveyor. It will take the full load of the shuttle car and then feed it out at the correct rate as a centralised load onto a belt conveyor. The ratio-feeder is more robust than the receiving end of a belt and withstands the normal wear and tear better.

The continuous miners can produce large lumps of coal: if these are fed onto a belt conveyor there is a tendency for them to roll off the edge of the belt as spillage. It is convenient to fit a breaker bar above the conveyor of a ratio-feeder to reduce the maximum lump size to around 15-20 cm. These lumps will self-centralise on a troughed belt and spillage will be minimised.

Management opinion is varied on ratio-feeders. Some think them essential; others feel that they can be a nuisance. It is necessary to consider the characteristics of the coal seam and the normal length of haul. If the coal is naturally lumpy and there is a long haul then a ratio-feeder will repay its capital investment. Most mines fit a breaker at some stage in the conveyor system before the coal is fed onto the main conveyor to the surface.

Continuous miner with shuttle car
A rare opportunity to see mining with the lid off. An area of shallow overburden has been stripped and the underground machinery is in use to mine and load the coal.

CONVEYORS

A full technical description of these is deferred to a more specialised book. However it is necessary to know the basic requirements of a conveyor. They can be divided into the two main types of a flexible chain conveyor used for short haul heavy duty work, and the belt conveyors used for bulk transport.

Armoured Flight Conveyor

The chain conveyor is frequently known as an *armoured flight conveyor,* or AFC, and is shown in Figure 7.6. They have been basically standardised in width, pan size, and chain sizes. The conveyors are made up of 1.5 m sections loosely linked together so that they can be snaked over bodily sideways, or used in a slight curve as a transfer conveyor. Some 2.4 m pans have been made for Australian longwall faces.

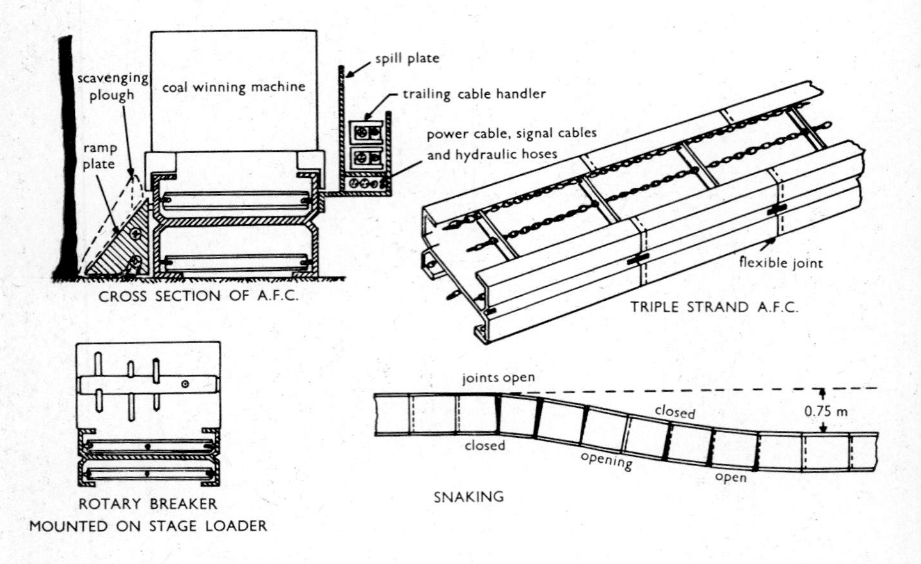

Fig. 7.6 Armoured flight conveyor

For longwall face work the conveyors may be up to about 200 m long. On high output faces a triple strand coal carrying chain with cross flights will be used. Two-strand chain is used for less arduous conditions. A spill plate and cable handling troughs and chains are fitted on the goaf side (the side away from the coal faces). In most longwall faces the conveyor also provides the travelling way for the coal winning machine which is captive on it and provides a further set of stresses to be resisted.

Triple strand AFCs are also used as *stage loaders* (or gate-end loaders). These accept coal at right angles from the face conveyor and deliver it on line to a gate belt conveyor. The flexibility of the transfer AFC both vertically and laterally allows it to correct for misalignment of the face and gate conveyors. As with ratio feeders a breaker bar or a crusher may be mounted on the stage loader to break large lumps to give better belt conveying.

The drive on an AFC is by engaging the chains on toothed sprockets so that tensioning is not a problem. However the ends of the conveyor should be kept parallel and the chains should be of equal length or else they will tend to jump off the drive sprockets.

Belt Conveyors

Whereas an AFC is built for brute force and rugged conditions a bulk carrying conveyor must have minimum friction over most of its length to provide economic operation. Coal from working areas (called panels or districts) is gathered by *belt conveyors*. Because the working area is expanding (or contracting) the length of the belt must be capable of being altered easily. Figure 7.7 shows the basic layout of a belt conveyor.

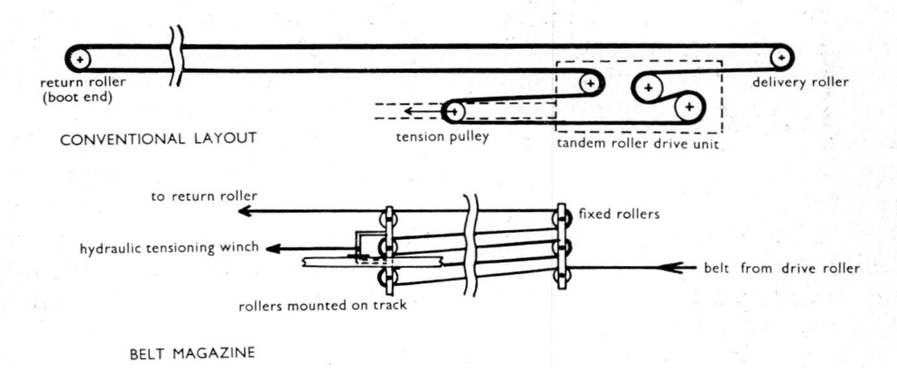

Fig. 7.7 Basic layout of a belt conveyor

The delivery head and drive units are semi-permanent and are firmly bolted to a concrete pad or braced against the roof. The delivery end feeds coal into mine cars or onto another belt in series. For delivery to another belt a chute should be fitted to give an in-line centralised feed. A belt is driven by friction from driving drums. The power that can be transmitted is proportional to the area of belt in contact. To increase power transmission the belt is lapped around two or more rollers, which may be rubber lagged to increase the coefficient of friction.

Power transmission is also proportional to the tension in the belt. The drive provides tension on the loaded side and tends to throw slack on the empty belt.

Empirically the limit of drive before slip is given by

$$\frac{T_1}{T_2} = e^{\mu \theta}$$

where T_1 = tension in loaded belt
T_2 = tension on empty side
μ = the coefficient of friction between belt and driving drum, and
θ = angle of contact between belt and driving drums.

To increase T_2 and to keep the belt in firm contact on the drive rollers a tensioning drum is provided to take up the slack. On large conveyors it is usually close to the drive drums but it is also possible to provide sufficient tension by pulling back on the return drum, particularly on uphill conveying where the gravity loading on the bottom belt helps to provide some tension.

The belt is made extensible by having the structure (its intermediate portion) made up in (conventionally) 2.74 m sections which each carry two troughed upper roller sets and one flat bottom roller. The tension end, or boot, is pinned to the floor or anchored to special supports. The boot can be pulled inbye with a shuttle car or a pulling device after the tension drum has been slacked-off. Further structure is inserted, then the boot is anchored and belt retensioned. The belt can be shortened by removing structure and using the belt tensioner to take up slack belt and pull the boot forward. For rapid-advance or retreat headings a belt magazine is used to replace the single tension roller. A complete roll of belt is inserted or removed as necessary and the magazine allows repeated changes of length without breaking the belt too frequently.

American practice is to have large lengths of structure based on tensioned wire ropes. These are also occasionally used in the U.K. but wire ropes in Australia are relatively expensive and rope structure is not used.

Where coal is to be carried over long distances out of a mine or overland the conventional belt conveyor, which is made up of a fabric carcase impregnated and protected with plastic or rubber, is not strong enough to provide long hauls. Instead a cable belt or a steel cord belt is used. These are different items but the Americans tend to call them both cable belts.

The *steel cord* belt consists of flexible wire ropes side by side in alternate lay embedded in a vulcanised rubber or a neoprene body.

One single layer of ropes is used and long lengths of belt are made in one operation. Joints are expensive to make and the belts are kept at a fixed length for as long as possible. Lengths of a single belt can be 6 or 7 km and about 1500 kw can be transmitted. Conveying practice is similar to Figure 7.7 but large diameter drive drums are used to avoid excessive flexing in the wire ropes. It may also be necessary to use two drive heads in series to transmit sufficient power.

Whereas in the steel cord belt the small diameter ropes which transmit the drive stress are embedded in the body of the belt to replace the fabric core, in the *cable belt* the wire ropes are thicker and are external to the belt itself. Two endless ropes run parallel to each other for the belt track, over rollers on supporting stands. They carry between themselves a neoprene belt located by means of moulded grooves. The ropes transmit the tension and the belt only has to carry the localised coal load. Consequently belt strength is not a limiting factor. At each end of the belt length the ropes are skewed out by sheaves. At the tail end they pass round rollers on a tensioning trolley. At the head end they are taken round a pair of coupled friction-drive wheels. The belt itself is taken round a delivery roller and a tail end roller and lowered back onto the ropes. Two existing cable belts have lengths of 8.8 and 8 km at about 1200-1600 tonnes per hour. A new installation in 1972 will have an unbroken length of 16 km at 1500 tonnes per hour.

LONGWALL MINING MACHINES

The history of longwall mining can be found in many English coal mining books. It was found more than 50 years ago as mine workings became deeper that the face-line of a coal mining panel had to become straight to provide good roof conditions. The earlier stepped longwall was virtually room and pillar working with the rib pillars replaced by packs built with roof stone. At about the same time as faces became straight early experiments were being made to replace with conveyors the mine cars (tubs) which formerly ran along the face.

Belt conveyors were used on the face for many years and there was a cyclic process of undercutting coal (or middle or top cutting) with a jib coal cutter. The coal was cut, fired and loaded on to the conveyor. (Figure 7.2). When the cycle had been completed once every 24 (or 48) hours the conveyor was dismantled and moved into the new track, the coal was cut and the process repeated.

The mining engineer was trapped in this cycle until the panzer-forderer conveyor was developed in Germany. This armoured conveyor was flexible and was intended to be pushed over bodily. It therefore had to be installed in front of the last row of supports. This was the *prop-free front* method. The support cantilever overhang had to be stronger and new types of link-bar roof supports were developed.

Previously the coal cutter had been pulled along the floor by an internal rope haulage. It was now logical to mount the conveyor on top of the armoured conveyor to reduce the exposed width of roof. Rope haulages were changed to captive chain haulage. A chain stretches the full length of the face and the cutter hauls itself along with a sprocket drive. Many experiments were made with complicated curved coal cutter jibs in England and Europe. The intention was to cut and load the coal onto the conveyor in one operation as far as possible.

Eventually James Anderton pioneered the shearer cutter loader named after him and developed through National Coal Board efforts in the U.K. The basic idea of the longwall shearer remains the same but the original 30 kW converted coal cutter has turned into a 300 kW giant with hydraulic servo-mechanisms to control cutting speeds, cutting horizon and tilt. Figure 7.8 shows the principles of a modern shearer.

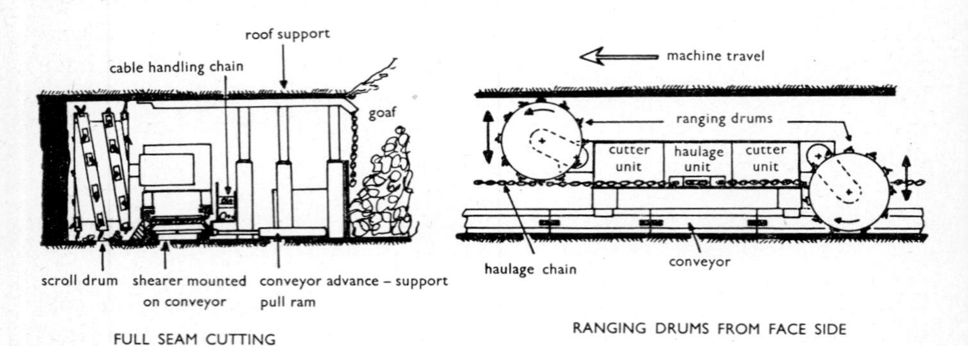

Fig. 7.8 Longwall shearer loader

The *Anderton shearer* is used on most modern high output longwall coal faces (a plough or a trepannner may also be used). A thin seam may be cut full section with one drum but roof convergence can present problems if the machine has to make a return passage without the drum cutting. The double-ended ranging drum shearer is very popular. It is mounted on the conveyor by trapping shoes and can cut in either direction. The leading drum is raised and trailing drum lowered and the machine pulls itself forward on the captive haulage chain. Currently experiments are taking place with rack and pinion haulage: the rack is fastened to the conveyor. The drums stay in their respective positions for the full length of a face cut. The cutting drums are about 0.6 m wide and 0.9 to 1.2 m diameter. The number of picks on the drum is kept to a minimum, even as few as 20. Experiments showed that when the picks were kept to a minimum there was less coal degradation, less dust produced, and a lower power consumption. The picks are fitted into a three-start scroll to load the coal on to the face conveyor. Loading action can be aided by a ramp fitted to the leading edge of the conveyor. The cutting speed of the machine is about 8-9 m per minute in soft coal with the rate automatically reduced in hard coal. Loading rate is dependent on the thickness of the seam. Details of the mining methods used are given later.

These machines are a.c. mains powered and have to tow an electrical cable along the face. Water hoses are also connected for dust suppression and motor cooling. The cables and hoses are led captive in a cable trough to the centre of the face and are then fed into a flexible cable-handling chain. This moves to and fro with the cutter and both protects the cables and prevents them from kinking and becoming tangled.

The shearer is kept within the coal seam by tilting the drum slightly to cut below or above conveyor bottom level. When the conveyor is pushed sideways it will drop or rise accordingly. In practice the whole body and drum are tilted together on an underframe. In a similar way the shearer is made to rise and fall along the length of the face. Radioactive probes are available to follow the seam horizon by measuring about 10-15 cm of coal left in the roof for level control along the face. Drum tilt can be controlled by a servo-mechanism set to follow the average grade normal to the face line.

Two other longwall cutter loaders are available. The *coal plough* planes coal off the face in a manner similar to a wood plane shaving down a wood surface. A narrow cut of 5 to 10 cm is taken by the plough which travels up and down the face guided by the face conveyor as in Figure 7.9. Several types of plough are available but none are used in Australia at present. Whereas the shearer cuts the whole seam to predetermined horizons a plough is not suitable for thick seams. It can be made to crop roof coal in a thin seam but becomes unstable if its height is increased unduly. Unless the top coal will fall easily a plough is not good in a thick seam.

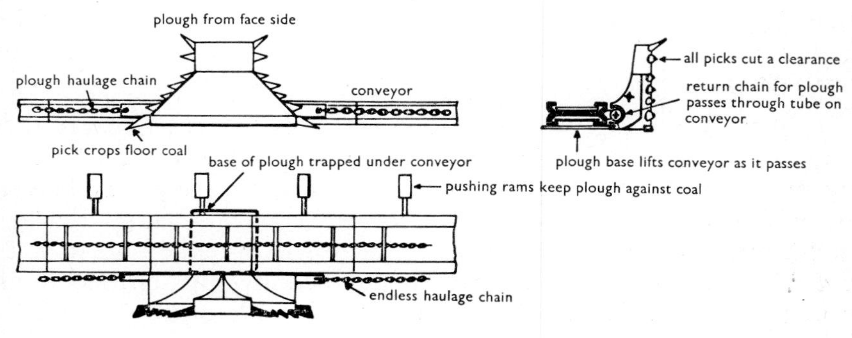

Fig. 7.9 Coal plough

Although a plough works excellently in soft friable seams it tends to lift up from hard floor coal or to ride out from hard bands in other seams. The plough is hauled along the face by a chain driven from motors mounted on one or both ends of the face conveyor. If the plough is forced too deeply into the coal face then hard pockets of coal can impose severe shocks onto the haulage system.

Considerable effort has been put into plough design and use to widen its range. Articles on variable geometry ploughs, activated ploughs and fixed increment ploughing can be found spread over the last ten to fifteen years in the *Mining Engineer,* and the *Colliery Guardian,* two of the U.K. coal mining journals.

A valid reason for persisting with the plough is that there is no moving machinery on the face that needs an electrical supply, and no rotating pick heads to produce large quantities of dust. Health and safety are thus at a premium. In addition the narrow web (width of cut) taken exposes a minimum area of roof before supports are advanced.

The *trepanner* is also used in Britain. It can be mounted alongside or on top of a conveyor and has a trepanning head which makes a tubular cut in the coal face. The circular shape then has to be squared off by extra coal cutting jibs at the back of the machine and on the top and floor. An alternative has been to fit one trepanning head and one shearer drum on opposite ends of a machine. The trepanner head was designed to give a larger coal size product than the shearer drum. Perhaps it is too soon to be prophetic but the trepanner will probably disappear with the market for large coal (above about 5 cm size).

Mine Development

As with metal mining it is essential to plan a mine carefully before operations are started. Maximum information must be obtained on the character and quality of the coal seam or seams, and the adjacent rock strata. Topographical and geological maps are a necessity to determine depth of cover and any geological hazards such as water-bearing faults or soft surface deposits.

Exploratory drilling is essential but the holes are wider spaced than for metal ore deposits. Down to 150-200 m one hole every 2-4 hectares may be sufficient, whereas at 300 m depth the company may be content with one hole every 400 hectares. Occasional large diameter exploratory holes are carefully sited. These will provide bulk samples for testing and details of coal cleat and bedding planes. The holes can also be used for auxiliary ventilation, cable and pipe entries, emergency shafts or pilot raises for full size shafts.

SHAFT OR DRIFT SITES

Bedded deposits stretch over wide areas and a lease can be several kilometres square. A mine may thus have several shafts, although the number of shafts or drifts decreases rapidly with depth of working.

The initial shaft or drift sites for deposits on a relatively flat grade may be determined by the necessity to place them close to the surface plant and possibly in proximity to a site for a power station. This shaft or drift may remain as the main coal haulage exit. Preferably one should choose the point where the seam is closest to the surface; This is not necessarily the highest point of the seam as a deep surface valley can be chosen. Care should then be taken that the mine entrance will not suffer from flooding, rock slides, or snow drift in cold climates.

For inclined seams, it is convenient to site the shaft or drift bottom at the lowest point so that both water and the mineral can gravitate to the shaft bottom, but this is not always possible.

Down to about 200 m vertical depth a drift is preferred and a high powered steel cord or cable belt can be used to bring the coal out at a gradient of up to 1 in 3 (20°). A vertical shaft is often favoured for ventilation and man-riding. Modern techniques favour pilot hole drilling and widening techniques to put a series of shafts across the lease for man-riding. Shift durations are now as short as 7 hours including underground travelling time. It is economical to use a small change house at a shaft top equipped with an automatic hoist and to ride the men close to their working face. Small diesel or battery electric man-riding cars are then used for underground transport.

Supplies are taken underground in the coal drift on rail track laid alongside the belt, or underneath a belt suspended from the roof. Alternatively a second entry is driven and rubber-tyred tractors and supply buggies may be used instead of rail transport.

UNDERGROUND DEVELOPMENT

In relatively flat seams main entry development is normally in the seam and for seams of about 1.5 m and above will be either of seam height or about 4 m maximum. Panel entries will be seam height from 1.2 m up to about 6 m above which thickness special techniques are used. Where seams are thinner than about 1.2-1.5 m then rock must be taken as well as coal to provide working height in the roadways.

Practice varies internationally. In the United States thin seams are mined with special low profile belts, supplies vehicles and man haulage vehicles, which will operate down to about 1.2 m. In Australia it has not been necessary to mine with these special techniques. Soft floor dirt has been taken to increase entry height to about 1.5 m in the infrequent thin seams.

In Britain and Europe roads are *ripped* or *dinted* (i.e. roof or floor is taken) to provide sufficient height. In Britain the Coal Mines Act requires a normal mimimum travelling height of 1.68 m. In addition the rock is weaker and only two entries are driven per longwall face. Road widths are kept to a minimum and the height increased to give an arched shape, which is stronger.

Where seams are steeply inclined they are mined in the seam but the workings are connected to the shafts by *horizons,* or *laterals,* as are metal mine stopes. An horizon can intersect and connect several steeply inclined seams which are mined in a descending sequence in vertical blocks. The mine workings can spread outwards and downwards contemporaneously. In modern coal mining these horizons are one of the few places where locomotive haulage is favoured. Mine car loading points with rail pass-byes, are installed at each seam intersection, and belt conveyor roads are taken out on the strike in the seam to develop longwall panels. In wet seams the 'strike' roads are driven at about 1 in 200 upgrade to help drainage.

Longwall panels are developed to the full dip up to about 1 in 1.5 (vertical to horizontal) grade to use conveyor-mounted machinery. If a face is developed off the full dip there are problems of machine stability.

ORIENTATION OF UNDERGROUND ROADWAYS

In the absence of geological problems underground roadways will be laid out on a grid suitable for pillar or longwall mining. New leases in Australia are now granted as square or rectangular blocks so that irregular boundaries do not have to be allowed for.

Major faults, dykes, washouts (old river valleys) or areas of thin coal, may determine the orientation of panels or main entries. A problem of more concern is the coal *cleat,* or *cleavage,* and the orientation of roof cleavages. Besides the bedding planes which break up coal and roof strata parallel to the seam, there is often a pattern of vertical or near vertical fractures in the coal and in the roof strata.

A pronounced coal cleat affects the stability of coal faces. Coal worked parallel to a cleat will fall off in large slabs. Working end on (i.e. the face is at 90° to the cleat, but advance is normal to the face line) needs more physical energy to break the coal off. The usual compromise for cleated coal is to work across the cleat at 30-45°.

A similar cleavage in the roof affects the stability of roadways. A common mining problem is that roadways in one direction give no trouble whereas other roadways driven at right angles to the good direction have frequent roof falls. This is particularly troublesome in the rectangular layout needed for pillar mining. As soon as possible a survey should be made of the roof joints. A paper by Connelly[2] describes the techniques and an application. Where the major joints are parallel to a roadway then roof conditions are bad: the roadway is strongest where the major joints cross the roadway. Consequently the mine workings should be orientated to allow all roadways to cross the breaks at about 45°. This is most important for in-seam roadways driven with a minimum of support. It may be necessary to have cut-throughs slightly angled to miss a minor joint system, but the angle must not be too acute.

ENTRY DRIVAGE

Entries, *bords,* or roadways, have three basic purposes, The mineral must come out, the men and materials must pass both ways, and sufficient ventilation must be provided to meet the statutory requirements of 19 per cent oxygen, and minimal noxious or inflammable gas values. Men and minerals do not affect the size greatly, it is the ventilation requirements and basic equipment size that is important.

Economic considerations and strata control requirements are the main controls. In good roof conditions where a heading is to be driven in coal a continuous miner will be used. Three passes of a 1.8 m width head, or two passes of a 3 m cutting head produces a near 6 m width roadway. (N.B. The present legal maximum width of roadway in New South Wales is 5.5 m.) This width is suitable for shuttle car operation and for ventilation with a wide and narrow side brattice cloth.

Ventilation

Whereas development tunnels in hard rock are driven individually and at relatively slow rates, advance rates in coal can be very high and multiple entries are driven and cross-connected. Single tunnels are ventilated with auxiliary fans but for coal mine development with continuous miners it is more convenient to let the main mine fan provide the ventilating pressure needed to move air. A system of brick stoppings and air crossings routes air into a panel, or district. In the development area curtains of brattice cloth (anti-static PVC sheeting) are hung across a roadway to restrict unwanted airflow. The main airflow is routed into headings as desired by stapling brattice cloth walls onto timber posts to split a heading into two as in Figure 7.10. Clean intake air passes up the wide side where the continuous miner works and the shuttle car travels. Dust-laden air is removed rapidly down the high velocity narrow side. Only one heading is worked at a time but all must be ventilated: some re-hanging of brattice cloth is necessary to allow free travel for the shuttle cars.

Figure 7.10 shows a five-entry layout with three central intakes and two outside returns. It is possible to have two adjacent returns and three adjacent intakes, but the outside returns ensure that there is a maximum of two intakes to feed gas and dust into a return, rather than three. To offset this it is necessary to build two rows of stoppings to isolate the returns.

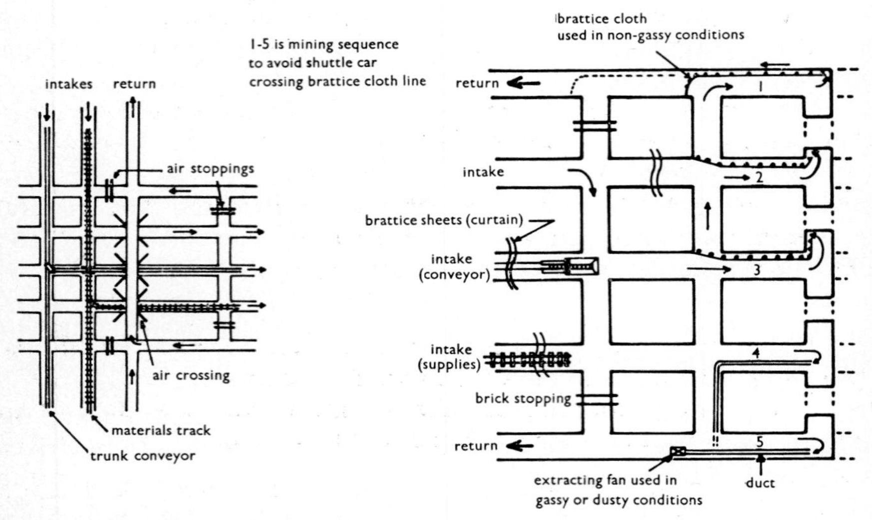

Fig. 7.10 Five-entry development

With borer type miners it may be necessary to mount a fan on the miner and to use flexible extending air duct to provide suitable working conditions in the heading. The fan will extract dust laden air which can either be exhausted downwind in a return for natural settling or be wet cycloned or filtered for dust removal. An auxiliary fan with dust removal facilities can of course be used in any heading.

Mining Details

The process of driving entries is often known as *first working*. There is no magic formula for the number of headings and the size of the pillar between them. In shallow mines with a good roof the decision can be an arbitrary one by the mine manager. In these good conditions the output per manshift from entry drivage is likely to be as high as the output from pillar or face work and there is no economic incentive to control the number of headings.

As the depth of seam increases the percentage of coal won by first working must decrease to maintain pillar stability. In New South Wales there are limits laid down in the Coal Mines Regulation Act.

To paraphrase the Act the amount of coal or shale to be left in the pillars shall be as under (N.B. depths from surface rounded down by author from values in feet).

Depth from surface	Percentage coal or shale left in pillars
Up to 60 m	not less than 50
60-150 m	between 50-60
150-300 m	between 60-70
300-600 m	between 70-85

The maximum road width laid down is 5.5 m and minimum sizes of pillars are also related to depth. The most critical effect is a jump in width of pillar from 16.5 m to 24 m when the depth exceeds 300 m, which is rather arbitrary. In any case a more general rule of thumb for long term pillar stability (for main entries) is that the width of the pillar shall be D/10 where D is the depth of the seam.

In practice for shallow working the pillars have been a convenient size for surveyors and workmen and have (in Australia) been one chain by two chains, or up to three chains square (one chain = 22 yd = 20.117 m). To continue the practice we would now use 20 x 40 m pillars. In Britain the arbitrary pillar was 40 yd width.

For deeper mines with weaker roofs research is currently in progress to determine safe minimum sizes for pillars. Even with pillars 40 m wide closure can be detected in roadways. If total extraction takes place against the pillar then a roadway 40 m away can show considerable crush at depths of over 300 m, but it is still usable. The mining engineers concerned at present use local practice to determine safe limits.

Whatever method of working is adopted the main layout for the mine is likely to be as in Figure 7.11 for development with continuous miners. For heading-type development the main entries will be driven in pairs with a minimum of cross-cuts. The layout of working districts is as given later for each mining method. Where thick seams are to be worked the initial development roadways will only be about 4 m high so that the sides of the roadway remain stable.

In Figure 7.11 the mine is shown with an upcast and downcast shaft plus a conveyor drift and a men and materials drift. A mine of this size may need between 250-500 m^3/sec of air. If this is passed through shafts the two drifts can be low volume air intakes which ensures increased comfort for men working or travelling in these roadways.

It should be noted that because of the continuity of a bedded deposit mining methods are much more predictable than in metal mining.

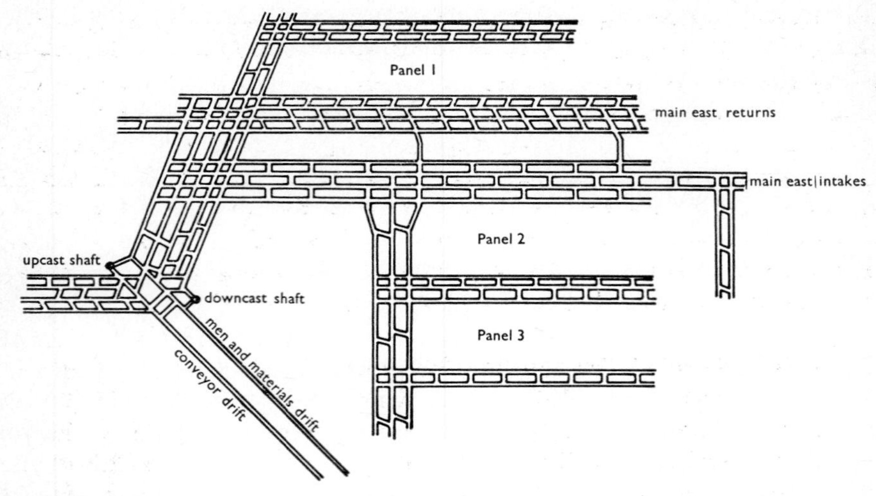

Fig. 7.11 Main entries plan

Layouts are more stylised; development of powerful cutter-loaders has mean that improvisation is now undesirable and that for at least the next decade it would appear that what coal cannot be worked by highly mechanised methods will be left for later generations. Apart from a choice of size of working unit and of orientation of panels one coal mine layout looks very much like another.

Methods of Working

PILLAR MINING

In different countries at various times this has been called Bord and Pillar or Room and Pillar mining. It could be argued that shortwall mining is also pillar mining.

In earlier days of handworking methods the essence of pillar mining was that casual pillars were left to provide support for underground workings. They were irregular and frequently were robbed to the point of failure. This of course gave a design criteria for the next pillar. In some cases pillars were also needed to prevent sudden collapse of the surface over shallow workings. There is at least one Australian house with its front floor joists supported on vertical RSJs resting on a coal seam floor a few metres below the surface. The back of the house stands over a coal pillar.

Increasing mechanisation and a better knowledge of strata control has required that pillars should have straight edges and a regular arrangement. The present descendant of the original pillar mining is probably the American room and pillar system for shallow workings. A layout such as Figure 7.12A could be considered typical. Main entries are driven and panels are blocked out with barrier pillars to protect the entries. Rooms are developed in isolation so that they can be sealed off against fire, or water, or after extraction to avoid ventilation problems. The room is developed by drives to the boundary and then coal is extracted on the retreat leaving small regular pillars for protection. Under good roofs all the coal except the pillars may be taken on the advance. Coal extraction is about 60 to 80 per cent. Mining can be carried out with continuous miners and shuttle cars as for entry drivage, or by using a cyclic process of undercutting, drilling, blasting and loading as described in shortwall mining.

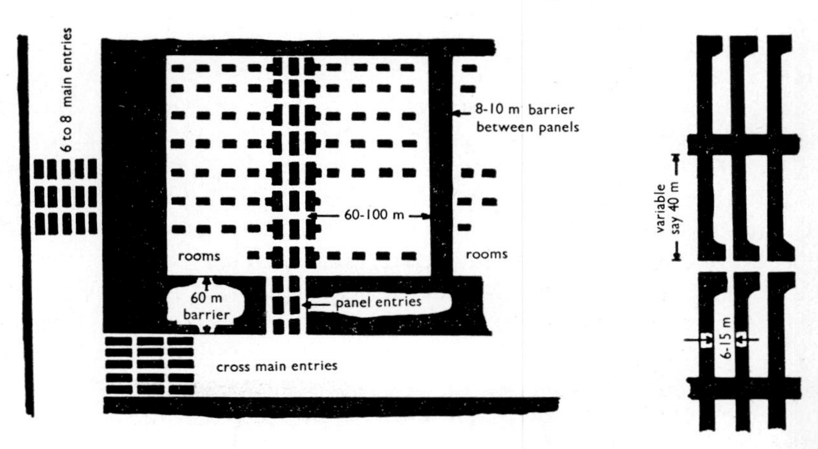

Fig. 7.12 Room and pillar layout

Figure 7.12B also shows the pillar layout used in the earlier bord and pillar mining. Coal mine workings progressed deeper in the U.K. before mining started in Australia or America. It was found that small remnant pillars at depths of 120-150 m were crushing and sometimes large areas of roof could collapse suddenly. Occasionally the roof would collapse around a pillar. Consequently a formal layout of working area (bord, or room) and pillar became necessary and the shape was frequently that of Figure 7.12B. Extraction rates were still about 50 to 70 per cent but pillars were more frequent and individually larger. Barrier pillars were not necessary because the room areas did not collapse completely.

This type of bord and pillar work was used first with hand-getting and then with shortwall cutters and conveyors. It was used in Britain with imported American gathering arm loaders and continuous miners in trials in the 1940s. The basic shape of pillar and bord is still suitable for cyclic shortwall mining under weak roofs. The short rooms give an opening which is stable for the limited extraction period.

In modern pillar mining in Australian conditions there is no attempt to form special rooms. Coal is blocked out into rectangular pillars with first working as already described, and then the pillars are extracted on the retreat. Although it is possible to cover the whole mine lease with first working before any pillars are extracted it is more usual to divide the mine into panels as in Figure 7.11 and completely extract each panel in turn. Pillars are left to protect main entries, but pillars between panel entries are extracted as the panel retreats. Some pillar remnants are unavoidably left and coal extraction is usually about 80-90 per cent. When working under water, or under urban areas, extraction may be limited to about 40 per cent to avoid subsidence. The mine area involved will then have first working only.

It is not necessary to block out a whole panel into small pillars. In fact if output from first working is low because of bad roof conditions it is useful to create large pillars and then *split and lift* them. Figure 7.13 shows a pillar extraction method for the Bulli seam at a colliery in New South Wales. (This system of working is known locally as the Wongawilli system after one of the first mines to use it.) Entry pillars are split and lifted off in sequence and then the solid large pillar between two panels is cross-cut before it is split and lifted. Ventilation is by bleeder leakage around the edge of the fallen waste. Entries 10 and 11 allow the shuttle cars to operate independently and facilitate ventilation. All work is by continuous miners and shuttle cars. It should be noted that timber supports are set across the splits adjacent to the goaf (fallen waste) to prevent the roof collapsing and running through the split. Timber cribs or hydraulic chocks may also be used.

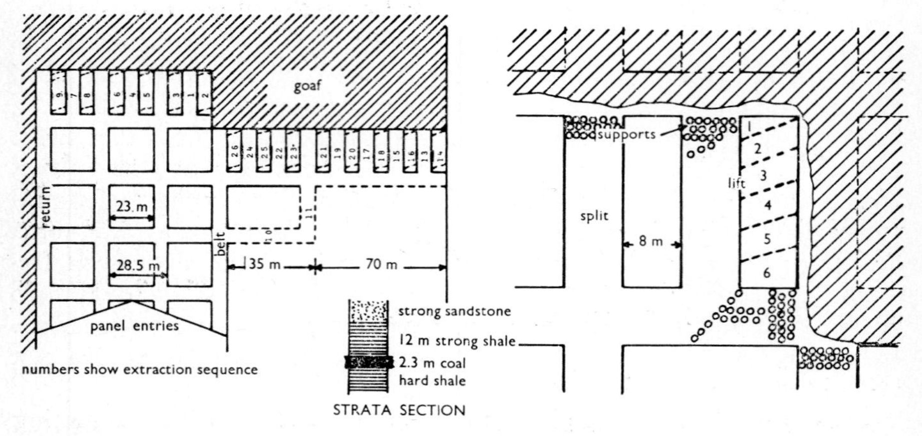

Fig. 7.13 Pillar mining in the Bulli seam

The split and lift system is commonly used in 'two chain' pillars because it gives good coal recovery. However small stumps of coal are left because recovery cannot be perfect. In some cases if the deputy in charge feels that the roof is weak, or the pillar shows signs of collapsing prematurely, the unmined portion of the pillar will be abandoned. However strong stumps must not be left because they prevent the proper caving action of the roof and cause further problems.

There are other methods of extraction which are used in small pillars at shallow depths. Figure 7.14 shows three of them. The pillar may be simply split through the middle and the remnants left. This is a common method of pillar robbing when retreating out of a room and pillar panel, or when abandoning entry pillars between mined areas. The pillar may also be worked by *pocket and fender*. The continuous miner enters the pillar diagonally, which is easier, particularly if it is a borer type. Pockets are cut out and fenders of coal are left to support the roof while adjacent pillars are mined. It may be necessary to blast the fenders subquently to ensure that the roof will cave. A pillar may also be worked by *open-ending*. Slices are taken off the pillar from the waste side towards the unmined area, or slices can be taken alternately at right angles from the two waste sides. Open-ending is convenient if shortwall coal cutters are used but has the major disadvantage that strong waste-edge support must be set for the full length of the pillar to protect the workmen.

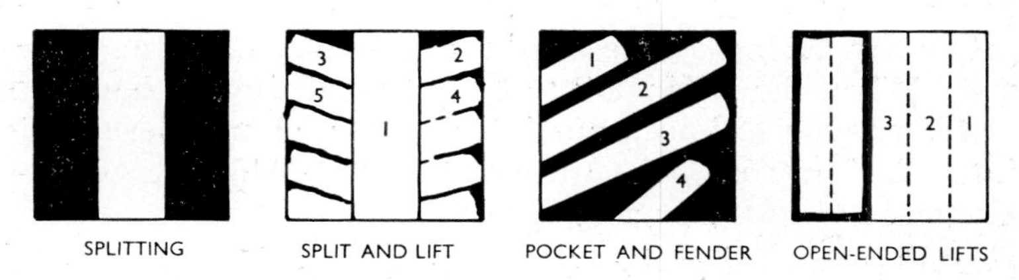

Fig. 7.14 Pillar extraction

The mechanised shortwall described later is an open-ending method but by using a self-advancing waste edge support system the disadvantage of the method has been overcome. Hand-set wooden supports were costly both in labour and materials. A continuous miner is expensive and the split and lift method keeps it at less risk than an open-ending method with weak supports.

Roof Control in Pillar Mining

The general theory of roof control will be dealt with in Chapter 8 but it is appropriate to look at its applications in pillar mining. There are two phases of roof control; the first occurs in heading out the roadways, the second when the pillars are being extracted. Mention has already been made of roadway orientation to avoid the roadways being parallel to the main joint system.

Roadway Drivage

Roof support in the roadways can be by rock bolts at regular spacings. The various types of bolts will be described in Chapter 8. It is essential to embed the bolts deeply enough to consolidate the lower roof layers, and to make sure that bolts cross roof breaks and joints so as to bind them together. In New South Wales mines bolts are usually used in conjunction with wooden split bars, or steel bars. As soon as the continuous miner has exposed either the area of roof laid down in support rules, or the workmen feel it is necessary, a bar is laid across the roof support jacks at the front of the miner. These bars have been pre-drilled with the hole spacing for the roof bolts. The bar is lifted to the roof and a drill bores through the holes at the ends so that rock bolts of 2-3 m length can be inserted. In some cases temporary hydraulic props may be set under the ends of the bars and replaced with wooden props further outbye.

If the roof is very weak the miner may have to be backed out to allow the men to insert the rest of the bolts. Generally however the miner continues forward and the rest of the bolts are inserted after the machine has passed. The bars may be staggered and fitted with one leg per bar as in Figure 7.15. A more usual pattern is to have the bars in line and to use two legs per bar, especially in the main entries. Bar spacing may vary from 0.75 m to about 3 m dependent on whether the roof is a weak mudstone or shale, or a hard sandstone or conglomerate. When the pillars are to be extracted the continuous miner is powerful enough to knock out legs as required and the roof is still held intact by the roof bolts through the bar. Alternatively a pillar can be entered from under a carrier bar.

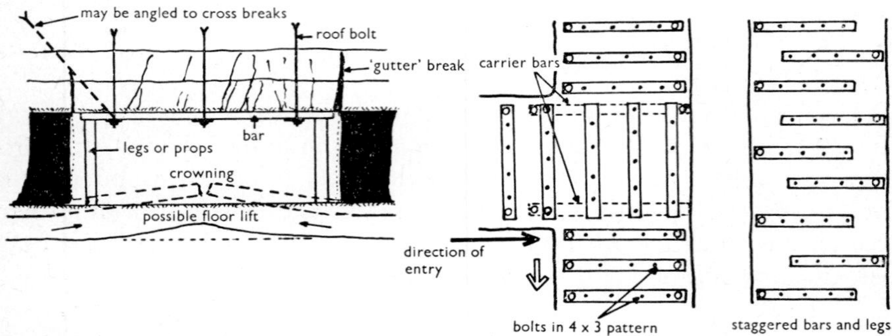

Fig. 7.15 Roof support in entries

The staggered bar and leg system is more suitable in wider roads. Cutting about one metre off the length of bars reduces their weight and cost. The single leg is set into the side of the road. There is a tendency for a roof to shear off along the coal edge and the leg set at the edge of the road will help to stabilise the break.

It is quite normal for a roadway to undergo closure[4,5] after mining takes place, even in the solid. There will be some roof sag and softer floors flow in from under the coal and lift upwards[6]. These effects are not marked until the mine reaches 200-300 m. Around this depth floor lift (or heave) can become quite noticeable and it is often this, rather than roof lowering, that tends to snap wooden legs or to bend steel legs.

In the deeper mines in New South Wales (300-400 m) roadway closure has sometimes occured to an extent sufficient to make it unusable by the continuous miner that drove it originally. However this closure de-stresses the area and if a new road is mined alongside the old one (within about 8-10 m) then the new road will not show any significant closure. This is the principle on which the split-and-lift pillar extraction method is based.

Even if a roadway does not close significantly in first workings it may close up rapidly under the influence of the *front abutment* (see Chapter 8) of the pressure arch created when pillars are extracted. The effects can sometimes be found about 150 m from the pillar extraction line, especially if dust suppression water has saturated the floor rocks. A more usual distance to be badly affected is about 20-30 m. Floor crowning can affect the shuttle car run, and conveyors may have to be hung from the roof to avoid them tipping sideways.

Pillar Extraction

Chapter 8 will show that for good strata control in extraction areas the span of the roof in a bedded deposit must be deliberately and systematically broken for several times the height of the seam. An area larger than a self-supporting beam in either transverse direction must be opened and the beam then breaks, releasing the abutment pressures on the edge of the coal or other mineral. These lower beams break up and 'swell' to give a similar support to the roof as is given by broken rock in a shrinkage stope. The main roof then squashes down onto this broken rock and is lowered gently as the broken rock crushes and is re-compressed to near its original volume. (Total extraction in longwall mining results in surface subsidence of about 90 per cent of the volume extracted.)

In older methods of pillar extraction there was ample time for the roof to break up slowly. The recommended method of pillar extraction was an average 45° line of diagonal pillars across the area. The pillars gave mutual protection and the roof broke around them. There seems to be little scientific evidence in favour of this method and it was only practiced at shallow depths.

When pillars were extracted more rapidly, and depths increased, it became obvious that the same theory and practice involved in longwall mining must be applied to pillar extraction. The edge of an extraction area must be snapped off cleanly over a sufficiently wide span to release any tendency to form beams. A cantilever of roof strata rests on pillar edges with some small extensions held by relatively weak supports. If the breaking load of a rock bolt is 10 tonnes, and of a timber prop about 20 tonnes, then this load is given by a piece of rock about 1 m square and 3 to 6 m high. The superincumbent load of the main roof beds must be held well back into the solid pillars and on the broken rock in the waste.

The recommended method of extraction of pillars is to present as straight a transverse line as possible across the waste. The method in Figure 7.13 shows systematic extraction across a 200 m front which will cause quite strong roofs to collapse. It is significant that as the New South Wales mines became deeper a pillar extraction width of around 60 m as a minimum was necessary to induce good caving. Neither strong remnants of pillars, nor strong artificial supports, should be left at the waste edge because they will create a new small beam and throw extra weight back onto the pillar edge. This can cause spalling of the coal, shear breaks in the roof over the coal edge, and possible guttered breaks into the roof which release roof joints and cause bad falls.

Supports must however be set at the waste edge as shown to prevent falls of the lower roof running into the roadways or the splits. These supports should be withdrawn or broken if they show signs of causing the waste edge to hang up instead of breaking. A small hang-up of roof has a cumulative effect because it then often needs a larger pillar to protect the continuous miner and workmen, and this in turn can cause further roof troubles during extraction. A bad sequence can cause coal extraction to drop as low as 50 per cent in the pillar areas. Roof bolts will not prevent the waste from caving.

The U.S. Bureau of Mines has studied pillar extraction and Stahl[7] produced recommendations for pillar extraction. These were basically the same as those given below which are up-dated and apply to Australian conditions.

Recommendations for Pillar Extraction

1. After extraction of a pillar is begun speed in completing the operation is essential. If necessary finish a shift off in another pillar, leaving a substantial block, not a small stump.
2. Complete extraction is preferable to only splitting pillars. Strong remnants should be blasted.
3. If floor lift tends to close roadways take a new skirting with the continuous miner to make sure that it is not trapped by floor lift. Make sure that it can be backed out quickly at all times.

4. Do not leave small pillars when first working. Form substantial pillars and split and lift (pocket and fender) them for extraction.
5. Do not rely only on roof bolts. They give no warning of impending collapse and can be badly weakened during extraction which causes roof joints and breaks to open.
6. Hydraulic props and chocks are good for waste edge support; they give a strong breaking-off edge and can be remotely withdrawn and pulled forward.

Case Studies for Pillar Mining

Two mines are briefly described to show the contrasts possible. Further details of both can be found in 'Mine and Quarry Mechanisation, 1969 (Magazine Associates, N.S.W., Australia).

Munmorah State Mine

This mine is situated in the northern coalfield between Sydney and Newcastle, and works the Great Northern seam which is a hard tough coal with an average thickness of 2.3 m. Coal is brought out of the mine by conveyor and passes through an automated crusher straight into the local power station. The workings are shallow and the immediate roof is a strong conglomerate. The only problem it causes is high pick wear if the continuous miner jib is raised too often into it, although delayed falling of the goaf in a new pillar area may cause a wind blast. The manager's support rules specified one bar and two props every 3 m. Occasional 1.5 m roof bolts are used where necessary. There is a small amount of floor lift because dust suppression water attacks the seat earth of the seam and causes it to swell.

Approximately $1\frac{1}{4}$ million tonnes of coal a year was won with a work force of 250 men (including 25 staff) who operate on a tonnage bonus. The coal was extracted with Lee Norse oscillating head continuous miners. The choice was partly dictated by the resistance of the coal to cutting. The head has to be dragged down through the seam and the extra dead-weight of the Lee Norse enabled the cutting force to be better utilised. There are seven continuous miners at the mine and they operated for 12 unit-shifts per day in 1970. The normal work force per unit-shift was

1 continuous miner operator
2 shuttle car drivers
2 skilled miners (roof supports, cable moving, ventilation, etc.)
1 fitter
1 electrician
1 deputy (staff supervisor)

although the full work force was allocated at 21 men per day per continuous miner unit shift.

The continuous miner loaded at about 130 t/hr into Joy SC15 cable reel shuttle cars, taking about 2 minutes to load 10-11 tonnes into a car. Production from first working was as easy and as efficient as pillar extraction. Dust and gas production was very low and ventilation was easy. Main development was with 8 entries (5 intakes and 3 returns) and panel development was either by 4 or 3 entry drivage.

Coal Cliff Colliery

This mine is in the southern coalfield of New South Wales. It is working the Bulli seam which is softer and easier to mine than the Great Northern seam. After treatment it yields a high quality coking coal with an average analysis of about 10.5 per cent ash, 69 per cent fixed carbon, 19 per cent volatiles, 0.3 per cent sulphur, 0.03 per cent phosphorous and a swelling index of 5.7. The seam is dusty and good spray systems are necessary. This in conjunction with careful control of cutting speeds and the provision of sharp picks keeps dust production within the maximum allowable concentration figures laid down in the Coal Mines Regulation Act.

The seam is also moderately gassy and care is taken to adequately ventilate all working places. In the north east corner of the lease the gas emitted was the Illawarra Bottom Gas. This was unusual and could contain up to 70 per cent of carbon dioxide with 30 per cent of methane. It is also found in Metropolitan Colliery to the north of Coal Cliff and is often associated with igneous sills and occasional gas outbursts. The working areas of Coal Cliff now have a normal seam gas which consists principally of methane. It is only given off when coal is worked and does not bleed through the strata.

The seam height varies between 1.4 m and 3.3 m, but is usually at least 2 m thick. The roof consists of about 0.9-2.4 m of shale underlying a massive sandstone, about 10-23 m thick. The seam was followed in under an escarpment from the outcrop and the present depth of working is between 420 m to just over 500 m.

Comparing the 1970 figures with Munmorah the output was about the same at 1 250 000 tonnes per year but the more difficult mining conditions necessitated a work force of 550 men which included the washery operatives and other surface employees. The proceeds earned for coking coal are fortunately higher than those for power station fuel. Since 1970 the mine has been reorganised to increase efficiency and to improve management control. It is now officially two mines; Coal Cliff Mine and Darkes Forest Mine, operated by The Coal Cliff Collieries Pty Ltd. Both mines share the same washery and supply service and the present total output of nearly 2 000 000 tonnes per year is brought out on a common 3.6 km cable belt with a 210 m vertical lift. A man-riding shaft with an automatic push button controlled friction winder has been sunk to serve the Darkes Forest Mine to reduce underground travelling time to the current working areas.

The depth of working is such that a mining engineer is faced with problems in pillar mining and considers other methods. The Coal Cliff experience shows how trials are made. In 1950 the New South Wales Coal Mines Regulations Act was amended to permit pillar recovery by mechanised methods instead of by hand. At this time the mine extracted 30 per cent of the coal in first working and 30 per cent on the retreat, leaving the rest. Dust suppression water from continuous miners softened the floor, remnant pillars both broke up themselves and caused floor heave, up to 2 m in some cases, trapping the continuous miners in the open-ended lifts used. Pillar sizes were 2 chains by 1 chain (40 m x 20 m). The mining methods at that time were aligned with contemporaneous American practice and a staggered diagonal pillar extraction line was used. Experiments and experience at Coal Cliff showed that pocket and fender methods had to replace open-ended lifts, and that the diagonal pillar extraction line had to be replaced by a straight extraction line.

In 1963 Coal Cliff had its first attempt at mechanised longwall mining. Westfalia-Lünen powered supports and an AFC, and a coal plough, were imported from Germany. The supports which had been designed for the easily-caved shale roofs in Europe were not able to cope with the stronger roofs in Australia and the longwall face experiment was terminated. The mine's progressive attitude was probably ahead of the availability of suitable equipment. In addition there were problems from excessive dust and spalling of large lumps of coal from the face.

The mine was meanwhile improving its pillar recovery methods and bought improved continuous mining equipment. The number of entries in first working were reduced and pillar widths were increased. The pillars were extracted by a skirt heading alongside the old entry followed by lifting the pillar.

A second attempt was made at longwall mining in 1964. Gullick 6-leg chocks were used with a ranging drum shearer on a 155 m long face. Again the chocks had been designed for weaker British shale roofs and were not strong enough for Australian conditions. The waste roof hung-up, then broke forward to the face line. The coal face spalled and there were roof falls on the face track. A coal roof was left to control the falls and only 2.6 m was taken from 3.2 m height of coal. Along with the heading pillars which have to be abandoned in longwall this only gave 60 per cent coal recovery and there was still a dust problem so the second longwall was stopped after eight months.

The method of coal recovery is now consolidated into pillar extraction on a split and lift basis. The first working output rate is about 250-500 tonnes per continuous miner shift from roadways, dependent on mining conditions and the machines used. Roadway conditions are relatively bad but when the entries have been driven they release the strata stress and subsequent pillar extraction has a productivity of over 500 tonnes per shift.

The method of support in first workings is

(a) heavy section half-round wooden bars are set at one metre centres as close as possible to the coal face with two 1.8 m bolts through them
(b) temporary hydraulic props are set immediately and are replaced with wooden props as the machine advances
(c) centre bolting with 2.4 m bolts is carried out on maintenance shifts
(d) intersections are pattern bolted with extra bolts and plates.

When pillars are extracted the support pattern is modified,

(a) wooden supports only are used in the splits
(b) bolts are set on the 'lift' side only (see Figure 7.16). The props on that side can then be taken out without using a carrier
(c) the intersection of split and entry may need extra bolts set.

Coal Cliff and Darkes Forest use Joy 8CM and 10CM continuous miners and Jeffrey Heliminers with Joy 15SC cable reel shuttle cars. The 8CM units are gradually being replaced with larger 10CM miners.

A unit team for one shift consists of

1 continuous miner operator
2 shuttle car drivers
3 or 4 skilled miners (timber, ventilation, etc.)
1 fitter
1 (shared) electrician
1 deputy (staff supervisor)

but the workforce allocated to each continuous miner for one day can total 50 to 60 men.

(Note: For Australian coal mines in general when unit manning is considered it must be remembered that the working shift is seven hours less travelling times. If the oncoming shift relieves the offgoing shift at the face then two or three production shifts can be worked plus a maintenance shift; i.e. four shifts within 24 hours. Personnel transport is now often organised on a basis of unit team cars. A diesel or battery driven personnel car will hold 12-14 men. These cars can be run down a drift as a coupled train, or in visual contact, and then dispersed to each working panel. The offgoing shift drive them back to the surface or to the shaft bottom.)

Figure 7.16 shows the current panel development at the Coal Cliff collieries. Because of their lower productivity the number of entries is kept to a minimum. Development has been with two headings designed to create panels of 1000 yd by 300 yd (914 m x 275 m) as shown. This requires a cross belt for pillar removal. Currently (1972) a 1 000 000 tonne block of coal has been programmed with a 500 ft (150 m) width panel. This will be worked without the cross-belt and the productivity of the two panel widths will be compared.

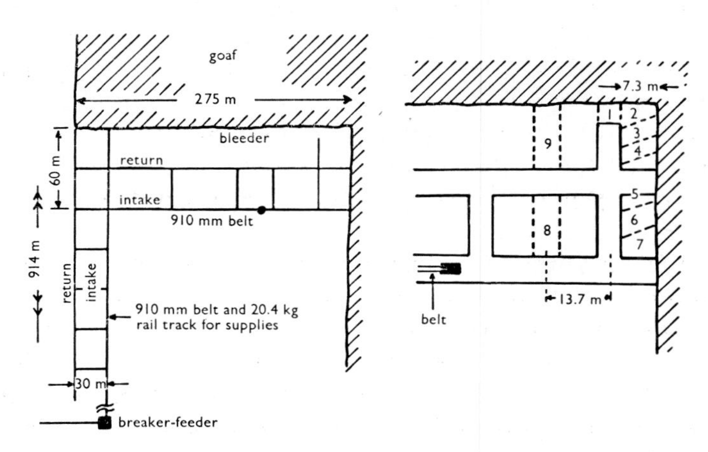

Fig. 7.16 Panel development at Coal Cliff Colliery

Panels are split into pillars for removal. Two entries are driven across the end of the panel. Ventilation cross cuts are driven between them as necessary: the cross cuts are kept to a minimum consistent with brattice ventilation which is efficient for about 80 m. Through ventilation for the lifts is provided by bleeder ventilation around the goaf edge.

Extraction pillars are formed 240 m by 30 m either side of the central entry as shown. The new pillars will be only half this length. The extraction pillars are then split at 13.7 m centres and lifted off. The splits are in an area de-stressed by caved goaf and this enables safe extraction to take place.

Coal Cliff considers the advantages of this method to be

(a) the continuous miner and operator are adequately protected from waste falls
(b) sub-panel conditions are good because of de-stressing by adjacent panels
(c) ventilation is simple and efficient
(d) mining operations are concentrated and flexible
(e) coal recovery is 90 per cent.

SHORTWALL MINING

Cyclic Shortwall

The original shortwall working probably developed from pillar and stall workings. (Note:— a stall is a working place similar to the bord in Figure 7.12). With cyclic mining which involves each working place in turn being drilled, fired, and loaded out there is an obvious need to achieve the maximum output per cycle. Whereas longwall mining gives a high output per cycle the pillar and stall method of working gives more flexibility between working places. If there is a breakdown of machinery or roof in one stall then there is not a complete loss of output because adjacent stalls continue to operate. In addition pillar and stall mining is more suitable to shallow depths.

In operation either mine cars were brought in to a stall for loading or a short cross-conveyor was laid to feed onto a stall conveyor. Both were preferably chain-type conveyors (these used roller type chains and preceded the armoured flight chain conveyors). The coal was undercut by a shortwall cutter similar to that shown in Figure 7.1, but which slid on the floor on a base plate. The whole length of the shortwall was bored by hand and then fired. Loading was at first by hand, and then later with gathering-arm loaders.

Some mines still operate basically on this principle. A group of shortwalls are mined adjacent to each other with rib-pillars left between (similar to the bord and pillar layout shown in Figure 7.12). A shortwall chain-jib cutter mounted on rubber tyres trams to each room in turn and cuts the coal to about 1.8-4.0 m depth. It is followed by a rubber-tyred mobile drill carriage which mounts a rotary drill on a hydraulic boom. The rooms are bored and then fired in turn. To complete the cycle a gathering-arm loader fills the coal out into shuttle cars or onto a chain conveyor. With a suitable grouping of three or more shortwalls all the machinery can be fully utilised. The machinery can develop headings as well as mine shortwalls. A particular advantage of this type of mining is its suitability for thin seams (say 0.9-1.2 m), and thick seams (over 4 m), where continuous miners are at their limits of operational height.

Awaba State Coal Mine[8] for example uses cyclic mining in a thick seam. This mine is close to Munmorah State Coal Mine and works the same Great Northern seam, but at heights up to 4.6 m. It was opened in 1947 several years before Munmorah and has continued with conventional cyclic equipment, buying newly developed shortwall machines as they become available.

The seam at Awaba is less than 67 m in depth and is worked under several water courses, roads, power transmission lines, small dams, swampy areas and a railway. Consequently the mine uses first working only. In this hard coal it results in 'tight' mining with no easy pillar extraction. Grades are up to 1 in 4. The coal is mined from panels made up of nine entries with crosscuts around square pillars on one chain (20 m) centres. This gives a satisfactory number of places for cyclic operation of one unit.

A mining unit consists of one Joy 15RU cutter (as in Figure 7.1). The jib length is 4 m and the machine has been modified to four wheel drive with dual wheels to cope with the soft floor and local grades up to 1 in 4. Two horizontal cuts are made at 1 m and 1.4 m above the floor; there are no vertical shears made although the jib can be used to trim loose coal from the sides of the entries. Drilling is carried out with a Joy CD71A face drill with four wheel drive and with the normal horizontal cable reel replaced by a 15SC shuttle car vertical cable reel. It is operated by one man and can drill 4 m long rotary drill holes in 35-45 seconds. It drills a full face of 18 holes and flits in less than 25 minutes so that it can service 12 faces per shift and still have spare time for odd jobs. The drilling range is to 5.8 m width and from 0.1 m to 4.4 m in height. It cannot drill the hard conglomerate roof and percussive drills are used for occasional roof bolts in weak patches of roof.

The coal is fired with full face multiple shot firing with permissible milli-second delay detonators under special conditions from the Mines Department. Joy 14BU gathering-arm loaders fill the broken coal into 15SC shuttle cars which then travel to the breaker-feeder and panel belt. The panel belts deliver to a central 150 tonne surge bin and coal is hauled to the surface by electric trolley loco. Total manning from coal cutter to panel belt is 16 men including tradesmen and supervisors. For seam heights above 3.6 m the normal shift output is in excess of 900 tonnes and has peaked to 1300 tonnes for a nominal 7 hour shift.

Shortwall Continuous Mining

Although cyclic shortwall is now mechanised it is a potentially complicated method with several pieces of machinery to maintain, drive around, and program. In the northern coalfield of New South Wales the Broken Hill Proprietary (BHP) Collieries have pioneered a shortwall mining method with powered supports at John Darling and at Burwood Collieries.[9,10]

The original concept was developed in 1959 as a method of open-ending long pillars (see Figures 7.14 and 7.17) to avoid troubles with crosscuts. Panels were developed by three main entries and pairs of roadways were driven off at right angles from the outer entries at 42 m centres. These roadways were at 10 m centres with a 4.6 m pillar between them. They advanced for 165 m to create large pillars 165 m by 27 m. The pillars were then retreated by open-ending across the 27 m width with continuous miners. Timber props were used as a waste edge support. The thin pillars between entries were abandoned.

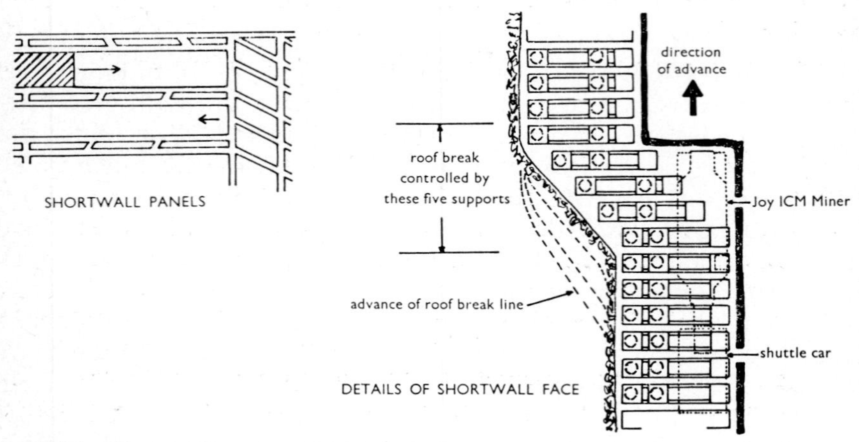

Fig. 7.17 Shortwall continuous mining

Through the 1950s powered supports of the self-advancing type were being developed in Europe and by the 1960s very robust types were available. Eventually in 1967 BHP ordered some Wild 134/134 self-advancing chock supports to use on the shortwall. The depth of operation was 107 m at Burwood Colliery and 259 m at John Darling Colliery. The supports, as their name indicates, consisted of two major vertical members each of 134 tons (136 tonnes) capacity. They are outlined in Figure 7.18.

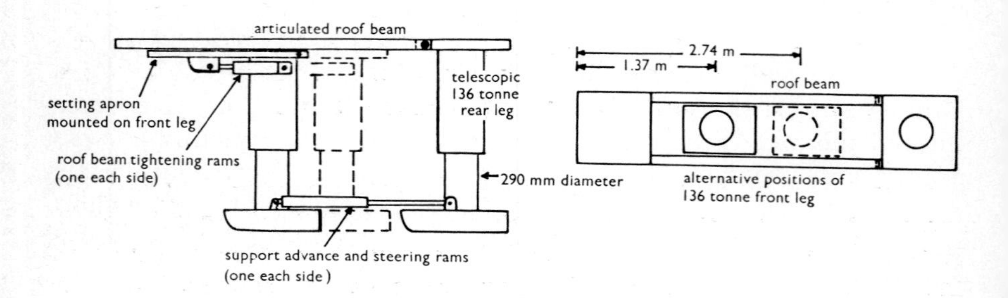

Fig. 7.18 Wild pillar extraction support

There is an articulated roof beam, its rear canopy is fixed onto the rear leg which has a load capacity of 136 tonnes. The front canopy and side arms are supported by a frame fastened to the front leg also of

136 tonne capacity. The front leg has its own separate contact with the roof and can be lowered and moved backwards or forwards within the side arms of the roof beam. After the front leg has been set a pair of rams can be operated to lift the front canopy of the beam to the roof. The front leg is moved forward and steered by a pair of rams which link the chock bases and which can be operated independently or together. The rear leg is pulled forward by the same double-acting rams.

The shortwall extraction method is shown in Figure 7.17. It can be divided into steps:

(a) start the cycle with the front support leg forward to give maximum support to the roof

(b) the continuous miner (Joy 1CM, a ripper-chain type) takes the buttock of an open-ended lift and loads into a shuttle car. Two shuttle cars may operate in tandem with the car behind the miner acting as a surge bin, or two cars may be used alternately by shunting into a convenient junction

(c) the back leg of the support is moved up over about a five support length to extend the roof beam forward above the machine track. This also allows the roof to break off gradually

(d) the legs are now close together giving a strong waste edge breaking-off support and cantilever support over most of the width of the miner and shuttle car track. It should be noted that both the continuous miner and the shuttle cars are handed to keep the operator away from the coal face, i.e. under the protection of the roof supports and safe from spalling coal

(e) after a full lift has been taken the machines are reversed to the roadway and all front legs are marched forward to finish with the cantilever tips close to the face and the legs apart to give maximum support to the freshly-mined strip.

Some points of interest from Burwood Colliery are: seam height 1.8 m, average width of lift 4 m, top production per shift 107 shuttle cars at average 5 tonne capacity: average production 335 tonnes per shift for 129 shifts. At John Darling Colliery the pillars were from 46 m by 320 m to 46 m by 387 m with a seam height of 2.2 m. Average production was 550 tonnes per shift, with a maximum of 870 tonnes per shift. A useful operating point was that all electrical switchgear, hydraulic pumps, etc., was mounted in an old converted shuttle car so that it could be easily retreated along the roadway in conjunction with the face.

The authors of the articles on shortwall continuous mining (Kay and Marsden) give the advantages of the method over normal pillar extraction as

1. large saving in timber costs
2. saving in brattice (for ventilation)
3. reduction in labour costs in handling these materials
4. reduction in pick costs and machine downtime because cutting is easier on these open ended lifts
5. flexibility of supports. Lifts may be opened out or tailed off (i.e. narrowed or widened) to suit the circumstances. The face is straightened out on the next lift. (It should be noted however that fenders cannot be left except at the tail end of the face.)
6. complete extraction and consequent good caving characteristics so that roof control is improved
7. there is easier prediction of subsidence effects from complete extraction
8. improved ventilation. The absence of crosscuts avoids leakage and there are less stoppings to build
9. less intersections so that there are less weak spots in the roof
10. increased safety because the men are working always under the cover of strong supports
11. the improved ventilation reduces dust concentrations in the working area
12. saving in purchase and installation of roof bolts; there is no bolting in the lifts
13. less cable damage on machines because of the simpler coal haulage system and the reduced materials handling off-shift.

It is suggested finally that alternate faces should be extracted in opposite directions to minimise the distance for support transfer. If this is done the belt conveyor must be laid in the entry against the adjacent panel for advancing faces because the entry against the working face will be collapsed. Coal is conveyed forwards to a crosscut and through it to the gate conveyor. Some mines could also suffer severe floor lift behind the advancing face.

A recent paper by Kay and Duncan[11] gives the latest developments in mechanised shortwall mining. It describes the mechanical and electrical details of the equipment and gives an account of trials with the method under stronger sandstone roofs. The chock was given a second rear leg to withstand the higher loads. Attempts were also made to use higher capacity continuous miners and shuttle cars but further modifications are needed.

It is interesting to note that before the development of the Anderton shearer attempts were made to use a ripper chain continuous miner (the Dosco) to work longwall faces. The machine was run up the face to take a buttock as for an open-ended lift and the coal was fed sideways onto a face belt conveyor. It was used both in Nova Scotia and in the Midlands coalfields of the U.K., but was not very successful on the 150 m long faces. At the time it was considered that it caused too much coal degradation, also there were no powered supports to give a rapid advance and a large area of roof was exposed.

LONGWALL MINING

In this section it is intended to describe methods in general, although a few case studies are mentioned. In general Australian and American coal mining methods are similar because the structural geology is similar. A useful reference to American methods is a paper[3] in *Coal Age*. Modern longwall mining owes a lot to development in Europe and by the National Coal Board in Britain in particular. However techniques continually improve and the rate of improvement is such that it is improvident to describe any detail as the 'latest', or the 'best'. The general methods are relatively stable.

To keep up with improvements in longwall mining (particularly longwall advancing) the U.K. mining journals *The Mining Engineer* and the *Colliery Guardian* are useful sources of information. It is stressed that any mining engineer must of course try to keep abreast of current literature.

The earlier longwall methods of the 1900s were needed so that a continuous stream of air could be coursed through faces to remove high gas concentrations. It was probably for this reason that hand-worked stepped longwall was introduced in Britain. Stepped longwall was a series of shortwall faces or stalls coupled end to end in a staggered line. All the coal was extracted and stone packs were used to provide permanent support for the roof. The interconnected stalls gave a lower ventilation resistance for the mine than did a network of roads and stalls.

It was found by trial and error that as the mines became deeper it was expedient to straighten the face line to obtain better roof control. This straight line also helped the introduction of early coal-cutting machinery and face conveyors. In its heyday some Yorkshire mines had a continuous length of 0.8 km of straight hand-worked longwall face served by numerous coal haulage gates.

As more machinery was introduced it became necessary in Europe to divide longwall faces up into units of about 1-200 m to provide mutual independence in the cut, fire and load cycle. Whilst belt conveyors were still in use along the face most mines used pairs of face belts to feed on to one central gate (roadway) conveyor. In steep seams the coal had to be conveyed downhill. When armoured flight conveyors were introduced along with hand-set supports the original thinking was to have a face length of about 200 m from which 1, 2 or 3 complete strips could be taken in one shift.

Present thinking and new powered supports have reduced the 'complete strip' psychology and face lengths have been adjusted to about 80-200 m dependent on roof conditions, seam inclination, seam thickness, and the coal winning machine used. A decisive factor is the maximum load which a face conveyor can carry. A fast-cutting machine in a thick seam will load a conveyor to capacity over a short length of face. In thick seams the roadways are driven completely in coal, and are profitable, so a short length of face is acceptable. In a thin seam the load of coal on the conveyor per metre cut is low, and roadways are expensive to develop, therefore face lengths tend to be longer.

Reasons for Use of Longwall Method

An abstract list of the relative merits of longwall mining is given below.

Advantages

1. Coal recovery can be increased by 10-15 per cent because no pillar remnants are left and very little loose coal is left on the floor.
2. Reduction in panel costs, particularly in support consumption and supplies manpower.
3. Concentration of supervision and service from tradesmen.
4. Men easily trained.
5. Face equipment and support is adequate in weak roof seams.
6. Roof action is good on the face with a clean break-off line, although the face must be kept straight (Federal No. 1 mine in the U.S.A. used a helium-neon laser to keep face and development drives on line).
7. Longwall mining generally produces a cleaner and coarser product than pillar mining.
8. Ventilation is more concentrated and is usually better upwind of the machine than in pillar mining methods.

Disadvantages

1. A breakdown in supervision, or machinery, can result in large tonnage variations.
2. Even minor troubles of roof control can cause slow working with high tonnage losses.
3. Dust control is usually much more difficult.

4. High output from a single unit can cause gas emission problems.
5. The high continuous rate of loading (Continuous Miner loads are usually intermittent and dispersed over 3 or 4 sections) can cause problems of belt overload, and much larger belts may have to be installed in a mine which changes over to longwall.

A more general discussion of why longwall methods are used follows.

Pillar mining essentially needs rectangular shaped roadways with easy side access for lifting the pillars. It is unfortunate that as bedded deposits are worked deeper in the ground the pressure on the rocks around the roadway increases. The successive stages in lessening consequent roadway closure are to reduce the width of the road and then to change its roof shape into an arch. Supports are changed from roof bolts and wooden bars to the much stronger steel arches. Neither arched roadways nor steel arches are convenient for pillar working and they are a nuisance if they are used in conjunction with retreating longwall.

In thin seams roadways are expensive to drive because they produce more rock than coal, and the rock is more difficult to mine. If, in addition, the depth of working and the roof strata are such that roadways are difficult to maintain, then roadways will be kept to a minimum and as much coal as possible will be worked from each road. This is one of the combinations of factors that induces a mine to adopt longwall methods. Even with thick seams, if the roadways are expensive to drive and to maintain, then longwall mining will probably be chosen.

Roof support on a longwall face is more systematic and is usually stronger than in pillar extraction methods. This is particularly so since the advent of powered supports with their pre-determined setting loads, and high yield loads which are controlled to avoid support breakage. In weak roof areas the narrow extraction web of a longwall face has marked advantages. It should be noted that when coal is extracted there is a marked roof lowering: it can be 5 to 15 cm per metre of face advance. A rigid support can fail but yielding supports accept this closure and keep the roof reasonably intact whilst it flexes downwards.

Britain and Europe in general, with deeper seams and weaker strata than Australia and the United States, have adopted longwall mining almost exclusively. There are other factors which must be considered; steep seams can be easier to work by longwall. If there are several seams in a vertical sequence then longwall mining gives almost complete extraction and avoids leaving small pillars of coal which will act as stress raisers on superjacent or subjacent seams. If the stress on mine strata is raised then bad roof conditions on faces results, and there is increased local closure in roadways.

The relatively low outputs per manshift in Britain for longwall mining have already been mentioned but this should not be taken as a reason for not adopting longwall mining. The better mines, particularly in thicker seams, far exceed the national average. A longwall face in a seam one metre thick will have to advance at twice the speed of a two metre thick face to give the same bulk output.

A target to aim at may be one set by Federal No. 1 Mine in the United States[12]. In early 1970 they managed to win 5400 tonnes of coal on 3 shifts from a longwall face in a 2.19 m thick seam at 213 m depth. There were 24 men which gives an output of 225 tonnes per face manshift. This is more than double the output per man-shift obtained when splitting pillars with a continuous miner under favourable conditions. The Federal mine averaged about 1100 tonnes per shift, 3 shifts per day on its first test panel with no previous longwall experience. Longwall mining gave an extra U.S.$1.50 per ton profit over the use of continuous miners, which were not in trouble but were taken out after an economic feasibility study.

It should be noted that the labour force is generally lower on longwall mining but the capital costs are higher. An Australian longwall unit may cost from $750 000 to $1 000 000 with the powered supports responsible for half to three-quarters of that amount. A continuous miner section may be set up for about $400 000-$500 000 on 1972 costs.

British longwall faces are currently aiming at targets of about 1000 tonnes per shift but many are below this output. A recent paper[13] gives a summary of faces as

Year	Number of Mechanised Faces	Output per day (tonnes)	Average Seam Thickness m	Average Shifts Worked per Day
1965	1660	487	1.22	1.96
1966	1561	506	1.27	2.05
1967	1518	454	1.29	2.19
1968	1224	549	1.32	2.17
1969	1057	600	1.34	2.19
1970	996	607	1.34	2.08

Australian longwall faces with thicker seams and retreat working should manage 1500 to 2000 tonnes per shift under settled conditions. The early longwall faces were mainly working panels originally developed for pillar extraction and some of the roadways gave troublesome conditions. An early installation at Kemira Colliery managed 6-800 tonnes a shift but peaked to 2000 tonnes per shift with a best daily output of 5800 tonnes.

Choice of Longwall Advancing or Longwall Retreating

For more than thirty years the theoretical advantages have all lain with longwall retreating so that practical disadvantages of retreat mining must be examined to see why advancing longwall is still necessary.

In longwall advancing the coal is worked away from main entries and roadways are made in the mined-out area behind the face. Figure 7.19 shows a typical longwall advancing layout from the U.K. In the absence of faults, etc., the layout would be a regular grid with faces developed on both sides of main entries.

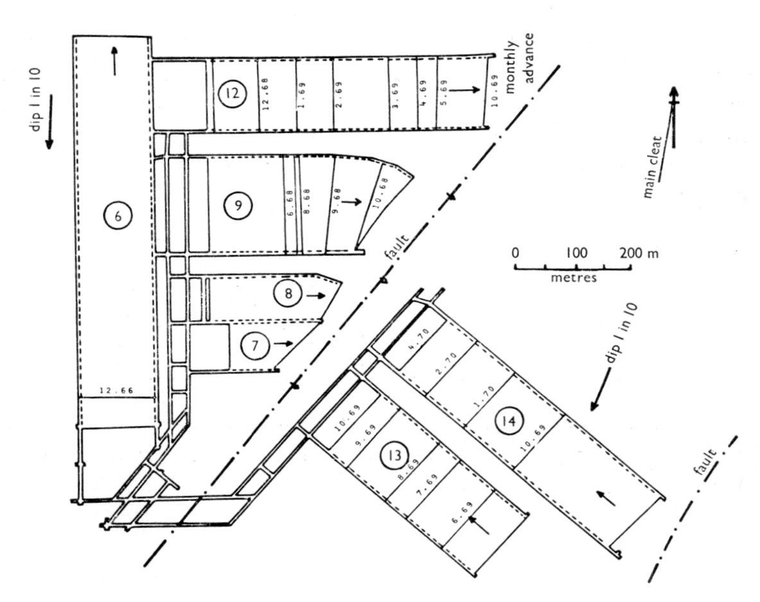

Fig. 7.19 Longwall advancing layout

For longwall retreating the panels are completely blocked out on all four sides and then the faces are retreated back along the roadways. Panels 2 and 3 of Figure 7.11 are in fact being developed for longwall retreat mining. Some earlier definitions of retreat mining imply or state that a mine is developed to the boundary of the lease before extraction begins, but it is more usual to block out panels as close to the shaft or drift as is safe and then to extract them in sequence spreading out through the lease. Large pillars are left to protect main entries as in Figure 7.11.

The advantages given for longwall retreating are:

1. the area to be worked is 'proved' in the development stage and there should be no hold-ups when the face is extracted

2. roadways in the solid are cheaper to maintain than goaf or rib-side roadways behind the face
3. entry drivage operations are kept separate from face extraction operations. Each can operate at their best speed. There is no interference between entry machinery and face machinery, so that entry drivage can be more efficient
4. seams liable to spontaneous combustion can have panels sealed off quicker in case of trouble. Potential fires are left behind in the goaf.
5. wet seams can be retreated to the rise so that water runs away from the face
6. recovery of face equipment when the face finishes is through a short distance of good roadways and is much easier
7. ventilation of retreat panels is easier. Any leakage through the goaf is designed to bleed off methane, rather than accidental.

The disadvantages of retreat mining are:

1. in deep seams, especially those with weak floors, there can be large amounts of floor lift in roadways in the solid. When the face retreats the pressure wave in front of it can almost close a roadway, trapping any machinery in it. In the event of holidays, or strikes, such items as stage loaders have been squeezed between roof and floor. In roadways made behind the face extra height can be taken to cater for closure which occurs at a slower rate than it does close to a face[5]
2. roadway drivage in thin seams is more difficult to mechanise than in thick seams. With longwall retreating the rock from the initial drivage has to be sent out with the coal. With longwall advancing it is possible to pack roadway dirt into the mined-out area
3. the T-junction between face and roadway is difficult to support, particularly where legs have to be removed from under bars. The worst case is where steel arches have been used in weak strata at depth and special arrangements are necessary. (These are described later.)

Because of these disadvantages the majority of European faces are still mined by longwall advancing, although some of the traditional conservatism of mining engineers has been overcome recently and fresh experiments are being made with retreat faces. This new trend has been helped by the fact that roadway-making operations behind the face have been difficult to mechanise and some potentially fast-moving faces are delayed by slow roadways.

It must be remembered too, that before the era of major long-term contracts, coal mining was a poor relation to metal mining. Seasonally fluctuating markets did not encourage long term development and a coal mine worked longwall advancing when it had a market, and then worked short time or closed down until the next sales took place. Modern capitalism and large companies have eased this situation, but there is still a seasonal demand for industrial and power station coal.

Longwall Retreating

High production longwall retreat faces exist in America, Australia and Britain. It is proposed to describe these in general terms as a shearer face equipped with self-advancing supports, although it is possible to use different combinations. The practice in Australia and America is to buy a 'turnkey' installation in which a main contractor supplies all equipment as an integrated system on a payment by results basis. Individual manufacturers may produce supports with different yield loads, or different shaped valve levers, but basically the equipment does the same job. Modifications are often necessary to meet more onerous conditions. An hydraulic chock with six-20 or 25 tonne legs suitable for English mines must be made with six-100 tonne legs for massive sandstone roofs in Australia. In most cases American and Australian mines prefer over-design for long life and reliability. The less in-service maintenance that is needed then the more operating time is available.

This is not to imply that the National Coal Board's engineers are not as efficient in Britain. However they operate under different circumstances. Equipment is not bought by a mine but is hired from headquarters plant pools at a fixed daily charge to cover capital costs and major maintenance. This covers all mining machinery and supports. Equipment is operated for a period of hours, or tons mined, as laid down in a planned preventative maintenance scheme. After this period it is exchanged and the replaced equipment is taken to a regional workshop for major overhaul. Routine maintenance is a responsibility for the mine operating the equipment. All supplies and machinery are bought in bulk by a national Purchasing and Stores Department. Requirements for coming years are forecast ahead by mines and must be provided for in advance. Under these conditions all equipment is as standardised as possible. There is maximum interchangeability of parts which are made to N.C.B. specifications. Mining equipment is often bought internationally to these specifications, which are frequently the only formal ones available, both for coal and metal mines. A paper by Spanton[14] describes the current N.C.B. use of these methods of bulk supply and use.

Early Australian and American longwall equipment was bought from within the specialised European market. Development in Britain was often by the National Coal Board's own research and engineering establishments, and by the Board's mining engineers. Overseas development was consequently a major responsibility for the mining company which bought imported equipment and found its operational limitations. The manufacturers then produced equipment to meet the new specifications.

The following description of longwall retreat mining is based mainly on Australian practice with general operating experience given. Details of individual mine's experiences, and of a United States installation, were given at a Symposium on Longwall Mining held by the Illawarra Branch of the Australasian Institute of Mining and Metallurgy at Wollongong in 1970. Generally the longwall mining faces were started up to cope with worsening roof and roadway conditions in the Bulli seam as it was worked at depths approaching 300 m or more. One installation is operating in the 1.2 m thick Balgownie seam that was considered too thin for continuous miner pillar extraction.

Layout

A general layout for a longwall retreat panel is given in Figure 7.20.

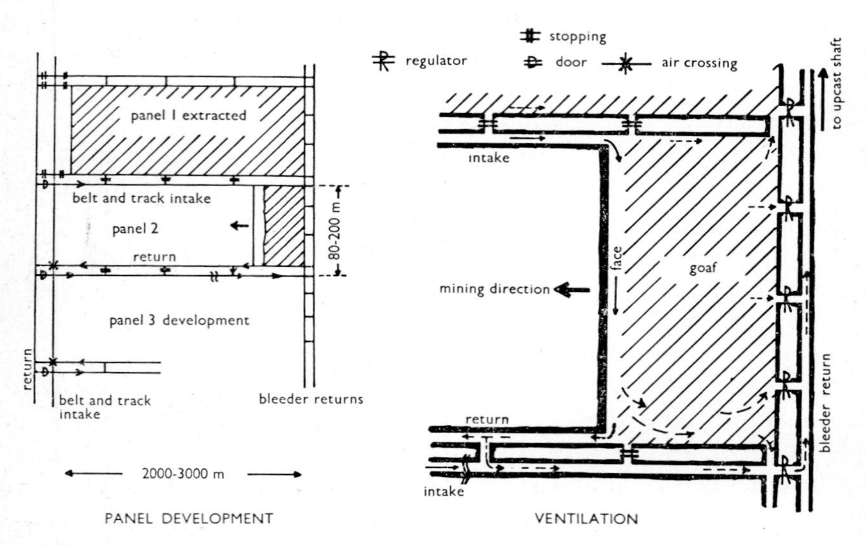

Fig. 7.20 Longwall retreating layout

A problem of coal mining is a build up of gas in fallen unconsolidated rock. When the barometric pressure falls it is possible for methane-rich gas to expand from waste areas and old roadways. In retreat mining it is convenient to establish *bleeder ventilation* to keep the goaf drained. A row of timber cribs (chocks, bulks, or pigsties) is built along the line of the abandoned roadway to keep it open for leakage ventilation. The intact return has its airflow restricted by sheets of anti-static (semi-conducting) PVC called brattice, or brattice cloth, to force the air back through the fringes of the waste. Bleeder ventilation is also useful because it can keep most of the desorbed methane gas from the coal in special return roadways away from new developments.

Powered Supports

The support system in a longwall face plays an important part in its success or failure. There is no longer time available to build stone packs to control the roof and to resist its tendency to move out from the coal face and towards the fallen waste. A powered support must:—

(a) hold up the lower roof beds
(b) stop tension cracks from opening in the roof between the goaf and coal face
(c) act as an anchor for the armoured face conveyor and provide the ram force needed to push it forward
(d) pull itself forward after the conveyor has been advanced (adjacent supports anchor the conveyor whilst this is done).

A typical support consists of four or six yielding hydraulic legs (like an hydraulic jack) in chock form for stability and comprises two main parts: a strong rear portion which gives a good breaking-off line for the waste and a front portion which moves forward to provide support over the new web of coal cut. Sometimes the front part forms an inner core between a pair of outer legs. Figure 7.21 shows the operation of one type of powered support. Each support in turn on a face goes through a cycle; set to roof; lower the front legs and advance to cover new track; raise the front legs; push conveyor forward; lower the rear legs and pull rear portion forward; raise the rear legs. Ideally at this stage the goaf collapses behind the support removing some of the deadweight on the rear legs. For economy only one chock in five to eight is a 'pusher' chock but all chocks have pulling rams for self-advance.

Support control is usually from the safety of the adjacent chock, or a small block of supports may be operated in semi-automatic bank control from one set of valves. The fully automated R.O.L.F. (remotely-operated longwall faces) installed in Britain as test-beds for new ideas were uneconomic but helped to develop a new generation of support controls to repay the initial investments.

Powered supports are fitted with a plate across the rear of the leg bases, and a chain mat at the rear of the upper part of the support. These prevent broken rock rolling forward from the waste between the legs of the supports. Loose rock will affect the pulling-forward action of the chocks and can cause them to tip sideways or to become fast. The Russians have developed complete shield supports that enclose a longwall face in a three-sided iron tunnel, but except in very weak strata this appears to be an 'overkill' operation.

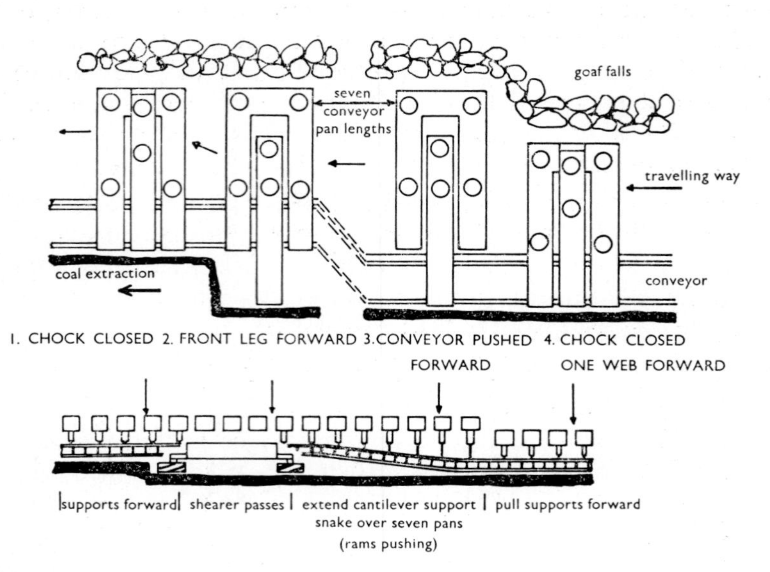

Fig. 7.21 Powered support operation

Face Operations

The equipment on a longwall face is generally disposed as in Figure 7.22. An armoured flight conveyor runs along the face with the coal winning machine mounted on it. The powered supports are pinned to the conveyor and are kept in location by it. There is often a tendency for a conveyor to creep along a face during snaking operations. This is particularly prevalent where the conveyor is always snaked (i.e. pushed over into the new track) from the same direction. Unidirectional snaking is from the head-end going downwind and a rule-of-thumb method of counteracting creep is to keep the tail-gate end of the face lagging behind the main-gate end by about 3-10 m. The supports travel parallel to the gates (roadways) and thus are slightly angled to the face conveyor. The conveyor advance rams will then counteract the tendency to creep. On low gradients conveyor creep can again be countered by angled faces. In steep seams the coal is conveyed down-hill and a strong anchor station must be positioned at the tail end to stop the conveyor from sliding down the face. Drive-head anchor stations have been used but tend to be a nuisance in an already congested roadway. Powerful chocks used as an anchor station which repeatedly stress and unload the roof can also cause deterioration of the roof strata in advancing roadways.

The face conveyor delivers on to an armoured flight conveyor stage loader whose flexibility is essential in aligning the face conveyor delivery onto the belt conveyor line in the gate. A lump breaker may be mounted across the stage loader to reduce lumps to a size where they will not cause spillage on the belt (about 10-15 cm is generally suitable). The stage loader changes the coal flow direction through right angles and delivers it centrally on to the gate conveyor in an even flow. Its delivery either straddles or runs alongside the belt so that there is an overlap to take care of advance or retreat for a period of several shifts. The layout of Figure 7.22 is for retreat working but advancing longwall has a similar layout with the roads turned through 180°.

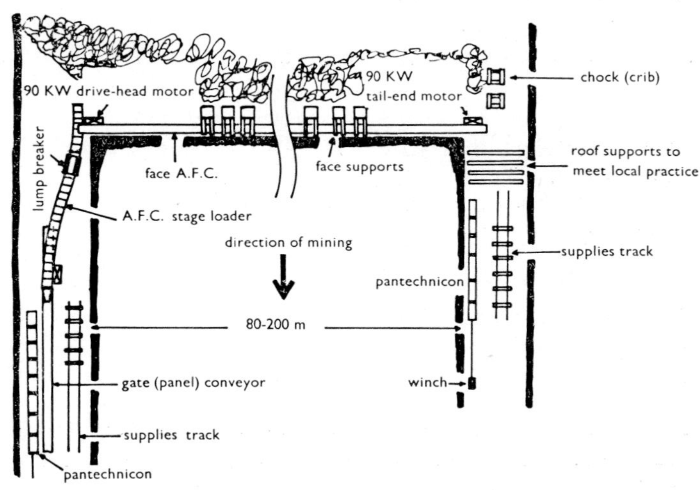

Fig. 7.22 Longwall face equipment layout

The face has to be supplied with electrical power for motors, and with hydraulic power for supports and, in some cases, hydraulic power for conveyor or plough motors. A transformer, hydraulic pumps, and electrical switchgear are installed near the face and must advance or retreat with it. These items are bolted together on wheels or skids to form a 'pantechnicon' which is pulled along as necessary by a winch or other device. Although transformers are not permitted in return airways in Australian and British coal mines the hydraulic pumps and motor drive switchgear are duplicated in the tail gate to cut down transmission losses onto the face. The trend is towards using 1100 volt operation (replacing 415 or 440 V) to reduce cable sizes for a given power of machine.

Each pantechnicon will contain two or three high pressure pumps, usually of the three-throw plunger type. Hydraulic legs in the supports are pumped up quickly through large bore hoses with a pressure of up to 17 MPa. When contact has been made with the roof an intensified pressure of up to 40 MPa can be applied to provide setting loads of up to 250 tonnes per metre of face for Australian or American hard roofs. In the European mines, with softer plastic roofs, about 60 tonnes per metre may be adequate. Conveyor and support advance is with much smaller rams, and pressures are limited to 4-5.5 MPa to avoid buckling them.

A feature of longwall mining which contributes to safety is that because the conveyor stays intact it is possible to run signal and telephone cables along its full length. Loudspeaker telephones are provided at intervals along with push button signals which can also be 'locked-out' to immobilise the conveyor and face machine. Audible warning of conveyor start-up is given because the conveyor operator is in the gate and may be 200 m or more away from the machine operator. The machine operator has to start the shearer manually, but once it is operating he can stand back a few metres and operate it by radio control. He can engage and disengage the haulage mechanism (but only stop the motor), raise and lower the underframe, and operate forward and reverse by remote control.

The normal crew for a shearer face is one machine operator, two support operators, one man at the main gate end, one at the tail gate end, one conveyor operator, a supervisor and an electrician and fitter. However the number of miners may vary from five to ten or more, dependent on face and roadway conditions. Also the men may interchange jobs or help each other as necessary. In mines with a high labour turnover the face may be deliberately overstaffed to keep a reservoir of men in training. On a plough face the same number of men may be used but each will stay in his own section watching the plough travel and operating his own supports and conveyor rams as necessary.

The shearer may be operated bi-directionally (Bi-Di) or uni-directionally. Where a ranging drum shearer is used as in Figure 7.21 it cuts along the face and the cantilever portion of the roof support is advanced by the first support operator. The second operator then operates the conveyor advance rams with the other support rams in neutral. The conveyor moves over against the coal and then the powered supports can be brought forward. Fallen coal in the newly-cut track may hinder the conveyor snaking operation by preventing a full ram extension of 0.5 to 0.6 m (the shearer drum width). To help clear the coal a ramp plate as in Figure 7.6 may be fitted.

It is also possible to have an *activated ramp plate*. This is a set of three or four miniature ploughs, the same height as the conveyor and about 0.3 m long. They are fixed onto an endless chain running behind the static ramp plate. The hydraulic motors which drive the plough chain are mounted on the conveyor ends and the operator in control of snaking operates the motors by remote control. The ploughs operate along the length of conveyor to be moved over and lift the coal up over the edge of the conveyor as it moves forward.

Since the conveyor has been snaked over immediately behind the shearer then the shearer has to cut its way back along the face, reversing the operation: this is bi-directional shearing and usually provides the most efficient operation. However there are circumstances in which it is not possible. The Bulli seam in the southern coalfield of New South Wales is a coking coal and has a high dust content. When the coal is cut then workmen need to wear a respirator if they work down-wind of the machine. The machine itself is fitted with sprays which saturate the coal with water across the portion being cut. However large quantities of dust are produced when the coal collapses off the face behind the machine, and also when the roof falls behind the supports as they move forward. Consequently all operators stay up-wind of the machine. The machine cuts from the tailgate to the maingate and then 'flits' back ploughing up any fallen coal on the way with the lower drum rotating. The conveyor and supports are moved over as the machine flits back, ready for the next cut.

Various dust suppression methods are available. Wetting agents can be added to water sprays but have been found ineffective in use. Longhole water infusion has been used where holes are bored parallel to and in advance of the coal face. Water is injected into the holes under pressure to seep through the coal and wet the dust. Unfortunately it also saturates the roof and floor and weakens them so that bad roof conditions can result. Venturi-type water injection into the cutting drums, and water fed through hollow picks, are also possible but provide maintenance problems and do not prevent dust caused by falling coal or roof rock. Water spray curtains have been tried across a face but airborne dust particles are very difficult to capture and the water is a nuisance.

Face-end Operations

The T-junctions between face and roadways provide many of the operational problems. When a roadway is driven in the solid coal then stresses are set up at the roadway edges. As a retreat face is mined back alongside the roadway then the roof sags and further tensile stresses develop along the roadway edges. There is a strong possibility of cracks 'guttering' up into the roof and causing support problems (Figure 7.23). The crack which eventually gutters up is often formed in the initial roadway drivage and enlarges during face extraction.

In retreat longwall both ends of the face conveyor protrude into the roadway so that the shearer drum can cut clear from the face coal. It is necessary then to remove the face side legs of roadway supports to allow the conveyor to pass. It is unwise to rely on roof bolts to hold up the roof bar and the lip of the entry. The roof bars must be supported by carrier girders. In their simplest form these are a pair of RSJs side by side supported by two or more yielding hydraulic props at each end. It is necessary to use a pair of joists so that they can be advanced alternately leaving at least one to the roof. They are set across the face entry next to the last face support (see Figure 7.28 for comparison).

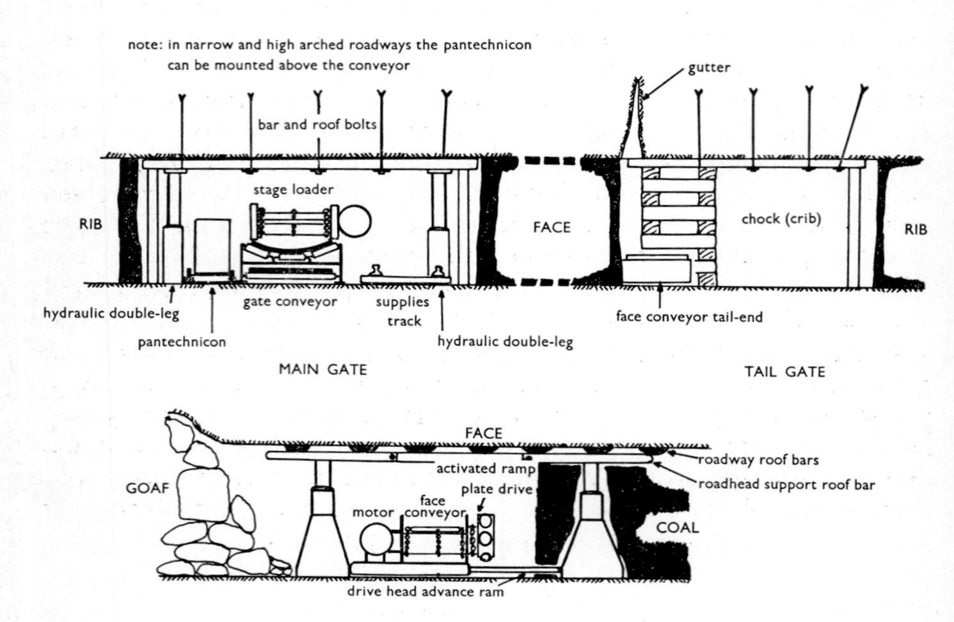

Fig. 7.23 Cross-sections of roadways and face

Under weaker roofs it may be necessary to use two pairs of special chock supports in parallel. Each support straddles the face conveyor as in Figure 7.23. They are linked at head and base and are fitted with rams to pull the gearhead over. They are interconnected to move themselves alternately on a mutual push-pull basis.

In front of the face hydraulic yielding supports are set under each roof bar to double leg it so that the effect of the front abutment can be resisted. Behind the face the main gate is allowed to collapse after recovering anything of value. Cribs are built in the tail gate to leave it as a bleeder. In some cases these cribs also are necessary to limit the amount of roof guttering. Occasional cribs may be built in the main gate, but they are undesirable if they prevent large blocks of roof from falling after the face has moved ahead.

Roof Falls

It will be noted that roof conditions are frequently worse in the tail gate than in the main gate. Large quantities of dust suppression water are used on the face and the bottom chain of the conveyor carries excess water back to the tail gate in flat seams. The floor there is frequently saturated, and occasionally even under water, and is consequently very soft. The roof supports sink into it and allow tension cracks to open in the roof, causing falls.

Falls can also occur when the face passes opposite a cross-cut junction because of the long diagonal spans of roof involved. For this reason cross-cuts should be kept to a minimum during development. In deeper mines the initial development roadways de-stress an area, but are damaged in the process. It can be beneficial to drive an initial twin-entry development, and then subsequently to mine flank roads for new face-end roadways. Provided that the roads are within a few metres (say 8-10) of the old roadways they are in a low-stress region and will give good roof conditions for the face ends.

Roof falls also occur on the face between the powered supports and the coal edge. They have to be filled in with timber cribs (or bulks) over the supports as in Figure 7.24. This is necessary to prevent the front of the roof beam rising into the cavity and the rear of the canopy becoming 'iron-bound', i.e. no further yield is possible in the supports. When soft patches of roof are encountered that are liable to fall it may prove economical to insert panels of square-mesh wire (concrete reinforcement or fence wire) over the tips of the supports to span them in threes or fours to stop the lower roof 'fretting' away.

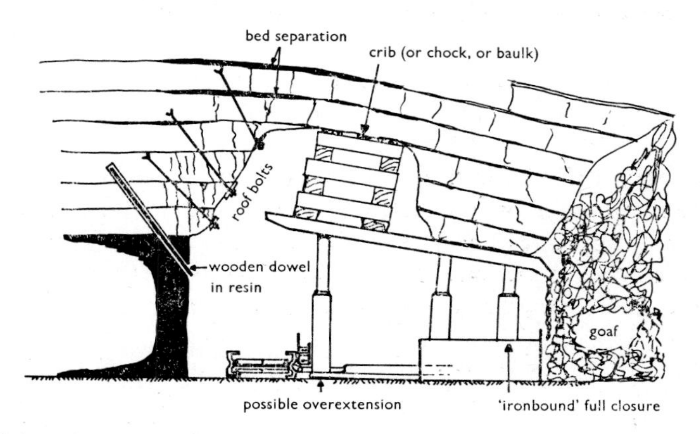

Fig. 7.24 Recovering a face fall

The edge of a face roof fall must be pinned over the coal by roof bolting forwards. These must be relatively short bolts to avoid waste hangups after the support has passed under them. They also have to be clear of the cutting drum to avoid damaging cutter picks. An elegant technique is to use wooden dowel bolts embedded in epoxy resin down through the coal. A shearer drum can cut through these wooden dowels without damage to picks.

Longwall Advancing

Face operations in longwall advancing are the same as in longwall retreating; it is the operations at the face ends which are different. The older methods of extracting coal by hand-working have virtually disappeared from developed countries with high labour costs. Hand-working was by the method shown in Figure 7.2. In stalls or in stepped longwall one team of men normally performed all the operations of coal extraction and roadway drivage as necessary.

For longwall mining teams of specialists each performed their own operations on a regular shift basis. There were the men who extracted ('got' or 'won') the coal, variously described as miners, colliers, strippers, hewers, etc. Their function was to load out cut-and-fired coal and to set roof supports. They were followed in sequence by the conveyor movers who advanced the belt conveyor one track, and the packers who removed the back row of supports and set chocks and stone packs as necessary to control waste breaking. The cutter men then came on shift to cut the next web of coal and a borer came to drill the coal ready for firing. During the conveyor advancing and coal cutting shifts the face roadways would be advanced an amount equal to the face advance.

When armoured conveyors were introduced longwall methods became continuous, which meant that the teams of men who made the roadways had to provide continuous advance of the order of 1 to 2 m per shift. In thick seams this presented few problems, but in thin seams at depth where (say) a 3 m high roadway was made after extracting 1 m of coal, there was a large volume of rock to move in a short time. Older methods called for hard physical work which was suitably well rewarded. Day wage payments do not encourage hard physical work and roadway drivage became an increasing problem for longwall advancing methods.

In fact the two major problems peculiar to longwall advancing are roadway drivage, and stables at face ends; the two are related. The other problems are those common to all mining such as falls of roof, floods, fires, faults and finance, to mention a few. The mining engineer can lead an exciting life.

A longwall face *stable* was necessary so that the face cutting and loading machinery could stand in it, or adjacent to it, while the face conveyor was pushed over, see Figure 7.25. This stable can be mined by hand, but a high productivity face then exists between two low productivity ends. Early attempts to mechanize stables involved the adoption of special equipment, the Dawson Miller for example, which mined the stable with a separate machine and conveyor in advance of the roadway and face. More recent attempts to improve efficiency have been concerned with eliminating the need for a stable.

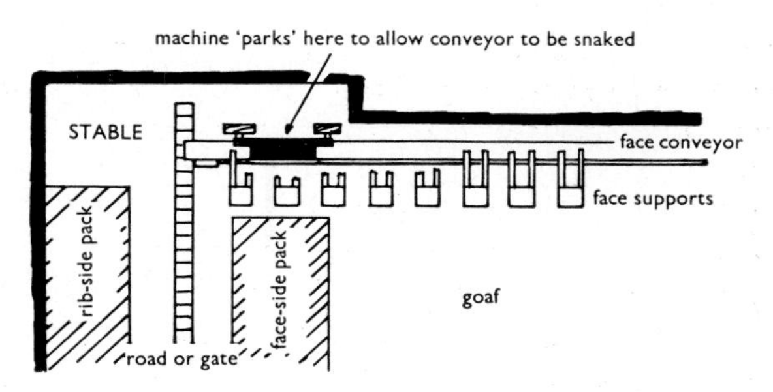

Fig. 7.25 Longwall advancing with stable

Roadways

It is necessary at this point to consider the construction of roadways for longwall advancing. Roadways can be made in front of the face as *advance headings*, or at the face line, or behind the face. They can also be alongside the rib, or have a ribside pack, and may be wholly in the seam, or enlarged into the roof or the floor as in Figure 7.26. Any combination of these factors may be used; the sketches are illustrative, not unique applications.

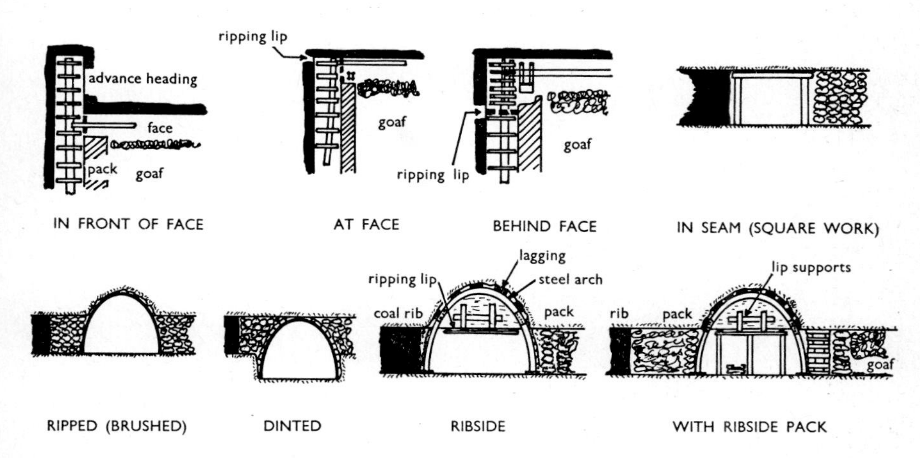

Fig. 7.26 Roadways for longwall advancing

The choice of where to make the road depends on many factors, the final choice should be the one which gives the most economical panel operation, and may not be the one which gives the cheapest roadway. Advancing longwall roadways may have to remain in existence for several months, although it is only the first 50 m behind the face which is likely to be heavily damaged by strata movement[4]. After the face has travelled beyond this distance then roadway closure and damage is small unless further excavations are made.

Under a weak roof at depths in excess of 2-300 m the ideal position for a roadway behind an advancing face is about 4 to 6 m from the coal ribside. This distance has been found to be the optimum value from over fifty years experience and has been confirmed in a series of field trials by the author.[4,5,6] If the roadway is positioned closer to the coal rib it is in a zone with a badly broken roof and is both difficult to construct if ripped behind a face, and can fall in readily. If it is moved further away it suffers heavy loading from the intermediate roof.

Under a strong roof (e.g. massive sandstone) the roadway is moved in against the rib to take advantage of the robust cantilever of the lower roof beds. Figure 7.27 illustrates this choice. With a road built away from the ribside it is necessary to have a pack built of rock, or concrete blocks, or a crib, to support the lower roof beds between the roadway and the coal rib. This is costly in labour, and sometimes in material, and attempts have been made to eliminate packing by cutting a horizontal slot in the coal rib so as to move the shear breaks away from the roadway. This is also expensive. On the face side of the roadway it is necessary to have a pack or a set of cribs to stop the waste flushing through. The strong roof-ribside roadway-crib combination is the cheapest, especially if it is possible to use rectangular supports. This is one case in which a floor may be dinted to gain height, rather than break a good roof.

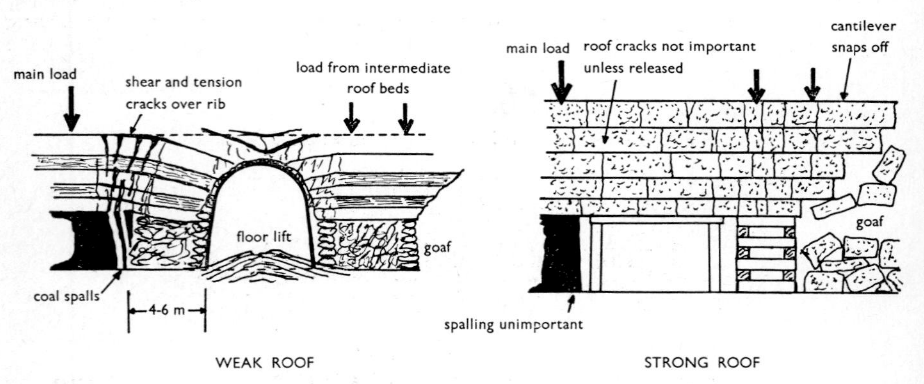

Fig. 7.27 Road position in relation to roof strength

In weak strata over about 150 m in depth it is possible to have as much as 1 to 1.5 m vertical closure in a roadway behind a longwall face. The roof lowers by one-quarter to one-half the seam height extracted, and the floor can heave up substantially. When closure of this type takes place it is often necessary to enlarge the roadway to near its original dimensions by removing the supports and blasting or machine-cutting the roof away. This operation is known as *back-ripping* or *back-brushing*. Similarly it is possible to regain height by *dinting* the floor. However if the floor is dug out too deeply, and uncovers the feet of the roadway supports, these may be displaced inwards bodily by a renewal of floor lift which is a result of extrusion from under the sides of the roadway[6].

Advance Headings

This problem of roadway closure is particularly prevalent with advance headings where the closure that takes place in front of the face line must be added to that which takes place behind the face[5]. Advance headings have been common in Europe for many years, particularly in the steeply inclined seams. They were not popular in Britain but extensive trials were made in the 1960s and some mines use them now. Where an advance heading can be driven in front of the face, and maintained behind the face, it provides one means of eliminating the longwall face stable. The coal loading machine can cut out into the roadway to enable the face conveyor to be pushed over.

In addition to roadway closure advance headings present a support problem at the T-junction with the face, especially in arched roadways. One solution is to provide 'false legs' on the arches coupled with a fish plate. Just in advance of the face conveyor these false legs are removed (Figure 7.28) and are replaced by special shoes which can rest on a pair of carrying girders or on a special face chock as in Figure 7.23. After the conveyor has travelled through, the false legs (or shorter ones to allow for the convergence that has taken place) are replaced under the arches.

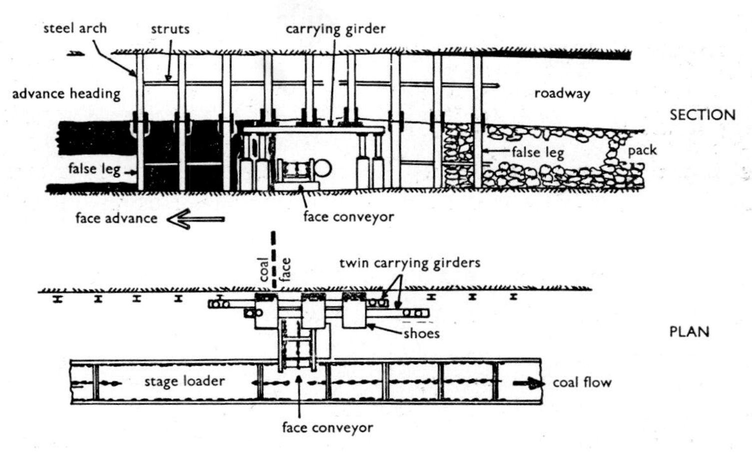

Fig. 7.28 False legs in advance heading

Stable Elimination

Another method of dispensing with stables is to use a sumping drum shearer as in Figure 7.29. The shearer drum is modified to carry picks on the side against the coal. After a normal cutting run to the end of the face the machine is withdrawn several metres and the conveyor is snaked over into the corner of the face. The shearer then makes a sumping cut to penetrate into a new web depth. After one or two cuts the shearer can remain in its new position in the face corner. The conveyor is then completely snaked over and the machine is ready for a complete face cut. To speed up production a single drum shearer may be kept at the return gate end especially for sumping and the face shearer will cut the rest of the face and the main gate stable.

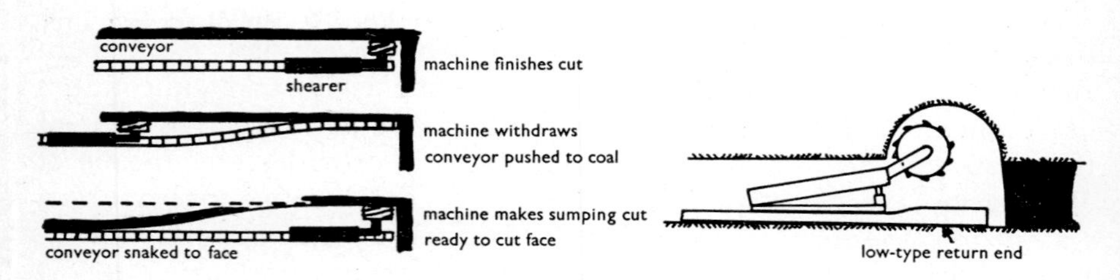

Fig. 7.29 Sumping shearers

Special modifications can be made to the machine and conveyor to allow the shearer to reach over the stage loader and into the main gate ribside. Another modification to provide a high lift underframe enables a shearer to cut the roadway at the face line. Any rock travels out of the mine with the coal, and cribs or special concrete blocks are used to support the roadway alongside the arches or squarework. This method of support is useful in shallow mines.

Ripping Roadways

Where the shearer cannot cut the roadway roof for strata control reasons then the roadway is made behind the face as in Figure 7.30.

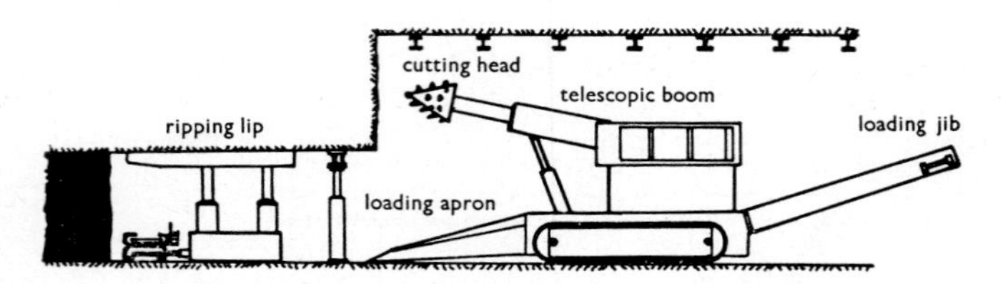

Fig. 7.30 Selective heading machine

The rock can be drilled and blasted, broken with a very large hydraulic percussive pick, cut with a conical head cutting machine as shown, or cut by a swinging arm machine. The cutting machine shown will cut an arch shaped roadway either in front of or behind the face. The head is fitted on a telescopic boom and broken rock falls on to a 'round-the-houses' conveyor to be loaded out over a tail boom. These road rippers may be skid-mounted or crawler-mounted.

The rock from roadway ripping can be loaded forward onto the face conveyor, loaded backward into mine cars or onto a conveyor, or filled sideways into packs. Hand packing is slowly being replaced by mechanisation. A slusher (scraper) can be fitted to the ripping machine to scrape rock into the packhole. A miniature hydraulically-powered bulldozer is available. Experiments are also being made with a special turnover armoured flight conveyor which will dump the rock into the packhole.

Conclusion

Much of the blame for a low output per manshift in longwall advancing can be laid onto the problems of face-ends, and this in turn is a result of poor strata conditions due to weak roofs and deep seams. Where face end work can be effectively mechanised then high outputs are obtained. It must be realised however that victory is not just around the corner: the majority of roadside packs are still put on by hand. Only the dirt from advance headings tends to be consistently sent out of the mine, especially where the roof rocks are weak shales. These tend to break down in the coal washeries and cause a slimes problem. About two-thirds of all rips still have their dirt packed in the face.

Mechanisation of ripping lips has its problems: road heading or ripping machines use tungsten carbide tips which will not cut hard or very abrasive rocks. Only a few swinging-arm ripping machines are in use because they will not generally cut rock above a strength of about 83 MPa uniaxial compressive strength (medium to hard shales). The selective heading type machine shown in Figure 7.30 can cut rocks up to a strength of 120 MPa if they are jointed, or laminated, by breaking the weaker bands first.

Quartz grains in rocks present a problem from incendive sparking if struck by cutter picks. A general classification is.

(a) silica (quartz) content higher than 50 per cent, strong probability of incendive sparking

(b) silica content between 30-50 per cent, possible chance of incendive sparking

(c) less than 30 per cent silica, low chance of incendive sparking.

An incendive spark is one that is capable of igniting a methane-air mixture. This mixture is lighter than air and can easily collect at a ripping lip.

The use of a full scale tunnelling machine which can excavate hard rocks cannot be economically justified for routine roadway drivage.

WORKING THICK SEAMS

A definition of a thick seam can be taken approximately as a seam in excess of 4-5 m where extraction in a single lift becomes difficult. It is not easy to standardise methods of working thick seams of coal or other minerals because a great deal depends on seam continuity and quality, and the depth of working. A seam about 30 m thick in central Queensland, away from habitation, might be economical to opencut down to 300 m. In an area where large scale open-pit mining is not permitted, then only a portion of the 30 m might be extracted by underground methods.

Restrictions on mining can be:—

(a) limitation of subsidence: for underground mining near built-up areas, or under water, or highways, etc., extraction may be limited to 40 to 45 per cent to leave substantial pillars to prevent subsidence

(b) limitation by quality: thick coal is often produced by many seams merging together on the edge of a coal basin. Some of the seams may be unworkable because of high ash, sulphur, chlorine, etc., and then only intermediate seams will be worked. In a bedded deposit of, say, potash, only the highest quality band may be worked, either by longwall or pillar methods, and the rest abandoned

(c) limitation by strata strength: most coals have a relatively high tensile strength in comparison with rock, and will absorb a fair amount of strain energy before cracking or failing. Thick seams of coal may be totally extracted in lifts but a thick bed of limestone may be partially extracted by first working only because it is highly jointed, and pillars must be left to support the incompetent roof.

Bedded Deposits other than Coal

Most thick deposits can be partially extracted by room and pillar working (*breast stoping*). Heights of 9 to 10 m in rooms are practicable using working platforms for drilling, and telescopic scaling platforms to insert rock bolts to support walls and roof, and for charging shot-holes. Modified methods of shrinkage stoping may be used where the rooms are worked from the top of an ore pile which is loaded from floor level by front-end loaders. Each blast from the roof of the room rills down to the loader. Heights approaching 30 m can of course be worked by any of the normal stoping methods described in Chapter 6. These methods are not generally suitable for coal or other highly-jointed deposits.

Thick Coal Deposits—Pillar Mining

The easiest method of working thick coal underground is by partial extraction, although this sterilises substantial reserves and can present problems in seams liable to spontaneous combustion (self-ignition). Really thick deposits (say above 10-15 m) can be stoped, paying due regard to possible problems from light methane gas in rise workings, and a high dust content in some coal seams. There are also frequently dramatic changes in quality in vertical sequence. The top portion of a seam may be low quality power station fuel and the bottom portion may be high quality coking coal worth three or four times more in sale price.

A mine near Ipswich in Queensland is working 9-12 m of coal under shallow cover at gradients of up to 1 in 1.8. Extraction is limited to 45 per cent. The entries to a new section are being driven by continuous miner. This is only possible by driving at an angle to the true dip. The floor of the main entry is partly in coal and partly in rock so that the cross-grade is difficult to maintain. In the extraction areas the top 3 m of coal is of lower quality than the rest and is left as a roof. Panel entries and first working entries are driven at convenient angles to the full dip to obtain reasonable working gradients (Figure 7.31).

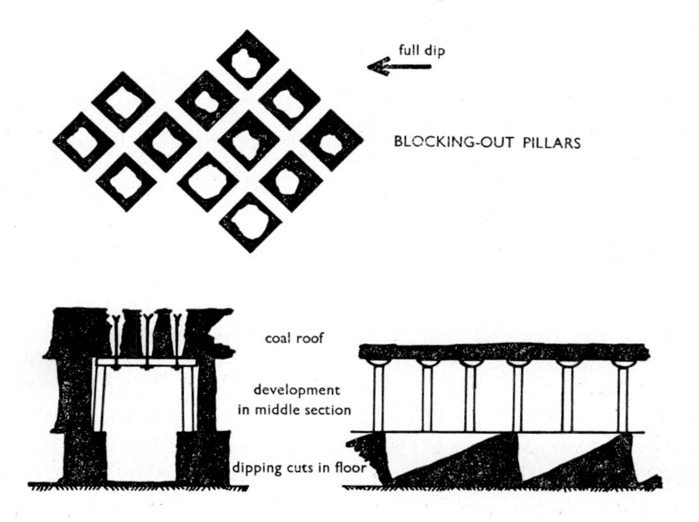

Fig. 7.31 First working in thick seam

All roads are developed in the middle section of the seam and when a panel has been fully blocked out the continuous miners retreat mine the roadways. In this process the miner retreats a few metres and then takes a dipping cut down into the coal floor. Where the coal sides are strong all the floor coal is taken. If the coal sides are friable, then when the machine contacts the rock under the seam it backs out to the level section of roadway and takes another dipping cut. The abandoned roadways thus have a level roof, supported by bars and legs resting on side ledges, and a narrower saw-toothed floor section as in Figure 7.31. The floor coal is in fact the highest quality but if it is all mined in bad conditions then the pillars are likely to be unstable and pillar collapse is not permitted.

In other mines where total extraction is permitted then first working will still be in the lower portion of the coal by continuous miner. This leaves a coal roof which is stronger than rock, and the intact coal roof will help to prevent spontaneous combustion in seams liable to this (see later). The mine, or a substantial part of the lease, will be formed into pillars before second working is started. Pillar extraction commences at the boundary. Each pillar is worked by pocket and fender to produce what is virtually a sub-level cave. Pockets are driven through the pillar as high as the machine will take them, say about 4 m. The roof of the pocket is then drilled on a wide spacing and shot down to retreat about 2 m at a time. The continuous miner is used to load fallen coal because its cutting head will pull lumps back and break them.

The fenders must be substantial enough to stop the roof caving too quickly and covering the fallen coal, but they must collapse, or be blasted or robbed, before the next pillar is mined. Pillars are retreated out in turn across a broad front and the waste sealed off in blocks to prevent coal remnants catching fire. Extraction can be about 60-80 per cent.

Pillar caving has been used for many years. In the 19th century the Staffordshire thick coal in Britain was blocked out into pillars on the bottom of the 10 m thick seam. 'Pikemen' cut up through the coal with long pointed bars to the roof of the seam around a pillar which was then undermined and supported by cribs (chocks). After the pillar had been completely undermined the cribs were withdrawn, and as the coal collapsed it was mined out by hand.

Thick Coal Deposits—Longwall Mining

This section also includes *contiguous seams* (i.e. seams in close proximity to each other) because they are worked by similar methods. Thick coal has been worked in Britain and India for many years although the amount left in Britain is declining rapidly, and most was extracted by older hand-worked longwall methods which left ample time for some of the more complicated practices.

All the methods of working thick coal rely on the fact that coal is relatively flexible and very resistant to breaking providing that joints are kept closed. A bed of coal can be lowered without major cracking taking place, provided that it is lowered gently. The usual method of longwalling thick coal is to take the bottom workable section first. A lift of 2-3 m will be taken to a natural parting by normal longwall methods, although advance has been preferred to retreat, probably for traditional reasons. The face must be stowed solid, either with sand-fill or with broken rock. In the Coventry colliery system in Britain strip packs were used but this gave an uneven floor and bad working conditions in the next lift.

The coal left intact as a roof settles behind the first face, consolidating the fill. The roads can then be roof-ripped to make entries into the next lift and the process of extraction and filling is repeated. The roads are again ripped and a third lift taken. Some mines have worked up to four lifts from 8-10 m of coal. The last lift was usually retreated, even if the other faces had been worked advancing. Coventry colliery in fact retreated three pairs of longwall faces concurrently, with only 15 m separation between each pair.

It can be seen that this system requires large quantities of fill; in fact it is a similar system to metal mine cut and fill. Some Indian mines have installed large aerial ropeways to carry river sand several kilometres to the mine to use as fill.

To avoid the necessity for solid stowing many mines have worked thick coal in descending sequence. Although Coventry colliery in the Warwickshire Thick Coal worked from the bottom upwards, the neighbouring mine, Newdigate colliery, worked the top lift first and then the rest in descending order. The Thick Coal was caused by a merging of several separate seams; Newdigate worked the top Two Yard seam first by longwall advancing and strip packed it because that was the current method. The mined area was left to consolidate for at least a year and then the middle leaf (the Ryder seam) was worked longwall advancing leaving a hard layer of coal (the Bare seam) about 0.6 m thick to stop the caved waste above from falling through between the roof supports. With careful mining practice roof falls on the face were rare.

The roadways in the middle lift were offset from the roadways in the top lift because the old roadways formed lines of weakness and could cause roof problems in the middle lift. However the roadways in the middle lift were dinted in the floor behind the advancing longwall down to the bottom of the coal sequence. When the middle lift reached the limit of travel (often about 1.5 to 2 km) then a new face was driven through in the bottom lift (the Nine Feet seam, actually about 2 m thick) between the dinted roadways and the Nine Feet face was retreated back. Again there was a hard band of coal, about 0.45 m thick, which was left to hold the caved waste above the working area. From personal experience there was more roof trouble in the initial Two Yard seam workings of the top lift than in working under the old wastes which had de-stressed the areas.

Older methods of longwalling thick coal deposits have been expensive on manpower. In some systems the need to keep to a cycle often wasted coal and extraction might only be 60-70 per cent. Modern systems of working by mechanised longwall require reliability of operation. A deliberate decision is often taken to sacrifice coal to reduce costs and where three lifts might have been taken with handworking only two lifts will be taken by machine. A coal roof of about one metre thickness will stand well above powered supports but will break off in the waste to permit succesful caving. Two to three metres of coal left to protect the extraction of six or seven metres still gives a reasonable percentage recovery.

One of the recurrent problems in thick seams is *spontaneous combustion*. The roof coal in particular, in Australia, Britain and America, seems liable to catch on fire once it is exposed and broken. For this reason initial development of roadways is advisably carried out in the lowest portion, or at least in the middle of the coal bed. Crosscuts between intakes and returns should be kept to a minimum and pressure differentials between intakes and returns should be as small as possible. Stoppings, ventilation doors and air crossings should be carefully sealed and the coal plastered over or gunited with cement/sand paste to prevent air leakage through cracks. As a final precaution a collar should be inserted into main panel entries and fireproof materials stocked nearby so that areas affected by self-heating can be quickly sealed off. Details of all these precautions and of fire detection can be found in a more advanced text.

WORKING STEEP SEAMS

Steep coal seams can be worked as for narrow metalliferous veins but coal is softer and machine working is preferable to a blasting and loading cycle. Steep seams have been worked by top slicing, or what is effectively square set mining or stull mining, but these are uneconomic if labour costs are high. A version of cut-and-fill mining is possible in steep workings. One mine in Spain uses an armoured flight conveyor mounted on buoyancy floats. The coal is shot down from overhead sequentially on to the conveyor. When a complete slice has been loaded out a layer of sand fill is run in. The conveyor floats up on the fill and when the water has drained away a new cycle can be started.

Longwall mining is possible up to grades of at least 1 in 1.5 as currently practised in Scotland. The longwall face is laid out to the full dip although it is good practice to have the bottom end of the face leading by a few metres so that falling rock in the waste slides backwards instead of bouncing between the support legs. High spill plates are fitted to conveyors to catch falling coal. The face and roadways advance on the strike. Handworked longwall on a steep grade was difficult and dangerous because anything dropped ran downhill, and supports were difficult to erect. Modern mechanised longwall is easier, but some dangers are inherent in the system.

The armoured conveyor is laid to the full dip (or slightly off) and must be held in the top gate by a firm anchor station. The conveyor acts as a coal retarder on gradients steeper than about 1 in 3 to 1 in 2 and may be fitted with high flights for extra control. Soft seams are ploughed and the plough is light and easy to control. When shearers are used they usually cut down-hill because with the drum in the coal the machine cannot break free and run away. Power consumption for haulage is also less of course. The particular moment of danger is when the drum breaks free at the bottom end; there is a sudden shock loading on the haulage chain which could snap it. It is advisable to attach a spare anchor chain or a rope from a hydraulic winch to act as a check on the last few metres into the stable or roadway. If the shearer cuts up-hill the conveyor must be snaked immediately to trap the machine into the coal.

Powered supports are linked with swinging arms, or are fitted with sideways-pushing rams to prevent the support sliding downhill when its legs are retracted. Special modifications may be needed to the hydraulic systems to allow for the vertical head of fluid. Particular care has to be taken to keep loose coal cleaned up because the centre of gravity of the support is often only just within its base, and if the rise side of the base is lifted the support tends to tip over.

Mechanised faces are not generally worked more than a few degrees off full dip because the machine will tend to topple off the conveyor, or to turn it over. Some cantilever supports have a tendency to nosedive if advanced to the dip.

In steep seams the immediate roof shares the common desire of everything else to slide down into the bottom roadway. It may be convenient to have restoring rams or springs to lean the legs of the powered supports uphill as they are raised to resist this movement, and to maintain stability. The bottom gate will be subject to higher strata pressures than in level seams and roadway maintenance costs for longwall advancing can be expensive. Longwall retreat roadways can close in advance of the face. For maximum roadway support a thick buttress pack, or two or three rows of wooden cribs, are usually necessary on the rise side of the roadway. A special powered support is trailed near the bottom of the face to protect the men erecting these cribs from rocks sliding down the face. Lightweight concrete blocks have been tried as buttress packs, but tend to be more expensive than wood, and their extra rigidity can cause bad shear breaks in the roof with consequent falls. The slight yield of hardwood cribs lowers the roof before it breaks.

References

1. Francis, W. *Fuels and Fuel Technology.* Vols. 1 and 2. Pergamon Press, London, 1965.
2. Connelly, M. A. *A geological structural assessment at Appin Colliery with reference to roof failure and directional mining.* Proc. Aust. Inst. of Mining and Metallurgy, No. 234, p. 17, June 1970.
3. Anon. *Mining methods; continuous, conventional, longwall.* Coal Age, p. 117, October 1969.
4. Thomas, L. J. *Rock movement around roadways at Babbington Colliery.* Colliery Guardian, **212**, pp. 475 and 507, April 1966.
5. Thomas, L. J. *Strata control investigations in advance headings.* Colliery Guardian, **216**, p. 473, June 1968.
6. Thomas, L. J. *Pack pressure measurements.* Colliery Guardian, **218**, p. 503, October 1970.
7. Stahl, R. W. *Extracting final stump in pillars and pillar lifts with continuous miners.* U.S. Bureau of Mines Report of Investigations, R.I. 5631, 1960.
8. Carrall, L. R. *Conventional mining of coal for power generation.* Paper in 1972 Newcastle Conference, Aust. Inst. of Mining and Metallurgy, Melbourne, 1972.
9. Kay, J. and Mowbray, G. J. *Shortwall experience in the Newcastle area.* Paper 8, Symposium on Longwall Mining in Australia. Illawarra Branch, Aust. Inst. Mining and Metallurgy, February 1970.
10. Kay, J. and Marsden, W. R. *Shortwall continuous mining.* B.H.P. Technical Bulletin, **13**, No. 3. Also in Mine and Quarry Mechanisation, 1969—Magazine Associates, N.S.W., Australia, and Mining Congress Journal, May 1970.
11. Kay, J. and Duncan, B. *Development of face machinery with particular reference to shortwall mining.* Paper in 1972 Newcastle Conference, Aust. Inst. of Mining and Metallurgy, Melbourne, 1972.
12. Katlic, J. E. *Longwall experience in the Pittsburgh seam.* Mining Congress Journal, p. 38, July 1970.
13. Bourne, W. J. W. *Ripping and packing: a mechanisation summary.* The Mining Engineer, No. 139, p. 323, April 1972.
14. Spanton, H. M. *The utilisation of material assets.* The Mining Engineer, No. 137, p. 221, February 1972.

SECTION C. ROCK MECHANICS

Chapter 8. Design and Support of Underground Openings

Introduction

There will be no attempt in this section to present a rigorous mathematical treatment of the subject of rock mechanics. The following material is an interpretation of practical experience built up over centuries of mining operations. A theoretical approach to mining rock mechanics can be obtained from the textbook by Obert and Duvall[1], based mainly on their experience at the U.S. Bureau of Mines. Jaeger and Cook have taken a mathematical approach[2] that is more applicable to shallow excavations and foundation work in civil engineering.

It should be noted that much of the published material on rock mechanics relates to civil engineering excavations at shallow depths with consequent low pressures, or to dam abutments, where even small movements can cause leaks, landslides or catastrophic failures. In underground mines relatively large rock movements can be tolerated, and indeed have to be. Most rock structures are basically self-supporting after a fall and can be repaired and re-used with varying degrees of production loss. Even in open pit mines, providing the rock slopes are monitored, a slip is rarely catastrophic if basic rules-of-thumb have not been broken.

Stability of Broken Rock

The stability of fractured, jointed, or broken rock depends mainly on:—

1. the ratio of the average size and shape of the rock fragments to the size and shape of the opening in the ground
2. the direction of fracture planes in the rock with respect to the boundary planes of the opening
3. the strength of individual rock fragments
4. the coefficient of friction between individual rock fragments
5. the shape of the opening.

The motive force for displacing the rock is the basic pressure of the superincumbent strata. The vertical load is strictly

$$\sum_{0}^{H} \rho h$$

where H is the vertical depth of working from the surface and ρ the rock density of the strata. In practice it is convenient to round off actual rock densities to a mean figure. If the rock is assumed to weigh about 2500 kgf/m^3 then the superincumbent load is approximately 25 kPa per metre of depth (0.25 kg/cm^2/metre). In Imperial units the rock was assumed to weigh 144 lbf/ft^3 to give load of 1 $lbf/in^2/ft$ of depth.

Much has been written about measuring virgin rock stresses, especially by overcoring. Internationally the measurers produce figures of horizontal stress about 2 to 3 times the vertical stress, although Hast measured horizontal stresses up to 14.5 times the vertical stresses in the irregular pillars and the walls of a lead ore stope. When the stresses in a concrete plat (shaft inset, or station) lining were measured in Britain the same ratio of two to three times horizontal stress to vertical stress was found to act on the lining.

Most stress measurements are made in the walls of excavations and are only a few metres from the free surface. However measurements in Holland and in Britain[3] showed that excavations of 9-11 m^2 cross-sectional area approaching instrumented zones 55-90 m distant produced measurable strains. To the sceptic this could suggest that there is no such thing as a measurable virgin stress. As soon as a hole is made in rock then its stress pattern is altered and only local elastic moduli can be measured. For convenience most stress measurements are made from drives (which tend to be similar in shape) rather than in production stopes. The ratio of horizontal to vertical stress is remarkably constant except for Hast's measurements which were made in an old production stope. The measurements made could be mining stresses. One would anticipate that the virgin stress ratios would not be so remarkably constant because geological conditions are not constant.

It is the stress in the rock around an opening that determines the behaviour of the excavation. The usual theoretical approach to stress distribution is to treat a tunnel as a two-dimensional hole in an infinite plate of an elastic homogeneous medium. These simplifying assumptions enable the theoretical stresses around a circular tunnel to be calculated and then the results are fitted to other shapes by photo-elastic techniques or calculations of local effects. Errors are not large for small openings at shallow depths in hard rock. A theoretical study by Hobbs[4, 5] based on measurements on broken rock in triaxial constraint, and on the deformation of rock asperities, also derived the general pattern of stress that probably exists around circular roadways.

The general form of the stress is given in Figure 8.1. The two curves shown represents the tangential and radial stresses on a circular roadway in solid rock in virgin ground, and on a roadway in broken or jointed rock. From actual stress measurements the more likely distribution is that shown in the right-hand set of curves where there is a change from non-elastic broken rock under increasing confinement, to broken rock behaving elastically.

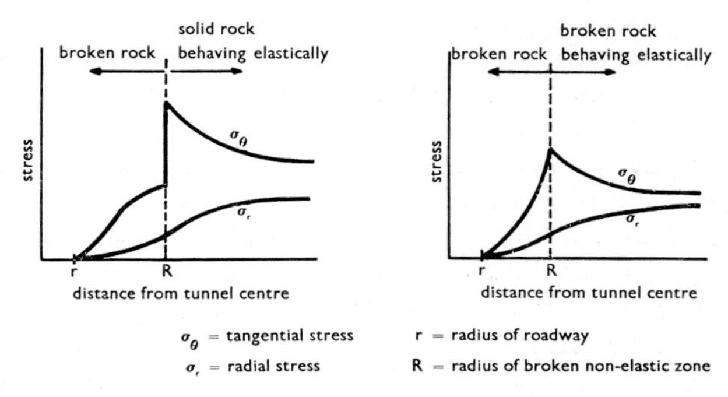

Fig. 8.1 Typical stress distribution around hypothetical roadways in solid rock and broken rock

The work by Hobbs (and that of other experimenters) has shown that:—

1. the higher the confining pressure at which rock is first broken the weaker it is in subsequent movement
2. increasing the confining pressure on broken rock increases its strength but at a decreasing rate
3. broken rock approaches the strength of solid rock as the confining pressure increases.

It is also easy to prove experimentally that the strength of solid rock can be greatly increased by confining it triaxially before it is broken. The increase is not linear. Hobbs[5] gives an expression

$$\sigma_1 = B\sigma_3^{b} + \sigma_3$$

where σ_1 is the fracture or yield stress, σ_3 is the confining pressure, and B and b are constants to be determined by experiment for a particular rock. In relation to 3. above it should be noted that confined broken rock can be stronger than unconfined solid rock of the same type.

If this theory is related to observation underground then the sequence of changes found can be understood. When a load is applied to a rock surface adjacent to an excavation, then if that load exceeds the unconfined compressive strength of the weakest layer of the rock, that layer will start to break up and extrude into the roadway.

The broken rock is not completely free to move. There is a frictional restraint between individual pieces, and interlocked asperities on joint and crack faces provide a shear-force restraint. As the thickness of broken rock builds up, then it increases the restraining force on the rock further away from the roadway, and consequently increases the load the layer can support. This is the practical equivalent to the curves shown in Figure 8.1. The actual shape of the curve is not important in the context of this chapter. What is important is that it is possible to build up stress in the rock above that which the edge of an excavation can withstand.

The increased confining pressure (triaxial pressure) on the remote rock increases its effective compressive strength so that eventually almost any superimposed load can be supported. Measurements have been made in both coal and metal mines of the stresses due to mining operations. They have shown that quite large stress changes are caused and that a high peak stress as shown in Figure 8.1 is normal in front of any advancing excavation. In relation to longwall coal mining the loads have been of the order of three to five times the cover load at about 3-5 m in front of the coal face. This peak load is known as the *front abutment*. It will be remembered that the cover load is depth $\times$ rock density.

Although the curves in Figure 8.1 are those for the stress distribution at any angle around a circular excavation under hydrostatic load they are similar in form to the vertical stresses observed adjacent to openings in mines. It has been mentioned that there is a peak pressure in front of longwall coal faces. There is however no similar peak behind longwall faces, but a stress gradient has been measured in the floor of packs behind longwall faces, both normal to the face and normal to the roadways[6]. Leeman[7] has determined the general shape and extent of the fractures produced ahead of a longwall gold reef in South Africa, and the stress changes, and they conform to the same pattern as longwall coal mining. Similarly measurements have been made in coal pillars in the United States and in Britain, and stress increases to a high peak are found in the walls of the pillars.

All these measurements have shown that no matter how broken the layer of rock adjacent to an excavation there is still a build up of stress within the rock. The zone of broken rock behaving non-elastically generally only extends some 4-5 m into the sides of an excavation, even at over 2500 m depth[7]. Laboratory tests have shown that in piles of broken rock the core of the pile can withstand loads in excess of the unconfined compressive strength of the pieces of rock. This indicates that even a fully-broken pillar or rib of coal, or of rock, can provide an effective support to prevent complete closure of workings.

There are however important exceptions. A pillar or a wall of broken rock can only support a strong roof above a strong floor. The processes of rock breakage and flow apply to roofs and floors. Small pillars of hard coal have been seen to penetrate a soft floor; the pillar remains unbroken and the floor flows up around it. Soft seat earths of a coal seam can flow into a roadway and close it completely within a few weeks, especially behind longwall faces. Under these circumstances, in order to control the movement, the floor may be deliberately buried by ripping into the roof of the roadway and dropping hard rock on to the floor. The dead load and frictional restraint of the fallen rock may then be sufficient to hold the softer material in place. Brittle roofs or soft roofs can shear off or flow downwards around pillars. In fact large scale collapse such as that at Coalbrook North are probably due to massive failure of the roof as well as of the pillars.

Effect of Tunnel Linings or Supports

Hobbs[5] gives a theoretical curve which indicates the distance from a roadway centre at which the transition from the non-elastic broken rock zone to the elastic broken rock zone is likely to take place. He also indicates on it the effect of a slight increase in boundary stress; in his case 1 lbf/in^2 (0.007 MPa). This stress is equivalent to that of arch or joist supports set at one metre centres and taking a load of about 2 tonnes. The curves are of the form shown in Figure 8.2 but units are deliberately not given. The theoretical reduction was 12 per cent off 0.15 m in a 3.66 m diameter roadway. A paper by Shepherd and Wilson[8] describes strain measurements in a shaft plat supported by 1.8 m square-section reinforced concrete ring beams of the shape shown in Figure 8.2. The side walls of these beams sheared away from the roof members and pushed inwards. In less strongly supported sections the roadways squeezed in to give a width reduction of 1.5 to 1.8 m on an original 6 m, despite cross-bracing with multiple RSJs. When forces of this order are involved it is unreasonable to expect any tunnel lining to prevent movements.

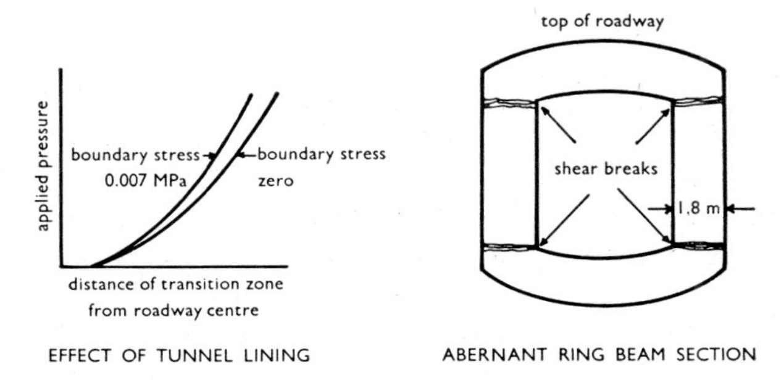

Fig. 8.2 Tunnel lining stresses

A general practice in deep mines or in soft ground has been to drive a pilot tunnel to de-stress the area, and then to enlarge the crushed tunnel to the desired cross-section and then line it. This process sets up the required de-stressed zone around the final excavation so that a steel or concrete lining can stay intact. Recently the practice has often been to drive a tunnel and to line it immediately. The horizontal stresses have then cracked the walls, or sheared construction joints, or caused compressional failures in roof or floor members. Generally these cracks and breaks are not structurally serious and the cracks are patched over with wire mesh and rock bolts, and then sprayed with cement or shotcrete (concrete mixture), to give a smooth surface.

Sometimes reports are written of how successful a change of tunnel lining has been. A recent technical article from America described how a concrete lining in a vertical shaft had broken up. The lining had been ripped out and the shaft was re-lined with shotcrete (sprayed concrete). The article then suggested that shotcrete would have prevented the movement in the first place. However, this type of article has been frequently written and the common message in each is

1. the first lining that was put in failed
2. the broken lining was removed and the opening enlarged to original size.
3. the new lining was put in, and held.

Sometimes a new lining is installed inside the old.

The essential point to realise is not that the new lining is better than the old, but that the original movement has set up a de-stressed zone to give the 'correct' stress gradient for the rock type, and that almost any second lining would be a success. There are some occasions, such as tunnels passing through fault zones, where small movements may occur continuously for years. However it is generally the first squeeze that is the worst and if a flexible lining is then installed that is easy to remove and replace, subsequent repair costs can be kept to a minimum.

Several coal mines have found to their cost that expensive treatments such as floor stitching with wire ropes and cement grout, designed to stop floors lifting, result in expensive repair costs when the ropes have to be cut out again because the floor could not be held down.

Design of Single Underground Openings

Holes in the ground can conveniently be divided into single relatively small openings, or multiple openings, or stopes and other working areas. The stability of single openings depends greatly on orientation.

SHAFTS

Vertical and near vertical openings in competent rock of depths to 6-800 m rarely give trouble in the absence of plats (i.e. openings off the shaft, also called insets, or stations) . Trouble may occasionally occur in clay beds that can swell if water trickles down the shaft and into the clay. However, if the shaft has been lined then the problem can be self-curing because expansion of the clay will seal off the water supply. Shaft sinking, lining, and furnishing is a subject for detailed study from another text. Generally a borehole should have been drilled through the strata to be sunk through and problems noted in advance.

Shafts in hard rock can be supported with a light timber lining down to 300 m or so and can be rectangular or circular. Shafts used for ventilation only may be left unlined, particularly if they have been raise-bored. The absence of shot-firing minimises the loose scabby pieces of rock that can weather and fall off the walls. An exception will probably have to be made for weak shales in damp conditions. Metal mines have favoured rectangular shafts. These give a convenient division into compartments for ore and rock skips, materials and man cages, ladderways to connect the levels, and pipe compartments.

In softer rocks, side squeeze can commence from 2-300 m downwards, but can easily be resisted by concrete linings to depths of 600 m or so except at plat intersections. Hard rocks can be stable to 6-800 m or more. In deep shafts it is advisable to use a circular cross-section for maximum strength because loading on straight lengths of shafts is usually uniform radially.

Coal mines working from one or a few seam intersections have used circular shafts for many years, and have installed skips in one shaft and service cages in the other, or used skip-cage combinations. Metal mines compromised at first by changing from a rectangular shaft to an elliptical shaft to eliminate sharp stress-raising corners. At Broken Hill North Mine the shaft extensions below 1000 m continued the elliptical shape. The original concrete lining cracked just off the major axis and was ripped out and replaced. The replaced concrete lining appears to be satisfactory with the shaft down to 1500 m in a sericite schist shear zone: the principle mentioned earlier of setting up a satisfactory de-stressed stress gradient in the rock still applies in vertical openings. Both Mount Isa and Cobar mines have already used circular concrete lined shafts.

Satisfactory theoretical stress design values for vertical shafts have not been produced to date. Trollope[9] has produced a theory to explain the slabbing mechanism of failure observed, and derives critical depths of excavation. Measurements in shafts[3] in shale coal measure strata to

1000 m have shown that strains in concrete linings away from plats are minimal. Shaft lining design has generally been empirical, based on the mining engineers' and contractors' experience. The National Coal Board in Britain standardised on shaft diameters, typically 6.1 m and 7.3 m finished inside diameter, to take standard cages. For the standard shafts it was easy to use a fairly standard concrete lining. A typical specification called for 0.38 m minimum thickness non-reinforced concrete to 880 m, then 0.46 m minimum reinforced concrete to 1000 m, followed by 0.53 m of reinforced concrete to shaft bottom at 1040 m. In the absence of other excavations the load on the shaft lining is uniform and reinforcement is not needed in the concrete because all stresses are compressive. If a shaft lining is elliptical then tensile stresses may exist in the inner surfaces of the long sides of the ellipse.

For a shaft sunk in liquid sand, or in porous rocks with free water, then the linings must be designed to withstand a hydraulic head and specialists will produce a composite steel and concrete or a cast iron (German) tubbing lining.

PLATS AND SKIP POCKETS

Plats are the short lengths of roadway which are immediately adjacent to the shaft. In Britain they are called insets and in the United States and South Africa they are called shaft stations. Skip loading pockets are hoppers (or bins) excavated into the shaft wall as in Figure 8.3 to enable coal or ore to be measured, and then loaded into a skip.

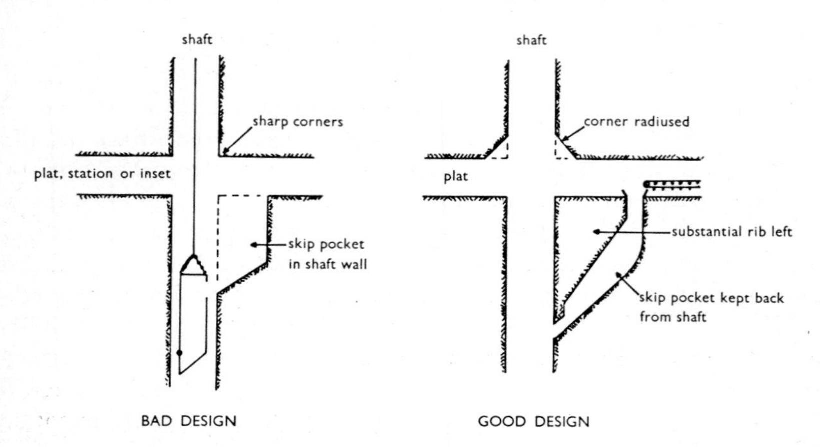

Fig. 8.3 Plat and skip pocket

The sizes and shapes of plats are usually dictated by the need to handle equipment in them, and in some mines by the necessity to pass large volumes of air. Usually the plat is the same width as the shaft, or slightly wider to provide a walkway past the shaft. The height of the plat may be the usual 3 or 4 m of a normal roadway unless a double-deck manriding cage is in use. For mines where a high volume air flow enters or leaves the shaft then the roof of the plat is given a transition curve to smooth air flow and raised to a height where the cage will not cause an obstruction. When the engineer has designed all these items then the rock mechanics expert may have some objections if the plat is going to be in weak ground or at depths in excess of a few hundred metres.

Early attempts to design plats to resist rock loads were based on an elasticity approach and on loads that were assumed to be vertical. A vertical ellipse is a good shape to withstand these loads but unfortunately measurements and experience showed that the horizontal load on a plat is greater than the vertical load. The best shape for a plat is roughly circular. Measurements of strains and closures in shafts and insets have been made in Holland and in Britain[3,8]. These show that the general types of closures are a sideways squeeze inwards of the wide expanse of shaft and plat. Figure 8.4 shows the squeeze in exaggerated form. The pattern is not unexpected but as a mine extends its workings deeper it is easy to overlook some elementary points.

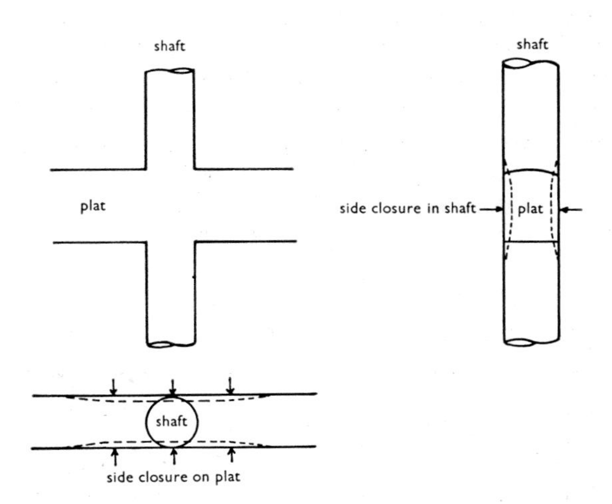

Fig. 8.4 Closure pattern in shafts and plats

The shaft buntons (cross girders) to which cage guides are fixed, and any compartment girders, should be set parallel to the axis of the plat. Even quite small amounts of side squeeze can buckle girders set transversely across the plat. Girders which must be set across a plat should be bolted onto stub carriers with slotted bolt holes, or should have pockets left behind their ends.

When plats are driven in weak rock, or in deep shafts, some precautions should be taken. The general need is to keep the dimensions of openings across the shaft to a minimum. The skip pocket must be moved away from the shaft wall and a substantial rock pillar left to stabilise the plat opening. The height of the plat must be kept to a minimum. If extra cross-sectional area is needed for air flow then a second plat should be opened about 30 m or so away from the first. It is important to remember that design should not be considered in terms of mining out rock and then replacing it by a strong concrete beam. The damage is done in proportion to the original excavation and a concrete beam will not necessarily support badly fractured rock. Two smaller openings with a strong rib between are stronger than one large opening braced across the middle.

A careful study of rock strengths should be made in the area where a plat is to be built and if necessary access roadways should be moved up or down to avoid weak beds. This is particularly important in coal mines. Soft shales can shear off very easily and large openings in them can be very weak; coal makes a better roof. In some cases it may help to drive the plat immediately below the coal seam to make use of its good load bearing characteristics.

LARGE SINGLE EXCAVATIONS

The earlier comments on stress gradients should be remembered. Many plat linings of concrete have cracked and spalled along the line of the roof and have had to be repaired. Some mines have gone to great trouble and expense to design underground shaft hoist chambers and crusher stations, which often must be of an irregular shape. Papers by Radmanovich and Friday[10], and by Ackhurst[11], describe photoelastic techniques used to study cross sections of the excavations. The stresses applied to the photoelastic models were determined from underground stress measurements. Tests on the models showed that the stress distribution could be improved by

1. further arching of the back (roof)
2. increasing the radii of the corners
3. removal of re-entrant sections at the bottom of the profile.

When the models were modified in this way there was still a high stress concentration on one corner of the excavation and this was reduced by cutting a slot along the angle. The slot was cut underground by using a percussive drill to bore 57 mm holes with 25 mm ribs between them to a depth of 2.4 m. It is unfortunate that production requirements prevented follow-up tests on the underground opening to see if the stresses behaved as predicted. The slot-drilling technique to reduce transverse stresses has also been used by the Hydro-Electric Commission of Tasmania, particularly in their Poatina Power Station. It is a useful method of accelerating the formation of an ideal stress gradient.

The photoelastic information obtained from tests can sometimes be predicted by examination of standard texts on photoelasticity. The solid objects tested by mechanical engineers indicate that sharp corners and notches, and protuberant ribs, all act as stress raisers. These stress-raising shapes should be avoided underground when designing equipment chambers.

For large chambers, such as turbine halls in hydro-electric schemes, crusher stations, pump houses, etc., it must be remembered that they are more likely to suffer from side squeeze than from vertical loading. Where possible horizontal equipment is to be preferred to vertical equipment in excavations at depth. For instance in a power station 2-300 m underground vertically (i.e. in a hillside) it may be preferable to accept a slight cost premium on tunnel drivage and on horizontal turbines to keep the height of excavations to a minimum. Excavations 25-30 m high can suffer badly from side squeeze effects, especially if they are long as well as high. If a chamber has a relatively small plan area, say 6-8 m square (or preferably rounded) then it is as stable as a shaft.

Wide flat excavations in weak bedded rocks will also give trouble and care should be taken to reduce widths. It would be preferable to align turbine pumps along a pump house, instead of across it. On a further note of caution the pump bases should not be cast integrally with a concrete floor. Some floor lift may be caused by side squeeze and the pump base may then crack or tip sideways. All underground equipment should be installed on a free base which can be levelled with packing shims.

This is not the place to provide details of perimeter blasting. However it should be noted that if a large excavation is to be made underground then the amount of explosives used near the edges of the excavation should be reduced to a minimum with techniques such as cushion blasting, or line (perimeter) blasting with low density explosives. The less shotfiring involved then the stronger the final surface will be. If the final surface is rock bolted, covered with wire mesh, and sprayed with concrete it can acquire extra structural strength.

ROADWAYS IN SOLID GROUND

Drivages at shallow depths in competent rock can be very large and will give little trouble in the absence of rock discontinuities (see later). As the rock becomes softer or the depth greater then the drivage becomes progressively more likely to fail. Drivages in hard solid rock usually fail by slabbing, spalling, sloughing, or brittle buckling of sides. This is one phenomenon with many names. In softer rocks, particularly laminated shales or mudstones, the rock will flow into the roadway and reduce its cross-sectional area. The two effects are shown in Figure 8.5.

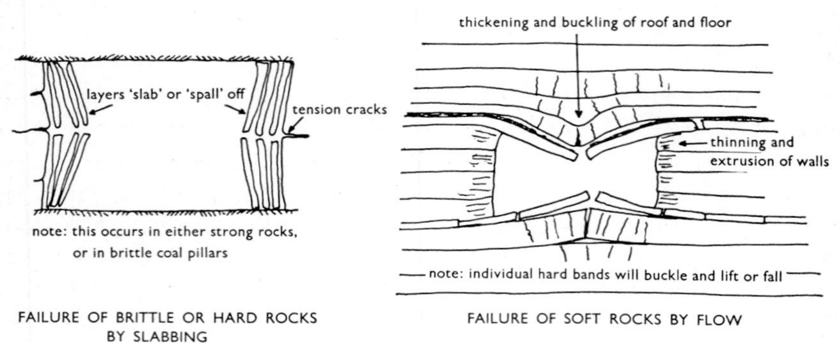

Fig. 8.5 Failure pattern of roadways

The quartzite rock of the South African gold mines provides excellent examples of slabbing in drives. These hard siliceous rocks with an unconfined compressive strength of around 200 MPa give little trouble to over 1500 m depth and can then start to fail violently if bad mining techniques are used, such as leaving small pillars at roadway junctions. In contrast, bedded sedimentary rocks such as shales with unconfined compressive strengths of around 25-50 MPa can start to fail at 120-150 m depth.

Practical experience, and model tests, show that the broad rectangular roadways driven by continuous miners in coal seams are the weakest shapes. The first change to be made with increasing depth has often been to use a borer type miner to give a narrower, smoother-walled, roadway. These changes may be made at 150 m under shale, and 300 m or more under conglomerate, but they will have to be made.

Figure 8.6 is based on a series of model roadways in laminated plaster-sand mixes tested by Hobbs[12] at the N.C.B. Mining Research Establishment. Failure of the roadways has been by sideways closure and the rectangular roadways were the weakest because the roof and floor strata buckled into the openings. In order of increasing strength came rectangular roadways, wide arches, narrow arches, circles of decreasing diameter, a pentagon and then a hexagon with a vertical apex. These latter derived their strength in laminated strata from the fact that the strata could not meet in the middle and buckle inwards readily. A trapezoidal roadway tested in another experiment has been included in Figure 8.6 for completeness. The trapezoidal roadway was the weakest.

The most practical shaped mine roadways have flat floors for transport, so a mining engineer does not readily adopt circular or invert arch shapes. However for civil engineering a circular tunnel is often acceptable and this profile is both strong and easy to drive with a tunnel boring machine. The machine carries with it the bonus that the absence of shotfiring gives a stronger unbroken rock to withstand higher stresses.

Although earlier it has been said that roadways in the solid suffer from sideways squeeze it must also be remembered that in weak laminated strata there is a considerable gravity loading on the roof beds. In addition roof buckling and floor lift is enhanced in wide roadways. Consequently for development work in weak strata a roadway of (say) 4 m high and 3 m wide may be preferable to a roadway 2 m high and 6 m wide. In a high narrow roadway a conveyor can be mounted above a rail track to give adequate transport facilities.

Although it is not primary rock mechanics the effect of rock weathering must be considered in semi-permanent mine roadways. Rocks with a clay or shale content will progressively deteriorate if they are wet. Humid air entering a mine shaft or drift in early summer can condense on rocks cooled down by winter air. Any argillaceous (clayey) rocks should be sprayed with gunite (cement-sand mixture) or with bitumastic compounds to prevent weathering and progressive deterioration. This precaution is usually only necessary close to intake shafts or drifts.

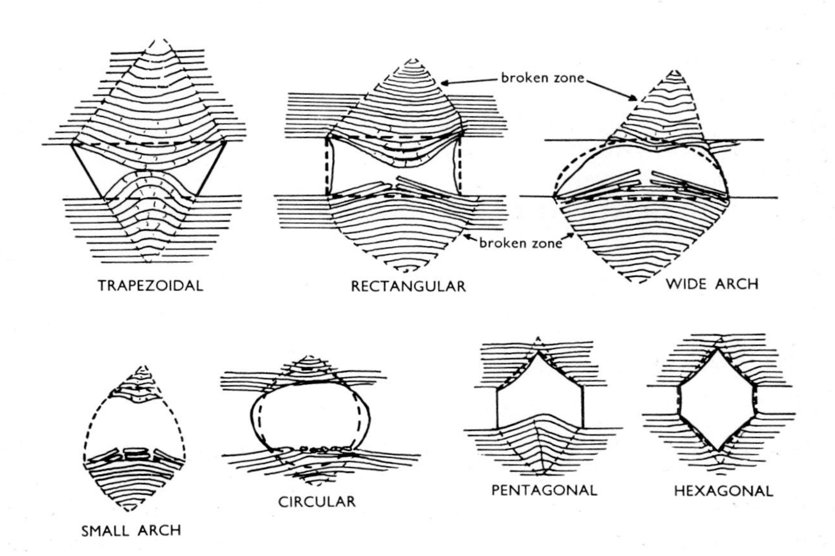

Fig. 8.6 Roadway shape tests in bedded strata (based on Hobbs[12])

Effect of Rock Discontinuities

Excavations are frequently designed on soil mechanics principles of flow of granular material. This is valid for shallow tunnels, and tunnels in clay or gravel, but is quickly invalidated at depth by the properties of jointed rock. To express the change crudely, if a large piece of rock bounded on five sides by joints or cracks exists in the roof or wall of an excavation it can fall into the excavation under the influence of gravity. Fortunately most joints have rough surfaces and remain interlocked so that a block will not fall out until the joints open in tension. However if there is a continuous series of blocks along a roadway or excavation then the combined effect if all these block are allowed to sag will be to open up tension cracks and allow the roof or wall to collapse.

The techniques of joint surveys will be dealt with in Chapter 9, and the stability of large jointed masses in Chapter 10. Problems of stability underground occur in two aspects; high pressure zones which will cause general breakdown of the rock, and excavations in areas where rock discontinuities may cause failure. The discontinuities can exist as:—

(a) normal jointing and bedding planes. Here the rock is divided into parallelepipeds and if kept in compression will remain stable

(b) geological faulting or uncomformities. Shear zones are essentially unstable and may cause continual movement and general support problems for years. There is an enhanced side squeeze effect in badly broken ground. Where a nest of faults or an unconformity intersect an excavation then there is a tendency for blocks of unstable ground to slide into the opening

(c) cleavages induced by mining. The act of driving a tunnel induces cleavages around the opening. Further mining operations produce more cracks parallel to their sides. (This mechanism is dealt with later.) The intersections of these cleavages, both with themselves and with pre-mining discontinuities, produces wedge shaped blocks that are particularly unstable. Induced cleavages produced in shear tend to have smoother surfaces than natural joints produced in tension, and because the wedges open downwards these blocks produce a dead weight under normal gravity which can load supports at a rate of over 2 tonnes per cubic metre of rock.

It is sometimes difficult to detect major discontinuities in rock because the normal viewing area is the exposed surface and an intensive drilling program is very expensive. There have been occasions where a major portion of the wall of a large excavation has started to slide bodily inwards. The cure is to bore long holes and to grout rope anchorages into solid ground to stitch the loose block back in. This is not to be confused with attempting to prevent floor heave, or sideways closure in 'swelling ground'.

Mention was made in Chapter 7 of the effect of regular jointing on the stability of roadways. In general, mine roadways and working faces should be aligned to cross a regular jointing system. This is easier in bedded deposits. In metal mines the roadways have to follow the direction of the orebody which itself may have been deposited along the line of geological discontinuities. The usual technique here is to drive the main development roadways in the footwall of the orebody, or below it, where rock conditions are usually better.

In civil engineering work the line of a tunnel may be relatively fixed in travelling from A to B. Where a tunnel goes across the strata there may little trouble, but the closer the line of a tunnel becomes to the strike of the strata then the more trouble can be expected. Tunnels in the strike direction should be carefully explored and it may be desirable in some cases to displace the projected line of tunnel sideways to a more competent bed of rock. An alternative technique is to drive a dogleg tunnel and offset the extra length against savings in mining cost, reduction in overbreak, and subsequent savings in maintenance.

Support of Narrow Excavations

From the foregoing it can be seen that there are many situations in which rock is ready to fall from the roof or sides of an excavation. It is perhaps superfluous to say that no artificial support can withstand the vertical loading of superincumbent strata at depth. Excavations in rock rely on the fact that the main load is redistributed on to both sides by a natural arching effect (Figure 8.7). If this arching effect does

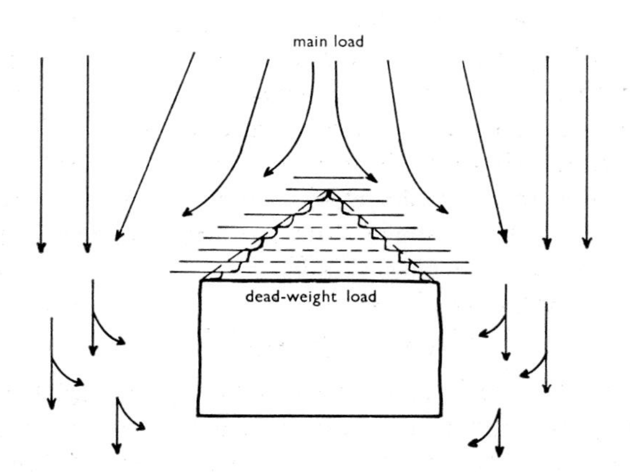

Fig. 8.7 Redistribution of stress around an opening

not exist, e.g. in a soft carbonaceous mudstone which shears off vertically, then an opening is extremely difficult to support. The arching effect is generally due to a progressive cantilever action from the sides which will eventually result in a natural arch or a beam over the top. The height of the dead-weight zone within the arch, or under the beam, can vary from zero in an extremely strong roof to a height of from two to three times the width of the excavation in shales or mudstones.

Supports for excavations have to be adopted in relation to this dead-weight arch. The higher the natural arch then the greater the load and the stronger the support which must be used; so that if a high narrow arch shape is adopted for the roadway it conforms to the natural shape, and better roof conditions result. A secondary principle must also be applied. If the joints and cracks within the dead-weight arch are kept closed, then the arch is 'competent' and will stay intact and bonded to the sides. The artificial support needed is minimised. If on the other hand the strata are allowed to relax, then the joints and cracks open and lose their frictional restraint. The strata become 'incompetent', and not only are stronger supports needed, but they have to be closely set to prevent the rock from falling between them.

Rock Bolting

Rock bolts, sometimes called roof bolts, are the first line of defence in many mining and civil engineering applications. Although it can be a matter of contention many engineers consider that bolts should only be considered a first line, and should not be used for permanent support. Provided that the bolts are installed immediately the rock surface is exposed they will provide good support for a few weeks, or perhaps some months. However, slow movements of a rock mass can open up joints, or extend cracks, and then large bolted-together pieces of rock can fall. It can be a feeble excuse to say, 'the bolts were all right, but initial tests did not reveal the hidden joints in the rock'. Where workmen are concerned an engineer should assume that joints are present and install a secondary line of defence as soon as possible in permanent or semi-permanent roadways. In extraction areas due to be abandoned in a few days or weeks then rock bolts are often all that is needed for support, but care should always be taken to inspect the roof on each shift before work starts.

TYPES OF ROCK BOLT

Figure 8.8 shows the basic types of rock bolt. The split rod and wedge bolt is the cheapest but requires a hole of accurate length and diameter. The bolt must be impacted against the bottom of the hole, usually with a percussive drill and non-rotating dolly, to drive the wedge down into the split rod and expand it. The expanded parts of the rod grip into the sides of the hole. These bolts have the weakest type of anchorage.

The expansion shell bolt has separate leaves fitted around a cone so that the rod is not weakened. The bolt is set by pulling the cone down into the shell either by rotating the bolt thread in the cone, or by using a direct pull technique. If the bolt is pulled with an hydraulic cylinder then the strata can be prestressed before any movement takes place. It will be seen later that the greater the tension in a bolt the better its action in laminated strata. The expansion shell bolt also has the advantage that the bolt hole does not need to be drilled accurately to length.

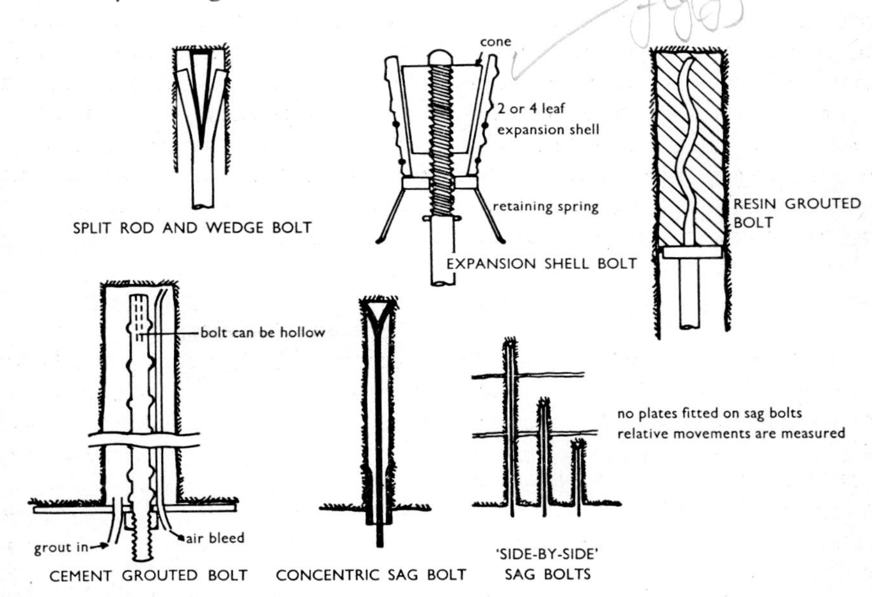

Fig. 8.8 Types of rock bolt

A resin-grouted bolt uses a two-part epoxy resin catalyst and hardener, usually with an inert filler to reduce the cost. A hole is drilled to the required depth and a cartridge of resin mix is pushed up the hole on top of the bolt. The bolt is forced into the cartridge and rotated for a few seconds to mix the separate compartments of catalyst and hardener. Within a minute the resin sets and a plate can be bolted on soon afterwards. There may be some limitations on the types of resin used in wet holes. Resin bolts have the advantage of giving a good anchorage in irregular holes where an expansion shell cannot grip properly. They are more expensive than other types of bolt but can be economical in some circumstances, when the total cost of a system is considered.

The usual application of a resin grouted bolt is to grout the head of the bolt only, but in weak ground the bolts have been grouted for their full length. In weak ground however a bolt is not acting as a support member, it is only holding fragmented rock from falling apart and there are possibly cheaper ways to do that. Full column grouted bolts have been used to reinforce draw points in metalliferous mines because the reinforcement still exists after the head and plate of the bolt have been worn away.

A further type of bolt is cement grouted into place. It can either be a reinforcement bar dowel, or a partially grouted bolt. However a cement grouted bolt is far more effective if a normal rock bolt is used and tensioned before the hole is grouted. In this case the primary purpose of the grout is to protect the bolt from corrosion. Cement grouting is frequently used in conjunction with long drill holes into which lengths of wire rope (rope bolts) are grouted. The rope bolt may have a proper anchor shell or it may be simply grouted into place.

A rope bolt may also have its anchor end grouted into place then, after a few days to cure the cement, the rope is tensioned with pulling jacks and the rest of the hole is grouted. The rock mass then behaves similarly to a tensioned concrete beam. The technique is expensive and is limited to major operations.

All these bolts may be used with a simple 0.15 m diameter or 0.15 m square steel plate bolted on their end, or may be used in groups to support a wooden or steel bar. Where a weak friable rock, or a large shot-fired chamber is supported, then wire mesh may be placed across the roof and walls before plates or bars are put on the bolts. If this wire mesh is subsequently sprayed with cement or concrete a structurally strong lining is obtained.

DESIGN OF ROCK BOLTING SYSTEMS

The use of rock bolts, or of full length grouted or resin-bonded dowels must be examined in conjunction with the material to be bolted. Three factors determine the type of support required in an opening underground, apart from the usual consideration that the total cost of the system must be assessed. If one is going to drill long bolt holes, install special expensive bolts, use expensive bearing washers and have men monitoring their performance, then it may be cheaper to install a simple standing support in the first place.

The physical factors which determine the use of bolts are:—

1. type of strata immediately surrounding the opening: this can vary from massive sandstone to plastic clay. Sandstone would probably be self-supporting except over wide areas. A thin layer of weak rock can be hung from a strong bed. Plastic clay would need standing supports

2. method of working: this can be considered as either systems in which the roof is deliberately allowed to collapse, e.g. longwall mining, pillar extraction or caving methods; or systems in which a roof beam is carefully preserved, e.g. development headings for room and pillar workings and overhand stoping methods. Where roof beds are collapsed then lateral movements usually develop in the roof strata and standing supports become necessary. Intact roofs respond well to bolting or dowelling
3. stress developed around the openings: the deeper the opening, the higher will be the stress. Adjacent workings create zones of high abutment pressure. Geological features such as faults can also increase stresses locally. Then the higher the stress the stronger the support must be.

These factors make it very difficult to lay down standards for rock bolting in terms of length of bolt, bolt spacing, rock tonnage per hole, etc. There are also practical limitations: for instance the time available for rock bolting may limit the length of the bolt, because longer holes would require jointed drill rods and jointed bolts.

There are however some general rules from observation and logic which always apply:

(a) bolts give better reinforcement if they are installed at the working face before strata relaxation starts

(b) for systematic support (as distinct from 'first-aid' applications) the bolts must be installed in a regular pattern

(c) bolts should always cross any planes of weakness, such as bedding planes, natural cleavage planes or cleats, faults and shear zones, and potential stress lines, at as great an angle as possible.

To check on the effectiveness of design some parameters must be monitored. The two that mainly affect mining operations are bed separation and total roof lowering; measurable movements are signs of incipient failure of the roof strata. Sag bolts (i.e. concentric bolts to different depths, or adjacent bolts anchored at different depths and not tensioned) are a convenient way of measuring separation of the roof strata. For general monitoring a dished or arched washer, such as the 'Brown' bearing plate may be used (Figure 8.9). The amount of flattening gives the load on the bolt, which tends to be the result of bed separation, i.e. sag of the lower beds.

As a final limitation it should be remembered that rock bolting does not really support an opening. The opening must be basically self supporting, or have firm columns. The principle function of rock bolting is to reinforce or stabilise the rock at the surface of the opening.

It holds loose rock, detached laminae etc., and prevents them falling due to the influence of gravity. Any small fall of unsupported rock is likely to build up into an increased failure until a natural supporting arch is achieved.

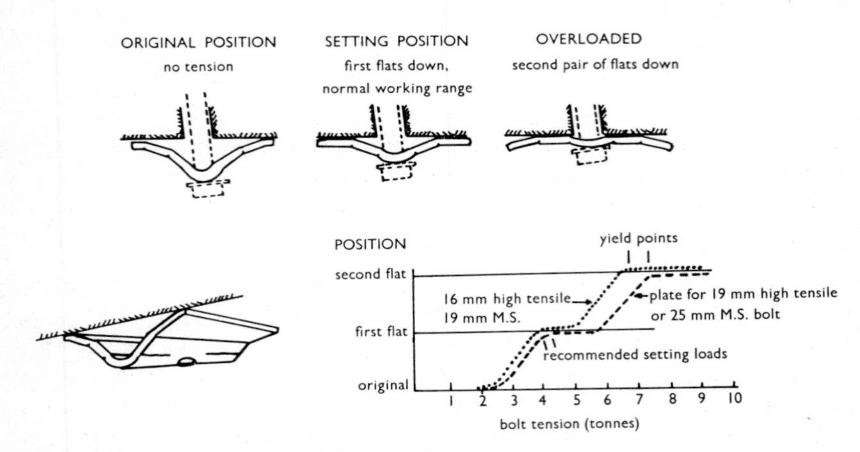

Fig. 8.9 Brown rockbolt bearing plate

DESIGN THEORIES

Although some distinction has to be made between different modes of action of rock bolts to be able to discuss them it must be realised that the behaviour of the bolts is usually a combination of these factors.

1. Laminated Beam

In thin to medium bedded rock two or more strata are bolted together to form a self-supporting laminated beam. Whether this beam will also support overlying rock is debatable. In coal mine strata, after slabbing of sidewall coal has taken place, the laminated beam has frequently been seen to fall, relatively intact, leaving the overlying roof also intact. The best laminated beams obviously include a strong stratum which will bridge over the abutments and stay intact. Obert and Duvall[1] report the results of theoretical and model tests on the laminated beam theory and give a design chart. They also state that their chart is based on continuous strata and that more than a few joints in a bed will invalidate the chart.

2. Support from Outside the Caving Arch

If there is a roof fall in a stope, roadway, or tunnel it usually finishes in a natural arch shape or at a strong overlying stratum (outside the 'dead weight' zone in Figure 8.7). Usually the height of an arch is proportional to the width of the opening and a series of measurements can determine the incipient failure arch, which will be influenced by the joint frequency and probable failure planes. If rock bolts are then anchored beyond the probable failure zone, and the roadway surface is supported by wire mesh and large bearing plates, then the loose ground is stabilised. Non-tensioned, full-column grouted bolts can only work in this way.

3. Consolidation of Fragments

Most rock around mine openings is broken into fairly regular slabs by bedding or jointing, and into irregular fragments as a result of the mining operations. The mining breakage is likely to be particularly bad if there has been uncontrolled heavy blasting, or if the opening is at the periphery of a large mined-out area.

In these cases a tensioned rock bolt has the effect of clamping or squeezing the rock fragments together in a cylindrical zone around itself to make a competent mass (Figure 8.10). If the bolts are spaced sufficiently close to each other then they will form the equivalent of a masonry arch and become self-supporting. This theory is possibly the widest held and has been the subject of considerable research by the Australian Snowy Mountains Hydro-electric Authority (referred to subsequently as S.M.A.), and their staff have published many papers; see for instance Paper 1 of the 'Symposium on Rock Bolting'[13]. The jointed and broken rock approach has also been used by the Mines Branch of Canada (M.B.C.), a government research institute[14].

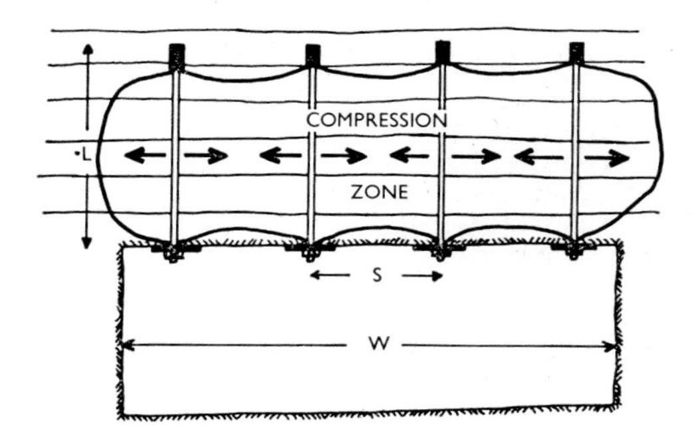

Fig. 8.10 Consolidation of fragments by rock bolting

General rule-of-thumb is that in competent rock a bolt should have a length equal to one-third of the width of the roadway and be spaced at two thirds the length of the bolt.

$$\text{i.e. } L = \tfrac{1}{3}W,\ S = \tfrac{2}{3}L = \tfrac{2}{9}W$$

Spacing can be rectangular or W-formation. In reasonably broken rock the bolt density should be increased to $L = \frac{1}{2}W$ and spacing approximately 0.9 metre: in badly broken rock increase bolts to $L = \frac{2}{3}W$ and extra bolts may be needed in the walls. The truss roof bolt (Figure

8.11) is particularly good for bonding fragmented rock as it is anchored over the solid sides of an opening and laterally tensioned to squeeze the fragments together. A more comprehensive code of practice is given below.

4. Surface Effects

A subsidiary advantage of roof bolting is that compression of the rock surface prevents opening of partings, cracks and joints which would be subject to attack by air and water. This also helps to prevent the crumbling away (ravelling) of rock from the surface. If left unchecked ravelling gradually removes rock from around the bolt until failure occurs. It can be checked by the use of mesh or gunite.

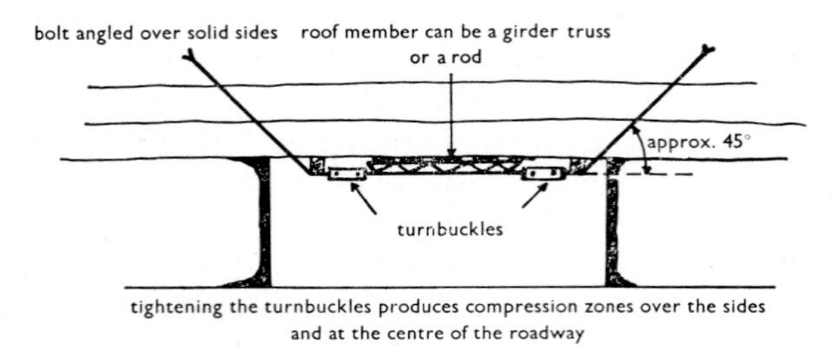

Fig. 8.11 Truss roof bolt

CODE OF PRACTICE FOR ROCK BOLTING

Both the Snowy Mountains Authority and the Canadian Mines Branch have produced rules of practice which are summarised below. Neither have been able to allow for the effects of time but suggest that continual monitoring is necessary. If a roadway is to stand for other than a short time the manpower costs of continuous monitoring may be high. For some tunnels, e.g. hydro-electric schemes, it may not be possible to monitor installations. The cost of permanent standing supports may then be worthwhile.

The least that should be done is to grout all bolts to stop them corroding and to use mesh and preferably a gunite or shotcrete coat to prevent weathering of the rock surface. Walls tend to slab off with time and this can produce an unstable beam by reason of failure of the abutments, or sagging of an increased span which will allow joints to open up and blocks of rock to drop out.

1. Strengths

Coates and Cochrane[14] recommend from Canadian results that before rock bolts are used for systematic roof support the rock be examined and classified, and the joint spacing and regularity be determined.

Then various items should be regulated as below. The symbols they use are

Q_y = representative yield load of system
Q_m = maximum yield load of system
Q_r = representative load capacity of the anchorage

and Q_r must be equal to Q_y or $0.7Q_m$ whichever is the lesser.

Bearing plates should conform to standard strength tests. If the collar of the hole is consistently broken and exceeds 10 cm then the bearing plate must be increased to a square of side (D + 5) cm (or (D + 2) in), where D is a representative collar size not exceeded by more than 10 per cent of the holes. Plates other than square shall have an equivalent bending strength. The standard plate is 15 cm square. The thickness of the plate is also adjusted to suit hole collar size. If t_4 is a thickness of plate found to be suitable for a 10 cm collar, then the desired thickness t is derived from

$$\frac{t}{t_4} \geqslant \left(\frac{D}{D_4}\right)^{4/3}$$

where D is the width of the plate and D_4 is 10 cm.

The load capacity Q_a of the roof bolt shall be equal to either the load capacity of the steel, Q_s, or the load capacity of the anchorage, Q_r, whichever is less.

2. Length of Bolts

The minimum length of bolt as recommended by the Canadians is L which should be either 1.2 m to avoid fractured ground, or more than the depth of a rock block containing the mouth of the hole. The maximum length of the bolt, if not determined by the size of the opening, is determined from

$$L \leqslant Q_a / (S^2P)$$

where S is the average spacing of the bolts and P is the average density of the rock involved. If the bolt cannot be installed with a tensile load equal to or greater than $0.5Q_a$ then the maximum length should be found from

$$L \leqslant Q_a / (2S^2P)$$

The Snowy Mountains Authority general practice is that

$$L = 6 + 0.004\,W^2$$

where W is the width of the opening (all dimensions in feet), with an over-riding consideration that

$$L \geqslant 3T$$

where T = the thickness of joint blocks. This is to ensure that the bolts are anchored behind the first two layers which can be loosened during excavations.

3. Bolt Spacing

The Canadians recommend that the maximum bolt spacing shall be three times the representative spacing of the joints, layers or fractures unless a membrane, (e.g. lagging, wiremesh or gunite) is provided between the bolts, in which case the maximum spacing would depend on the type of membrane. The maximum spacing in any case shall be limited to either 0.9L where L is greater than one quarter of the breadth or span of the opening, or 0.5L where L is less than one quarter of the span of the opening. At an intersection the span is the diagonal between abutments.

For the S.M.A. the spacing must not be greater than half the length of the bolts. This gives a compression zone of at least one third of the bolt length.

4. Bolt Tensions

The bolts should normally be tensioned to a load of between $0.5Q_a$ and $0.8Q_a$. If tightened by torque action then the maximum torque should be calculated from

$$T = 0.8r^3\sigma_y$$

where T = maximum torque, r is the radius of the bolt at maximum section and σ_y is the specified minimum stress at the yield point of the steel in tension.

In yielding rock, or where further excavation is to take place the bolts should be periodically checked with a torque device or dynamometers, or by pulling tests with a jack. If 10 per cent of the bolts sampled are shown to have a tension less than $0.5Q_a$, either the slack bolts must be retensioned or extra bolts added.

Monitoring may also be continuously carried out by indicator washers such as 'Brown' bearing plates. These have two bearing angles (Figure 8.9); the first degree of flattening shows correct tensioning. Full flattening indicates overloading on the bolts. Bolt slackening can also be detected. There is a bonus in that the plate gives a good bearing for angled bolts.

Steel and Wooden Supports

Rock bolts are very convenient to install in the face of headings, and as temporary support in chambers where columns would be a nuisance, but a more secure form of support should be installed as soon as possible under most roofs. In some instances these extra supports may never bear any heavy loads but in that case they can often be recovered and used elsewhere.

The choice of wooden or steel supports is based on cost as well as strength. The National Coal Board for example buys a standardised steel arch in bulk. These are the shape found to be best by experience and are cheaper than a timber support of equivalent strength. In contrast in a mine buying small quantities of steel, and not needing an arch shape, then timber will probably be cheaper. In general however, the steel support occupies less space for a given strength and if it is new (not embrittled second-hand rails for instance) the steel support yields slowly to failure. A wooden support will snap and fail abruptly.

POSTS AND BARS

The simplest form of support is the post and cap (or prop and lid). It consists of a single upright with a small wooden or steel plate above. However the post may be a simple wooden leg, or a rigid steel joist, or a yielding hydraulic prop. The hydraulic prop pumps up like a car jack and will lower at a pre-set load value, which may be from 5 tonnes for a light prop to over 100 tonnes for a 0.3 m diameter leg. Because roofs generally lower, even in first working headings or development roadways, a wooden crushing pad should be inserted over rigid legs to stop them buckling.

The next stage is to use a cross-bar and two legs (or one leg and rock bolts through the bar). These were described in Chapters 6 and 7. The cross bar and legs can again be wooden or steel. If both bar and legs are of steel then they must be fitted with positive locations to stop the steel surfaces skidding along each other. The principles involved are so simple that most mines prefer to make their own brackets. A wide cross-bar tends to bend quite easily in the centre and a change from wood to steel becomes necessary in deeper mines. If clearances permit the bar may be centre legged for extra stability. In some cases a fabricated engineering truss may be used as a roof member. Its extra cost is offset against retention of square work with a continuous miner, instead of adopting an arch shape.

Permanent installations such as shaft plats or drift-bottoms are usually given stronger support than ordinary cross-bars. They may be concrete lined but are often supported by straight or cambered RSJs which may be set on steel legs, cribs, or concrete or brick side walls. The points to note are that, again, some provision must be made for vertical crush and sideways squeeze. Steel legs are set clear of the sides and can generally be relied upon to penetrate the floor slightly before buckling. Concrete or brick side walls can generally push inwards because the roof members rest on wooden blocks on the side walls and are not keyed in. The joists are braced between their webs with wooden blocks, or a concrete bridge, to provide lateral stability. Figure 8.12 shows some of the possible combinations.

This type of support may also be used at large roadway junctions. In some cases carrying girders may be mounted on steel stanchions made of four vertical RSJs braced together for extra strength and to prevent column buckling.

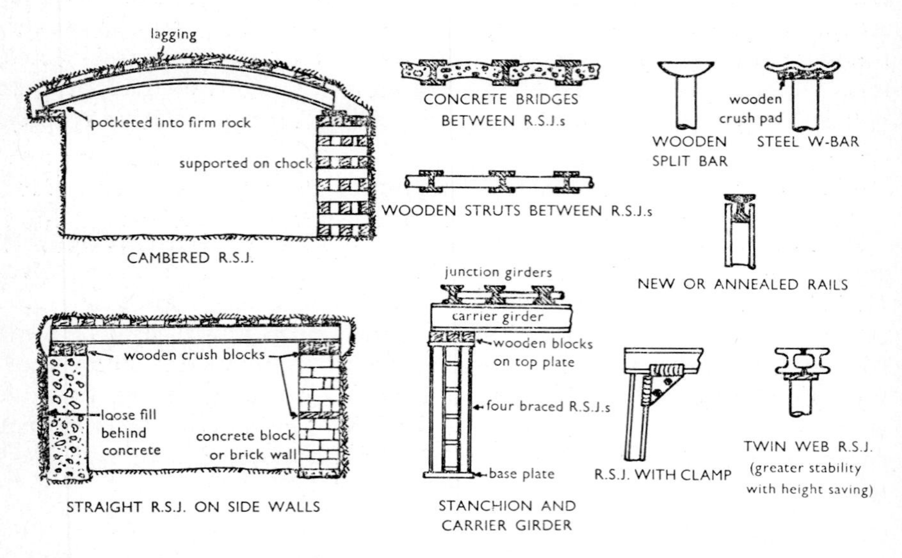

Fig. 8.12 Girder supports

When a change must be made to arched roadways then the supports are usually of steel, although both laminated wood and concrete arches have been tried. In Australia arches would be a special order and may be a multi-sectional polygonal shape for cheapness although extra fishplates increase the cost. The useful pentagon of Figure 8.6 may be the cheapest option. In Europe there is a wide choice of rolled steel sections available in cold-formed arches. They will probably be either German DIN standard or National Coal Board (NCB) standard profiles.

The basic choice is between a rigid or a yielding arch. The usual arch shapes (as distinct from full circle or invert special orders) are all splay legged; Figure 8.13 shows some of them. It has been found from experience that most roadways suffer from some side-squeeze and the splay legged arch is more fitted to resist this. The usual splay is to have the base of the arch about 0.3 m wider than the spring for widths of 3-5 m.

The rigid NCB section arches are produced in standard sizes to an NCB Specification. (Note: there are a large number of NCB specifications to cover most items of mining equipment and these can be purchased from the National Coal Board headquarters in London. They are used on an international basis where local standards are not available.) The RSJ cross-section is thicker than a normal structural steel RSJ.

Yielding arches are of two forms, those fitted with yielding stilts or boxes, and those of trough sections which slide over each other. The Usspurwies arch is shown in Figure 8.13. This is pin jointed to allow side squeeze and has yield boxes with wooden crush members which accept a load found from tests to be between 5 and 20 tonnes dependent on wood type, condition of arch and lubrication from coal dust. Other arches may have deformable steel yield pieces or hydraulic legs. These arches are expensive and are only justifiable in special cases. Frequently a rigid steel arch will penetrate a floor sufficiently to give the required yield. A 115 mm x 115 mm leg section will penetrate a shale of about 50 MPa compressive strength at a load of about 15-20 tonnes.

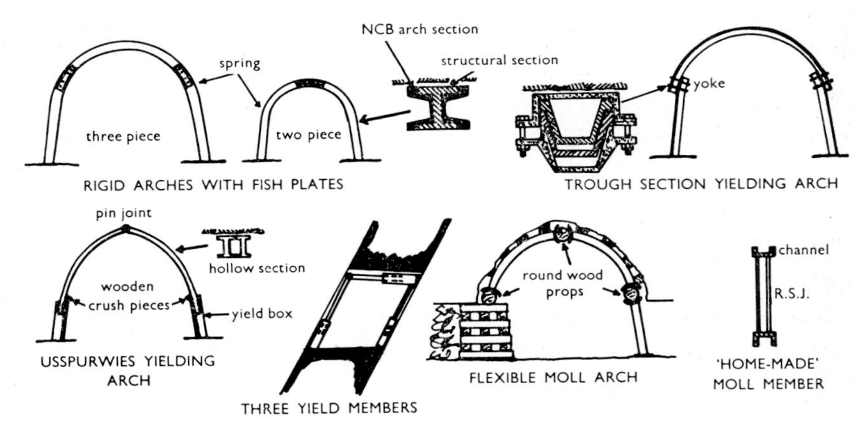

Fig. 8.13 Steel arches

Trough section arches are typified by the TH arch; the trough sections of a three piece arch overlap and fit into each other and are held by clamps. Steel to steel friction provides the yield load but the clamp action and different curvature of the crown and legs tend to give erratic yield values. A Moll-type arch relies on crush of the round wood members to give slight yield but has the great advantage that it can deform with a roadway without damage to the individual members, which can be straight or curved as necessary. These can be easily 'home-made' from channel and RSJ for emergency use in soft yielding ground.

Design of Multiple Openings

Multiple openings are usually taken to refer to coal mine or bedded deposit development roads and to development for pillar extraction.

Provided that the individual roadways are not excessively wide then the problem is one of pillar stability. From the amount of current research on this topic one must deduce that the theory is not fully understood and that practice is still based broadly on experience.

The openings considered in this section are small ones that by themselves would be relatively trouble free, but in combination they affect each other. The combination of roadways in metal mines rarely give trouble and thus there is very little literature or laboratory studies on them. However these roadways can be affected by stoping in which case roadway protection is necessary as described later. One frequent problem in metal mines is that of too small junction pillars. Sometimes a junction will be driven as in Figure 8.14 and to reduce the length of drivage a small pillar is left. Subsequent mining, or slip along bedding planes in the pillar, can cause it to start cracking. The pillar should have been larger in the first place but a quick first-aid measure is to wrap the pillar with an old wire rope held in place by rock bolts, and then spray it with cement or concrete to prevent it crumbling.

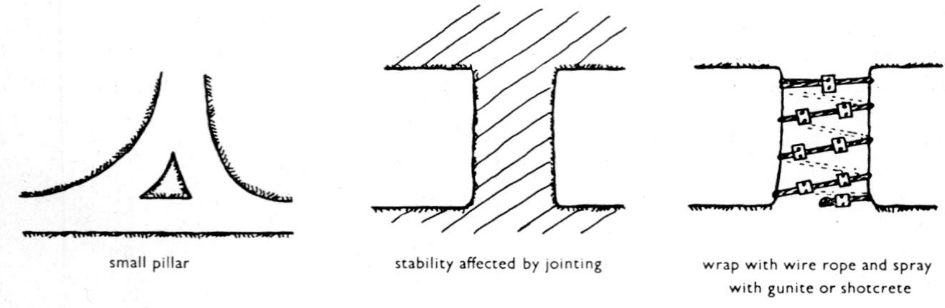

Fig. 8.14 Junction pillar protection

The principal study of multiple openings has been made by the U.S. Bureau of Mines. Duvall reported several photoelastic studies of mine openings and the results are summarised in his textbook[1]. However one must be careful to use photoelastic studies only as a guide to stress distributions in the early stages of mining. Small pillars in a mine will often yield beyond their elastic limit, and load shedding and load sharing takes place fairly quickly, sometimes catastrophically. Although many research centres have studied long term creep in rocks the deformations found have little significance other than in rock salt and potash deposits. These largely crystalline materials exhibit a strong tendency to continuous creep under load but large pillars still have a useful life of many years.

In the absence of valid mathematical theories for the three-dimensional behaviour of mine pillars the probable load is generally obtained by calculating the total load of the superincumbent strata over the mined area and placing this load on to the pillars left to protect mine openings, i.e. if the average unit vertical stress on a pillar is σ_p, and σ_v is the total vertical stress, then

$$\sigma_p = \frac{(\text{mined area}) + (\text{pillar area})}{(\text{pillar area})} \cdot \sigma_v$$

Thus if first working takes 40 per cent of the coal then the load on the remaining 60 per cent is increased to 100/60 per cent of the original. The load formerly taken by the empty spaces is shared between what is left. It should be realised that this is a square power law and for extractions of above 40-50 per cent the load on the remaining coal rises very rapidly.

This average load assumption is so simple that there must be many qualifications in practice. It is usual to make some allowance for pillar, floor and roof deformation, and a safe load on a pillar is expected to be about 50 per cent of the maximum calculated strength.

There are more complicated formulae to determine pillar loads; many of them introduce a factor to allow for pillar height because a high pillar is less stable than a low pillar. This is obviously true of small pillars, and of column buckling theory, but logic suggests that when a pillar is 15 m or more square then the effect of height changes between 3 to 5 m or so is not significant. Troubles only arise when joints as sketched in Figure 8.14 can provide a slip plane to seriously weaken the pillar but this can only happen with small plan area pillars. Metal mine experience of stope pillars 30 m or more in height (admittedly in rock or in ore, not coal) show that high pillars can be stable.

CLASSIFICATION BY USE

Average load theories and front abutment theories enable an engineer to determine the maximum load likely to be carried by pillars. When the physical strength of the pillar material is known then the desirable total area of pillar can be calculated. The next step is to determine the lineal dimensions needed to provide stability. Average stress theories assume an infinite area with no edge effects, whereas it is quite possible for a stable arch to be set up over five or more closely spaced entries in a virgin area, so that the pillars carry a load less than the original stress in the ground. Consequently before pillar dimensions are assessed the purpose of the pillar must be decided on. A possible classification is:—

1. permanent pillars; e.g. rib pillars in civil engineering water storage schemes

2. virtually permanent pillars; e.g. main entries for mining schemes
3. limited-extraction pillars (percentage mining); these are pillars to prevent surface subsidence and may be designed as yielding or non-yielding
4. temporary pillars; these pillars only have to support the roof between development and complete extraction.

PERMANENT PILLARS

For case 1 the engineer is probably working in barren ground with no economic value. There is no thought of maximum extraction of mineral. The excavations will probably be shallow with low vertical loads. In these cases long high openings with thin pillars between them will probably be best. This will keep the total span of undermined roof to a minimum. However as the rock becomes weaker, or more prominently bedded, or if the depth of excavation goes below 100 m or so, then the height/width ratio of the pillars must be reduced to maintain stability. The openings must approach a circular shape and must be spaced wider apart. The main load should be deliberately deployed onto solid rock on either side of the group of excavations.

MAIN ENTRIES FOR MINING

The basis for design here can be sharply opposed to civil engineering practice. There is a mineral which has to be recovered to make a profit. The area on either side of the entries will be mined therefore the pillars must be large enough to take the superposed loads and to protect the roadways for several years. It is also desirable that these pillars should provide an economic extraction area when the mining equipment makes its final retreat from the mine, but this is not always used as a design criterion.

The plan area of pillar needed can be calculated from the crushing strength of the coal plus a safety factor, but the answer could be satisfied in two ways; long thin pillars with few breaks, or wide rectangular or square pillars. The stress distribution on pillars will be discussed in detail later. At this point it is sufficient to say that long thin pillars are basically unstable, because the stability of a pillar depends on its minimum lateral dimension. Pressure arch theories will also be discussed later but there is every indication that a group of entries over a width approximately equal to the pressure arch may well suffer damage to the outside entries, even before adjacent coal is extracted. The lower roof will sag over the inner entries so that the heavier load of the intermediate and main roof is thrown outwards (Figure 8.15). It is not possible to be specific but as an indication care should be taken to minimise the lateral spread of entries under shale roofs at depths of 120-150 m, and at depths of 250-300 m under massive sandstone or conglomerate roofs. At these depths a barrier pillar of about 90-100 m may be necessary to prevent crush on the outside entries. The alternative often taken is to deliberately sacrifice the outer entries to destress the area.

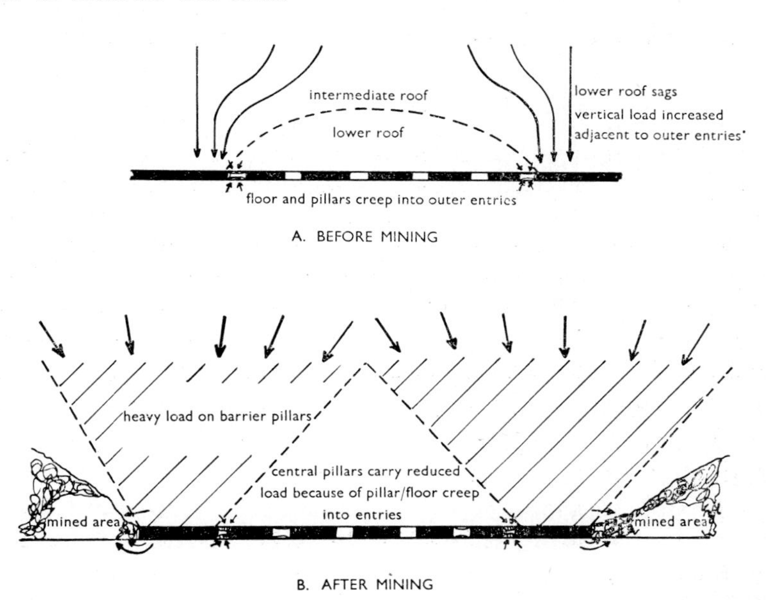

Fig. 8.15 Crush on outer main entries

PERCENTAGE MINING PILLARS

There have been suggestions made and attempts to calculate, or produce by trial and error, a size of pillar that would stand long enough to allow a proportion of coal to be won, and which would afterwards crush down and spread out to lower the roof gently. These are *yield pillars* and were nominally designed to give a fixed amount of subsidence at surface level. However, trial and error techniques which put expensive machinery at risk (the men learn to run fast) are not to be recommended.

Consequently percentage mining extraction for subsidence control is usually based on leaving large stable pillars in strong rows, and letting the roof sag in between. With pillars, say 15 to 20 m square, three rows may be extracted, leaving two rows behind, to give about 70 per cent coal extraction (including the initial entries) as in Figure 8.16. This course may be adopted where a surface structure, e.g. a railway, has to be protected. A substantial two or three pillar row must be left under the structure and each individual pillar must be stable. There will be some surface strains (see Chapter 11) but these should not be serious. A similar method can be used in mining shortwall panels leaving substantial blocks of coal between them as rib pillars, say 80-90 m shortwalls and 30-40 m ribs.

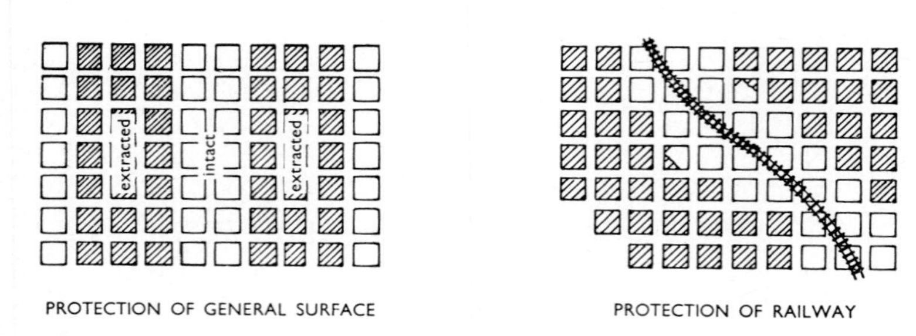

Fig. 8.16 Percentage mining pillars

TEMPORARY PILLARS

Until about 1950, while machine extraction of pillars was forbidden and many mines were hand-working coal, it was obviously a good plan to have drives and bords as wide as possible, to make as many crosscuts as possible, and to keep pillars small where roof and floor conditions permitted this. Pillar sizes could be determined by robbing until a few collapsed. Now with mechanised pillar extraction that can produce higher outputs per man-shift than roadway drivage the position is reversed. The rule is to have a minimum of entries and a maximum of pillars which automatically guarantees stability. In general the more recently a pillar has been exposed (i.e. it is 'greener') then the easier it is to mine. Where possible the mine is divided into small panels off main entries and pillars are mined rapidly. (Cross-reference should be made to pillar mining in Chapter 7.) Main entries into sections of the mine may have to stand from two to five years and the pillars will present poor mining conditions if they have crushed edges. Each mine must decide whether to use small pillars between entries and accept a poor recovery, or whether to have large pillars to maintain good conditions within the pillars.

Recent laboratory compressive strength tests on specimens of rock and coal have been made with *rigid machines*. These are presses which are specially built to prevent the stored elastic energy in the steel platens which load the specimen from unloading itself rapidly. Thus when a rock specimen starts to crack the load can be taken off it and a strength curve of the type shown in Figure 8.17 is obtained.

The conclusion to be drawn from this is that when a coal pillar is robbed to leave only a small stump or stook, and the compressive stress on the pillar starts to rise until it exceeds the maximum strength of the pillar, then when the pillar fails it will do so almost explosively because the effective strength of the material drops so quickly. The rapid reduction of support from the pillar stump can release a roof fall because of a sudden shock load onto the brittle rock.

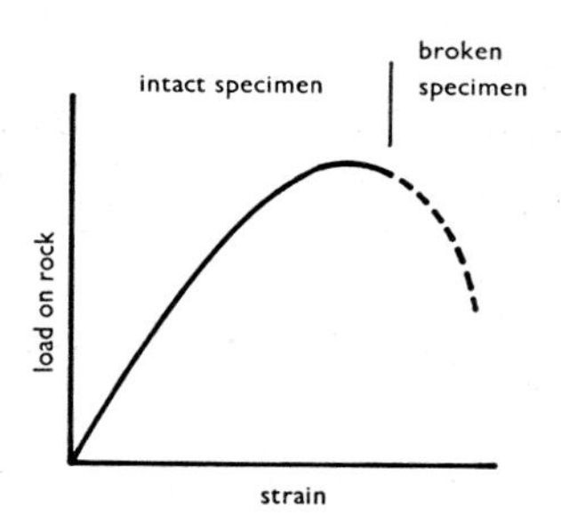

Fig. 8.17 Compressive strength of rock in a rigid machine test

MINING BEDDED DEPOSITS OTHER THAN COAL

Some valuable information on mining bedded deposits has been published about the White Pine copper-silver mine, Michigan[15]. Tests were conducted to determine the safe span of rooms in room and pillar mining. Roof bolted entries of various dimensions were constructed and then the bolt heads were blasted off one by one. None of the entries under test collapsed which shows either how useful bolts often are, or else how unlucky research workers are. Systematic roof bolting was continued because there were occasional falls in unbolted entries and it was not possible to differentiate good entries from bad.

Some roof failures occurred in rooms. One occurred where a diamond drill hole from the surface had been plugged at the base only. Surface entry water built up a hydrostatic head and spread out through the roof strata until it eventually caused a collapse. A second type of roof failure occurred through the directional nature of joints in east-west headings. White Pine's metal deposits occur in shales under mixed sandstone and shale roofs and the most troublesome roof failures were eventually attributed to lateral compression. This gave the most trouble in narrow headings in areas of the mine where the roof consisted of a massive sandstone bed overlying a thinner bed. The modes of roof failure are shown in Figure 8.18. One solution was to widen the rooms to allow the lateral stress to buckle the thin sandstone bed instead of breaking it. This of course is only possible because of the massive sandstone bed above which will span the extra width. It is interesting to note that similar problems occurred in the early 1900s in English bord and pillar mining and that the solution adopted there was to deliberately slot (nick) the roof at one side of the bord to destress the roof.

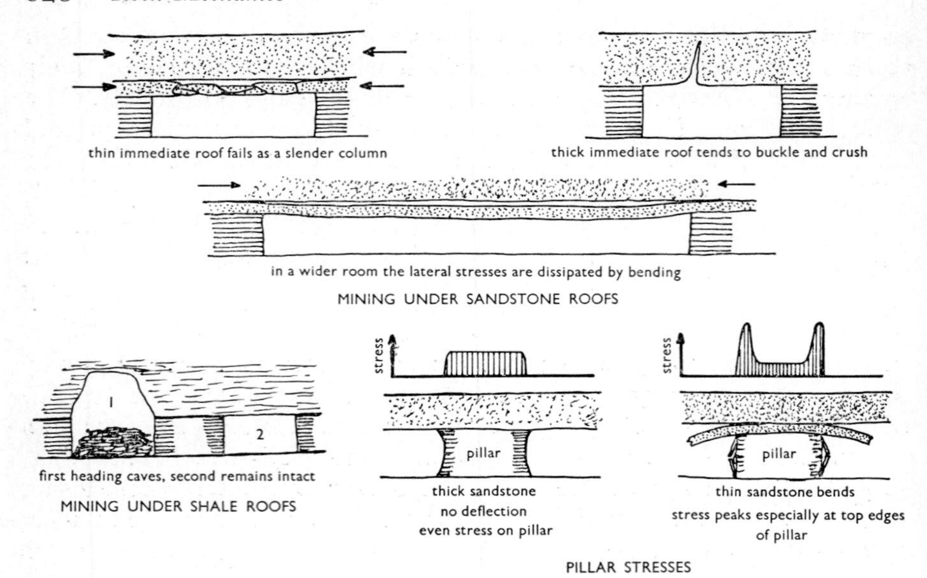

Fig. 8.18 Roof failures at White Pine mine

Research at White Pine showed that, as in coal mines, shales were affected by moisture changes and this affected the general stress pattern. Also the first heading in shale would cave and this then destressed the local area and subsequent headings gave no troubles. They also monitored an interesting pillar stress pattern. When a thick sandstone bed rested on a pillar the stress distribution was uniform. Thin sandstone beds deflected downwards and caused stress peaks.

Four cases of multiple pillar failure are also recorded in the article. One set was produced deliberately by robbing the pillars. A second case occurred when a flooded area was dewatered. There the pillars were tall, thin, and jointed. The third and fourth series occurred in close proximity to each other in an area which had stood perfectly well for eight years. The pillars originally had a safety factor of 5 but suddenly started crumbling away until the area collapsed.

Pillar Stability

Pillar stability has been a problem for many years but fortunately there has only been one disaster of the magnitude of that at Coalbrook North Colliery in the Orange Free State on 21st January 1960, when 437 men were trapped and died in a collapse of nearly 300 hectares (750 acres) of ground. The area had been entered by first working under a hard dolerite sill and was supported by pillars. For some reason the pillars suddenly collapsed in a chain reaction destroying a section of the mine. It is possible that the dolerite sill and faulted ground were the main contributing factors[16]. Accidents from pillar failures are fortunately few because failures are normally localised.

The accident in South Africa stimulated research. Salamon and Munro[17] carried out a statistical survey of size of pillar, and its stability or failure, in South African coal mines. The pillars were assessed on an average load basis using unconfined compressive strengths. Of the 125 pillars analysed 27 had collapsed in service. On the basis of a combination of empirical and statistical methods the authors deduce that 99 per cent of the collapses should occur at safety factors lower than 1.48. The safety factor is the excess of compressive strength over the calculated average stress load. All the pillars considered were at depths of less than 220 m and had a maximum width of 21 m (collapsed pillars were less than 16 m width) and heights less than 5.5 m. A general analysis of published work indicates that a safety factor of 2 is sufficient in most cases with a value of 2.5 to 3 for important protection requirements.

Currently many papers are being published on pillar strengths. To develop a mathematical argument an elastic analysis is used but this is not true in a practical sense. Current arguments also introduce a relationship between pillar size and strength which is primarily based on laboratory specimens. It is generally agreed, and proved experimentally, that the strength of a rock specimen decreases with increasing size because of the increased number of defects in the rock. The largest size of specimen tested so far is a 1.5 m cube block of coal[18], nominally 'in-situ' because it was carefully carved out underground. Bieniawski[18] gives the results simplified in Figure 8.19 and compares them with a theoretical formula developed by Protodyakonov, which gives a similar shaped curve.

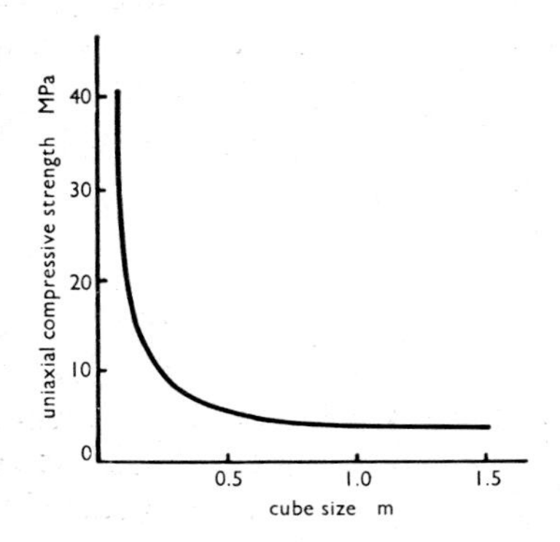

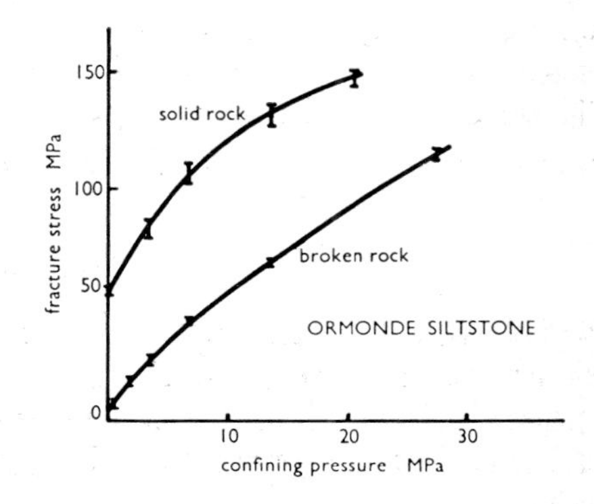

Fig. 8.19 Strength of coal specimen in relation to size (Bieniawski[18])

Fig. 8.20 Relationship between confining pressure and stress at failure (Hobbs[5])

The development of these small scale tests suggests that pillars of increasing size should then not increase in strength. However, Hobbs[5] has shown the relationship between confining pressure and stress at failure. Values for a siltstone are plotted in Figure 8.20. It has been mentioned earlier that the stress around an opening increases rapidly to a peak because of the restraint given by a layer of broken rock. The frictional restraint strength increase also applies to a pillar between roof and floor. It can then validly be argued that once the pillar size goes beyond the double layer of 3-5 m of broken rock on both sides, then its strength increases.

It is certainly true that measurements of changes in stress within pillars show that high stresses can be built up without the pillar failing. Stresses were measured in coal pillars at the Olga mine in the USA and despite the title of the report[19] none of the pillars instrumented failed before they were extracted. The pillar coal had strengths of 7-20 MPa, dependent on specimen size. The depth of working varied between 240 m and 430 m under hilly country. Pillars were roughly rectangular and measured about 18 m by 21 m. Hydraulic borehole plugs were used to monitor pressure changes in two pillars as mining retreated towards them. Peak pressures of 62-69 MPa were recorded 6-7.6 m inside the pillars. These peaks of 8-10 times the cover load were higher than the usual 3-5 times, but both roof and floor strata were hard. The significant point was that the weak coal was able to sustain an enormous increase in load, with only minor slabbing of the pillar sides parallel to the cleat, and one is justified in assuming that the strength of a pillar must increase with size, provided that the pillar exceeds a critical minimum. The minimum size is that needed to carry the peak stresses separately, or as one large central peak which will not exceed the constrained compressive strength.

The stress distribution across a pillar can be derived theoretically in the same way as the stress around a single opening. From general measurements and theory it is accepted that it is of the form shown in Figure 8.21 for the stress across a rib pillar between wastes, and in

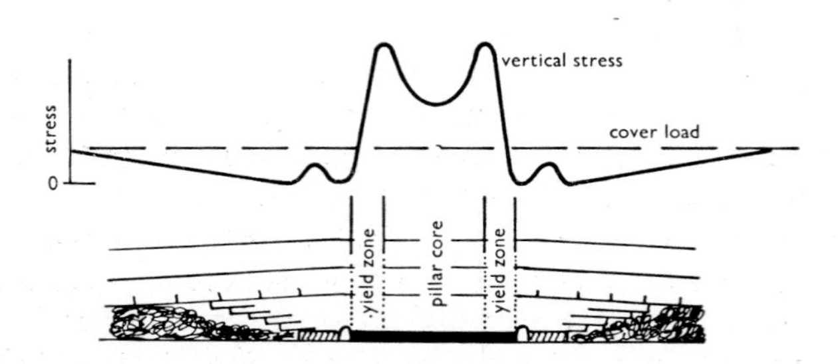

Fig. 8.21 Probable stress distribution across a rib pillar

Figure 8.22 for the stress across a series of pillars. The extra stress on the outer members of a series of pillars causes more damage to the outer roadways. It should be remembered that this stress redistribution also occurs every time a cross-cut is driven through a pillar, and every cross-cut locally de-stresses the pillar it is driven through, and therefore weakens it.

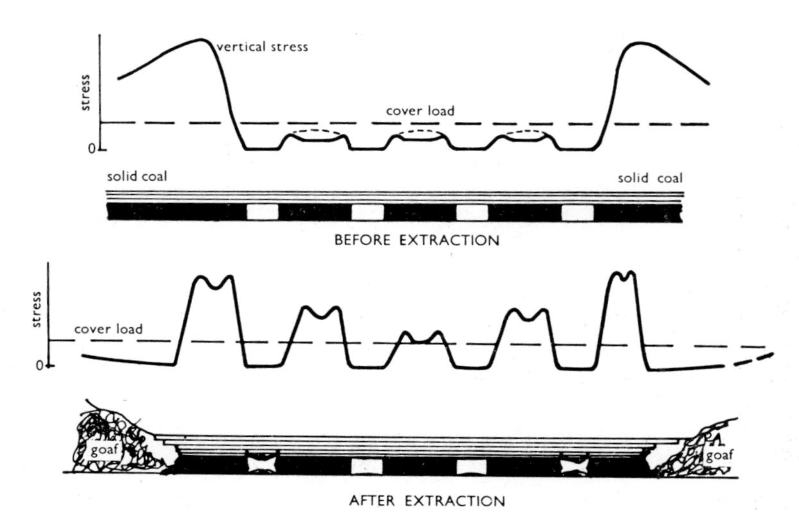

Fig. 8.22 Probable stress distribution across entry pillars

From measurements to date, particularly those collated by Orchard[20] in Britain, it appears that an empirical size for pillar stability is still the normal method of design. Theory has not yet produced a more acceptable alternative. The rule-of-thumb size for a pillar is that its minimum lateral dimension should be D/10 plus a few metres to allow for slabbing at pillar edges, where D is the depth of working. The rule generally accepted by the National Coal Board in Britain for rib pillar stability is (D/10) + 15 yards, so that (D/10) + 15 m should give a stable pillar to protect either main entries or the surface. When the stress distribution across a series of pillars is examined then it would seem reasonable to retain (D/10) + 15 for the outer pillars, but relax the values to D/10 for the inner pillars. Certainly the pillar strengths measured by Barry[19] indicated that a pillar within a group is remarkably strong. However these pillars were extracted so that the long term effects are not known. Subsidence measurements indicate that smaller pillars do collapse and spread. All these sizes assume of course that the depth of working will be sufficient to stress the area and provide loads above the unconfined compressive strength of the pillar material. For first working only, where regular pillars are left, then the average load method will probably provide satisfactory results.

The minimum size of pillars for one Australian coal mine was assessed by model tests at the Australian Coal Industries Research Laboratories. A width of 50 m was obtained for a depth of working of 488 m to 518 m. This is remarkably in agreement with the D/10 value.

Multiple Seam Pillars

Pillars in subjacent (below) or superjacent (above) flat seams should be vertically above each other to transmit the loads properly. Roadways under superjacent pillars can be expected to suffer more closure, particularly floor lift from soft floors, than would roadways not under pillars[21]. This is logical because pillars transmit load downwards whereas open workings destress an area. The worst effects on roadways or faces in adjacent seams are produced by pillar corners and pillar edges. These produce sharp stress peaks in their own seam (Figures 8.21 and 8.22) which induce fractures in the strata above and below the seam. Workings in adjacent seams are affected by these fractures which commonly hade about 10-20° off vertical sloping backwards over the mined area. The hade varies with the nature of the strata, being nearer vertical in weaker rocks.

When an opening is mined across these cracks above or below a seam then roof trouble can be expected, especially if the induced cleavages cross the natural geological cleat in the coal. The same effect will be found in metalliferous deposits. Conditions are expectedly worse in overlying seams because of the effect of gravity which has displaced the strata downwards. The effect of room and pillar workings can be measured in seams over 200 m above them, and were noticeable in workings nearly 100 m below. A longwall face worked 250 m below an existing roadway[21] produced 0.4 m of vertical roadway closure on an original height of about 2 m in addition to the expected general lowering of the roadway due to subsidence. The subjacent face passed under the roadway at right angles and caused marked diagonal distortion. The general symptoms were consistent with a pressure wave travelling ahead of the longwall face in accordance with the arch theory mentioned below.

The effect of mining adjacent seams is obviously greater the closer the seams are together. The induced fractures are more persistent and closer together. Subsidence measurements mentioned in Chapter 11 show a maximum damage angle of about 14° from the vertical that is probably also related to the induced cleavage angles of 10-20°, although at depths of around 300 m there is a general flow of strata into wide openings that spreads the angle of movement out to about 35° from the vertical.

The essential point in working adjacent seams is to place pillars above or below each other. If seams are close together, say 15-25 m, then the pillars can provisionally be equal sized. However it is generally best to taper the pillars outwards by 10-20° in each successive seam worked, as in Figure 8.23. An examination of the roof breaks in the first seam worked should give a more exact angle, but 14° can be taken as a design compromise. If one has the option the top seam is worked first and then lower seams are worked in descending sequence. However the best seam is usually taken first in this commercial world and the mining engineer visited with the sins of his fathers works either upwards or downwards.

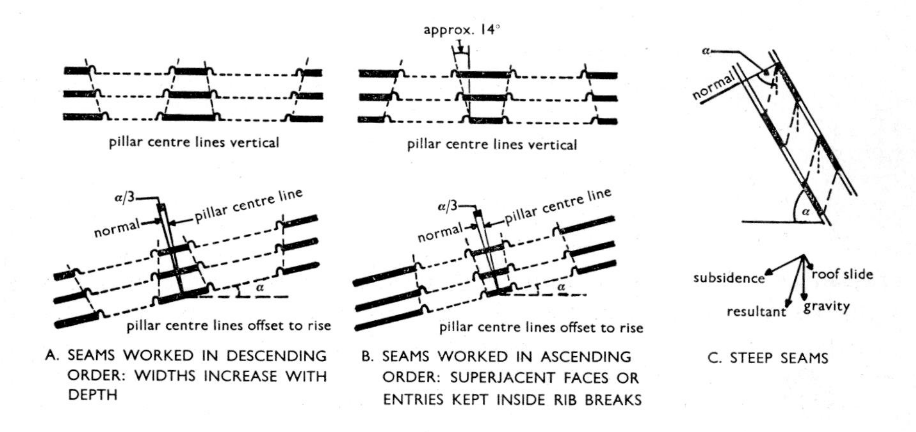

Fig. 8.23 Pillar positions in multiple seams

Opinion varies about practice in inclined seams. If the effect of gravity is ignored the pillars would have their centres on a line normal to the seam. However some concession is usually made to the tendency for roofs to slide downhill and pillars may be offset to the rise by about one third of the seam inclination for gradients up to about 1 in 3. Model tests made at the Australian Coal Industries Research Laboratories for one particular mine with two closely adjacent steep gradient seams indicated that an offset of this order was necessary. By the time the offset error is noticeable with increased seam separation then the influence of the pillars is correspondingly less.

At grades steeper than 1 in 3 the best principle is probably to revert proportionally to vertical superposition because the roof action is approaching the vertical and strata slumping alters the cleavage angles.

Stability of Wide Openings

There is a great similarity in the behaviour of rock in coal mines and in metal mines, especially if allowance is made for different rock strengths. The gold reef deposits worked by longwall methods in South Africa exhibit the same fracture patterns and closure behaviour as do longwall faces in coal mines. This must be qualified by the fact that the gold mines are working in hard quartzite at depths in excess of 2000 m, whereas coal mines are working in shales at depths generally less than 1000 m. There is a tendency to overlook similarities when mining steeply dipping orebodies, because the longwall face becomes the backs and the heavy gravitational loading of a relatively flat roof changes into squeezing and slabbing of the hanging wall. However the behavioural similarity still exists.

Theory and practice agree that when a large opening is made underground it creates a zone of relaxed pressure with zones of increased pressure around it. Attempts to calculate these peaks have not met with success and the engineer has to rely on measurements in similar situations to predict likely behaviour. Measurement and observation indicate that an excavation travelling forwards produces rock fractures of the form shown in Figure 8.24. They are particularly noticeable in longwall coal faces in soft sedimentary rocks and can be traced through workings in adjacent seams. Leeman[7] records their presence around the gold reefs in South African mines.

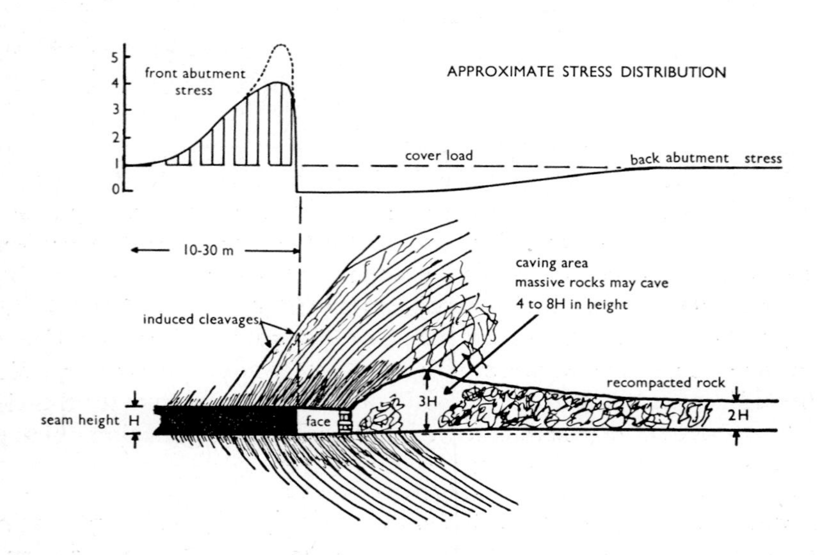

Fig. 8.24 Fracture patterns around a longwall face

It is possible to examine the edges of falls of roof, and the sides of ripped roadways, and trace the induced cleavages from mining. These cleavages follow parallel to the edges of the excavations and cut across the corners as in Figure 8.25. The most comprehensive description of them is given in a paper written in 1934[22], but they can be easily seen in most deeper mines. The spacing of the cleavages is proportional to the hardness of the rock. In shales at 500-1000 m depth it is commonly 12-50 mm. Spacings less than this occur only in weak mudstones or in zones of high stress. The fractures generally hade backwards over the opening at 5° to 25° from the vertical in conformity with Figure 8.24. Sandstones tend to have lower fracture angles than shales, and a wider spacing for the induced cleavages. Cleavage in sandstones may not be observed at less than 500-800 m in depth. In quartzite at 2700 m in South Africa the spacing is around 24-100 mm, and angles are about 15° to 20°. Evidence is not yet conclusive but it is possible that there is a relationship between the applied stress and the fracture angle: the higher the stress then the steeper the angle.

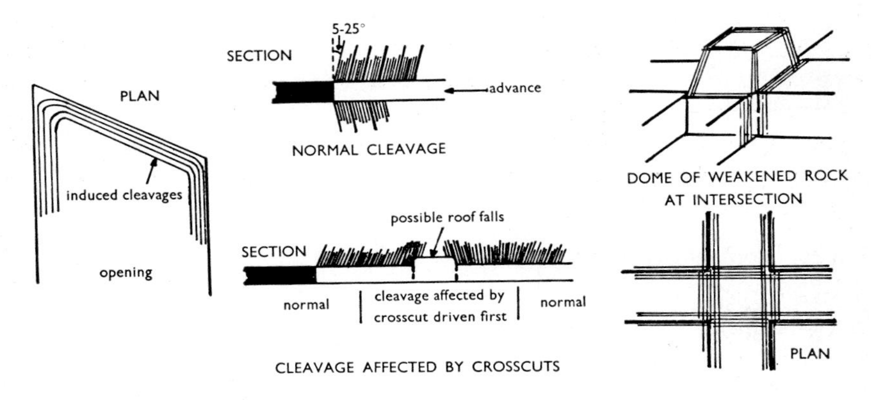

Fig. 8.25 Cleavage patterns around an excavation

This induced cleavage does not normally cause trouble but it is possible to have larger cleavages which will. If a face or a stope advances slowly with a regular increment then the roof can shear with a matching regular increment, and the induced cleavage is interspersed with more pronounced breaks which have varying degrees of settlement (the pronounced cleavages were known as *cutter breaks* in cyclic mining). These larger breaks have angles as low as 45° hading back from the face. In seams worked below old mining areas the new cutter breaks can intersect old weighting breaks sloping back from the seam above, and wedge shaped pieces of rock will fall out. Trouble will also be experienced if a longwall waste does not cave regularly. If the roof weight of an unbroken cantilever see-saws over the face supports then an induced cleavage sloping forwards over the face can be formed, and others will form sloping back from the face, as in Figure 8.26. Geological discontinuities can cause problems but early joint surveys should have obviated these. Serious trouble can be expected if strong supports such as cribs (chocks) are left in the waste. Roof weight is increased and thrown forward by the intact roof.

Although the concept of induced cleavage has been introduced in connection with wide openings the cleavage also exists in roadways, especially wide ones. They rarely give trouble, and are not particularly noticeable except at intersections, where they can help to cause falls. The edges of the fall at the lower portion are usually inclined inwards at the induced cleavage angle unless a geological discontinuity intervenes.

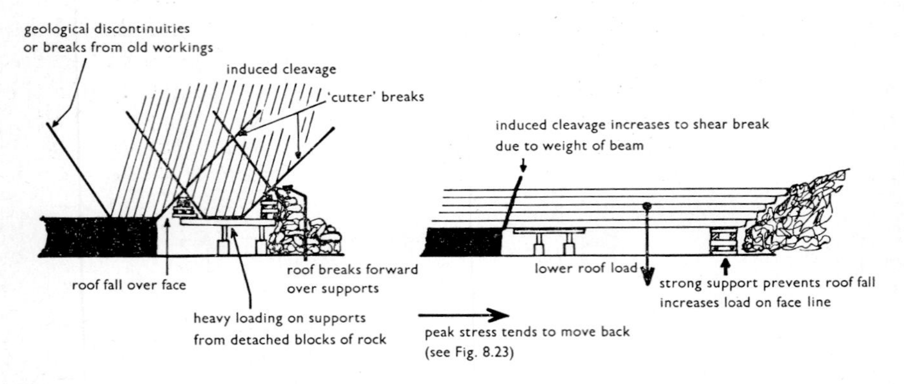

Fig. 8.26 Roof falls caused by induced cleavages

PRESSURE ARCHES

As an underground excavation is opened out it gradually forms an area of relaxed roof spanned by transverse arches. The inner surface of the arch is composed of stepped-out cantilevers held in place by friction between the beds (Figure 8.27). Eventually these cantilevers will meet to form the arch. Further widening of the excavation in both directions gives a situation in which the lower beds cave (fall) and the upper beds sag down on to them. The caving height depends on the ease of fragmentation of the roof beds, but can be about three to eight times the seam height extracted. As the extraction face moves forward, or adjacent areas are mined, the whole of the superincumbent strata sags down to apply sufficient load to re-compress the fallen strata to take the original cover load. There is then a stress pattern set up as in part 3 of the Figure 8.27 on all four sides of the excavation. One side of the excavation, the face, travels forward keeping its stress peak sharp. On the other three sides the pattern is static and as a result the stress peaks are dissipated by plastic flow of the strata.

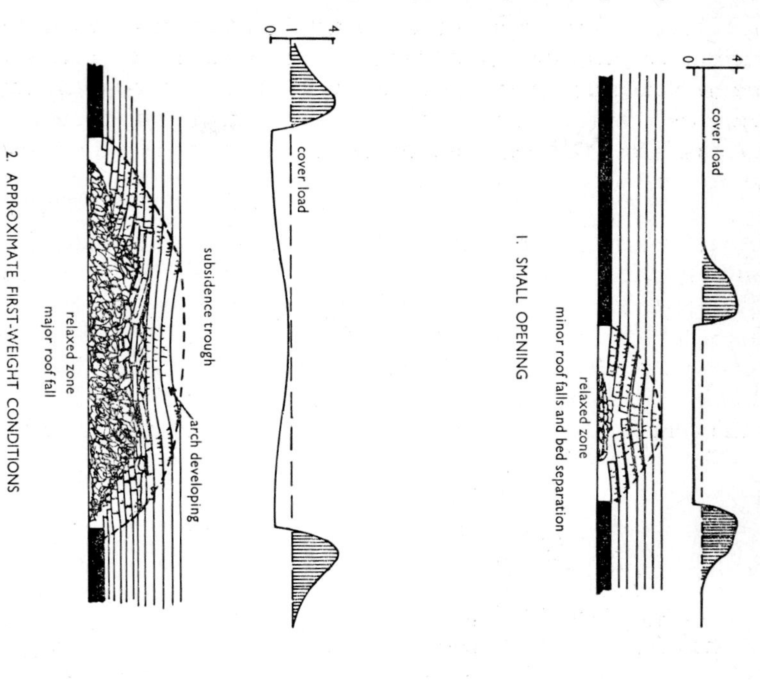

APPROXIMATE STRESS DISTRIBUTION

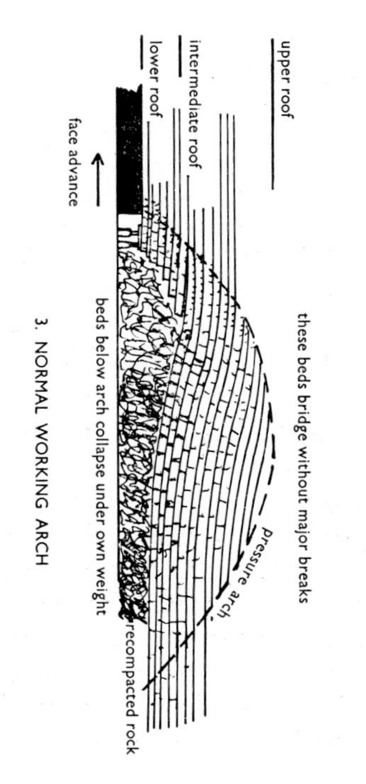

Fig. 8.27 Sequence of collapse in a mined area

Figure 8.28 shows how the stress patterns interfere around a longwall face or stope. There is one particular arrangement that gives trouble when a face is opened out. Before the first major fall of intermediate roof beds takes place there can be a condition known as the *first weight*. The pressure arch is at its maximum span with two concentrated abutments fore and aft and is carrying both its normal load of upper roof and some excess load from the rear cantilevers which have not travelled far enough to snap off. This position is that at which zones A and B^1 coincide. With slowly moving faces there can be considerable damage to roof supports, and local collapses of roof along the face line for a few metres of face advance. The condition does not persist for long and a fast moving face can travel through this period of maximum loading without much trouble. A similar condition of excess weighting can occur if a strong roof hangs up in the waste instead of breaking off.

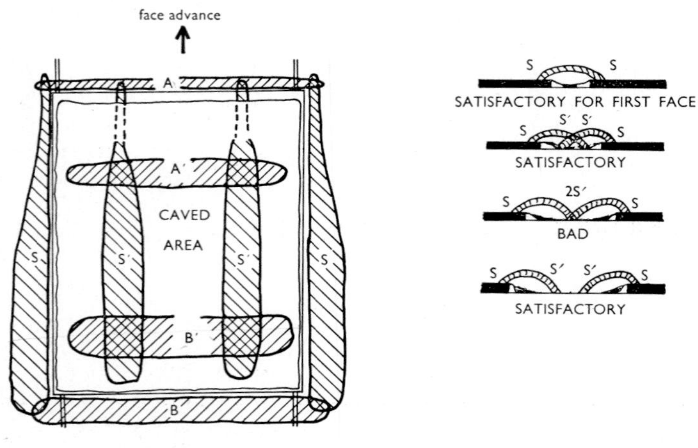

A = concentrated travelling stress zone in front abutment
B = distributed stress on rib where face was opened out
S = stress on side ribs redistributed in proportion to time and face advance
A′ = distributed stress of travelling rear abutment
B′ = distributed stress on caved waste after first weight collapses
S′ = stress on caved waste redistributed in proportion to time and face advance

Fig. 8.28 Interaction of stress patterns around an excavation

When the normal pattern of stress as in Figures 8.27 and 8.28 is set up, then there is only one concentrated stress abutment just in front of the face and the back abutment is a distributed load. Earlier investigators reported loads in excess of the cover load in back abutment areas but their methods are open to criticism because the load transducers were installed under, or close to, pack walls. Measurement of vertical stresses more recently (e.g. Thomas[6]) have shown only a slow rise to the cover load in European coal mines with longwall faces. This condition is probably true for all mined areas, although no investigator has specifically tried to measure stresses at the intersections of the high pressure zones shown in Figure 8.27.

Further points to note are that all of the area bounded by A^1, B^1, S^1, and S^1, and probably including those zones after a very short time, is relatively evenly stressed at cover load pressure. The zones A and B have significance in that they both cause heavy damage to roadways which pass through them, or through which they pass. The zone B would originally be more concentrated but disperses with time. Observations by the author on a road behind a longwall face, and running 5.5 m from the side rib (i.e. parallel to zone S but not in it), showed deformations as in Figure 8.29. The seam involved was 1.1 m thick and worked at a depth of 496 m. The two heavily crushed zones show the position of first weighting and thus the probable width of the pressure arch. Another seam had previously been worked 13 m above. The undamaged portion of the roadway lay inside the corner of the excavation which had never had a high pressure zone imposed on it.

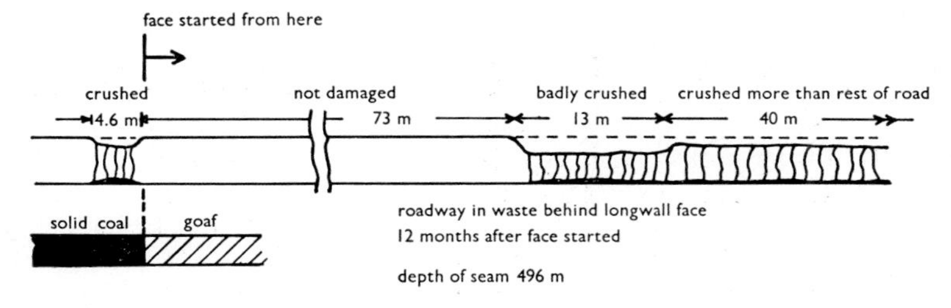

Fig. 8.29 Pressure arch damage to roadway

When the lateral arches are considered care must be taken in choosing a width for the working place. If a width of working is chosen less than the pressure arch (the actual figures are discussed below) then the zones S^1 both disappear and the whole of the working is spanned by an arch resting on solid ribs, S-S. However if an adjacent face is opened there may then either be a double loading 2S on the rib pillar, or an abutment of type S^1 may try to rest on the new workings. Hence either substantial rib pillars must be left or near-total extraction is advisable. In wide workings the effects of S^1 may be noticed near the face line: if support pressures are measured then peak loads may be detected at pressure arch width. Figure 8.28 is intended to show a wide working with adjacent pairs of S and S^1; if the working narrows then first the zones S^1 become coincident which can cause bad roof conditions at the face. After crossing each other trouble will not usually be found until the zones S^1 become nearly coincident with zones S on the opposite ribside.

The maximum stresses measured in these abutment zones are usually of the order of 3-5 times the cover load, although most investigators have been measuring loads in the front abutment. It is possible that a rough check on the magnitude of the stress can be made by observing the depth of working at which induced cleavage first appears. For instance if a sandstone with a compressive strength of 69 MPa first starts to break regularly at 620-650 m, where the cover load is about 16 MPa (at 25 kPa per metre), then the peak stress must be 4-5 times the cover load.

Widths of Pressure Arch

The most consistent attempts to determine the width of pressure arches has been made in the north of England[23]. The arch is usually determined by looking for the point where maximum damage appears in a roadway following an advancing face. The formula generally applied in the Durham and Northern divisions was

$$\text{width in yards} = \frac{\text{depth in feet}}{20} + 20$$

with a little rounding off this becomes

$$\text{width of arch in metres} = \frac{3}{20}\,(\text{depth of seam in metres}) + 18$$

Typical arches are then:

Depth of Seam m	Width of Arch m	Depth of Seam ft	Width of Arch ft
200	48	400	120
400	78	800	180
600	108	1200	240
800	138	1600	300
1000	168	2000	360

The recommendation of the committee was that stalls should be limited to three-quarters of the width of the pressure arch for stability, and that pillar width should be equal to stall width. However these stalls would have been stabilised with stone packs, whereas present practice requires caved areas because packing is too expensive. To be certain of caving an area it should presumably be wider than the pressure arch by a working margin, so that perhaps the minimum width of a caved area could be one and a quarter times the width of the pressure arch.

These arches were also primarily in carboniferous shales and would presumably be wider in stronger rocks such as conglomerate or quartzites.

Ground Stability in Metal Mines

When a stope is to be block caved it is necessary to open a critical area which will cave under its own weight. This area must contain a pressure arch which will collapse in the backs under the weight of the ore and keep on collapsing with the aid of cut-off raises. The fact that the cave area is greater than the pressure arch means that any roadway close under the cave may be subjected to high stresses because the natural arching action of a roof has been deliberately destroyed. If the basement rock is strong, and the finger raises and draw points are sufficiently long, then the haulage drifts may have a sufficient cover for protection (see Figure 6.28).

Fletcher[24] reports on rock behaviour in block caving and similarities with coal mining in bedded deposits can be seen. The boundaries of the block caves tended to slope inwards at 6° from the vertical; this can be compared to the induced cleavage. Pillars between block caves became fractured and presented poor working conditions when they were recovered. High stress zones are found around the edge of a cave, but when caves are filled with waste rock, and this has consolidated, then the stress in the adjacent rock is found to have been relieved.

The whole concept of pressure arches and pillar failures also applies to steeply bedded metalliferous deposits. Considering induced cleavages and abutment stress peaks of 4 to 5 times the cover load, and knowing the compressive strength of the rocks involved, one can predict the depth at which strata control problems will be experienced. It is noticeable that the hard South African quartzites with a strength of 140-200 MPa did not really start giving trouble until mined past 1500-1800 m. In a conglomerate copper mine in Michigan, with a medium grained igneous rock as a roof, rockbursts and other strata problems forced a change in mining methods at a depth of about 1000 m.

The United States Bureau of Mines has studied and measured closures in stopes, both open and sand filled.[25,26] Both these references refer to the Star Mine in the Coeur d'Alene silver/lead/zinc district of Idaho. The mine is working a 3 m thick steeply dipping vein in hard brittle quartzite at depths of around 1800 m. When measurements were taken in a sandfilled stope[25] it was found that lateral closure of about 0.5 m on the 3 m width (17-18 per cent) took place at a lessening rate over a period of two years. Pressure measurements indicated that the sandfill was bridging across at a height of about three times the stope width. This figure has also been determined experimentally on small scale models (unpublished work at Sydney University) but extrapolation to wider stopes would not be warranted. It is interesting to compare the closure figures to those for sandfill in horizontal coal seams where closures of 10-20 per cent of extracted seam height would be expected, i.e. movements of the same order.

The second report on the Star Mine[26], gives details of measurements taken in laterals and crosscuts while a crown pillar of a sand-filled stope was being extracted (Figure 8.30). When the depth of the crown pillar was reduced to 26 m then the movements observed in the adjacent roadways increased rapidly. When the pillar was only 7.6 m deep then lateral movements ceased. It was considered that at this stage the pillar remnant had crushed completely and that the sand fill had accepted the full load. However, roof falls now occurred in the laterals and cross cuts because there was no sideways constraint to keep joints and induced cleavages closed. When movements were assessed it was noted that at 18 m away from the crown pillar they were negligible, at 15 m there were noticeable movements, and at 6 m distance there was about 0.5 m of movement in towards the stope.

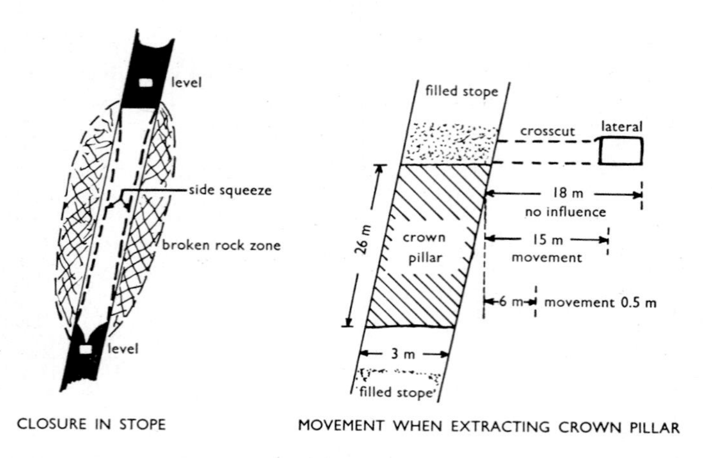

Fig. 8.30 Movements in a sandfilled stope

Although both these reports deal with sandfilled stopes other measurements and practical experience show them to be typical of all stopes. The walls of a stope will eventually collapse in broken fragments, or it is possible for a narrow stope to squeeze in until the walls meet; especially if the walls had been rock-bolted to prevent slabbing. In areas where high grade ore has been open stoped along a line of lode (e.g. at Kalgoorlie in the West Australia gold mining area) it is only possible to recover the lower grade ore using sand fill methods for stability. Earth tremors occur regularly as old stopes collapse and squeeze in together because new mining disturbs the pillars between them.

Protection of Roadways

Protection of pillar entries has been dealt with to some extent in Chapter 7, but there are some general comments to be made. When large openings are to be extracted underground then main roadways should be driven a safe distance away. The distance should increase with depth, and allowances should be made for faults or for major cleavages. Main entry pillars in coal mines are of the order of 90 m.

In metal mines major roadways are generally situated in the footwall, because the orebody must be mined, and the hanging wall will be affected by subsidence. In hard competent rocks the main drives should be at least 20-25 m into the hanging wall. In a soft rock with a pronounced joint system it may be necessary to either increase this distance or to set strong supports in the roadway. The cost of extra length of crosscuts must be balanced against the cost of extra supports used.

If mining operations involve working below underground roadways then either the roadway must be virtually abandoned or pillars must be left beneath the roadway. The layout is substantially as in Figure 8.23 but with the roadway or roadways at successive levels running through the centres of the pillars. The rules for pillar stability at each level must be followed: if there is to be only one pillar with a roadway through the centre it must be remembered that the pillar is in fact then two pillars, because the roadway and its adjacent zones of broken rock reduce the stability of the area.

If roadways have to pass through abutment zones, or through faulted areas, then it is wise to anticipate trouble. The roadways should be supported with flexible or yielding supports to allow for deformation. The alternative is to use cheap expendable supports and to make provision for a repair program.

Rockbursts

Rockbursts, as distinct from gas outbursts, are usually a result of mining in such a way that large stresses are set up in remnants of rock or mineral, or in a way that produces high stresses on a front abutment. It is also necessary to have a hard brittle rock, because a soft rock deforms easily under pressure and relieves the stresses produced in mining.

Gas outbursts tend to be associated with hard brittle coals and geologically disturbed areas, such as the anthracite area of West Wales in Britain, or with strong roofs and igneous intrusions as in localised areas of the southern coalfields of New South Wales. Metropolitan colliery has been liable to outbursts of mixed carbon dioxide and methane, usually in association with weathered dolerite sills. The colliery currently prevents outbursts by augering large diameter holes in advance of first working in areas liable to outbursts.

The presence of hard intrusive igneous rock sills has often been associated with rockbursts at shallow depths, e.g. that at Coalbrook North mentioned earlier as a multiple pillar collapse. There are probably two reasons involved: the rock sills are usually medium grained, tough rocks with an absence of regular jointing. Consequently they will not cave readily and can span several hectares of ground before collapsing. Secondly the intrusions themselves may be associated with large scale tectonic disturbances. Hard massive conglomerate roofs and sandstone roofs have also been associated with rockbursts in bedded mineral deposits. These roofs do not sag or shear off, thus relieving the load. They stay intact over large areas and as more and more mineral is extracted then the load on remnant pillars, or on abutment zones, increases past the safe compressive strength of the mineral or rock. Finally a geological weakness or a heavy round of shotfiring will trigger off a roof fall or pillar collapse. This can happen in a mined out area without major destruction except that caused by an *air-blast*. An air-blast is produced by the falling roof acting as a piston to compress air. The high pressure air escapes through adjacent workings with explosive violence and can displace heavy machinery and roof supports.

The true rock outburst is more likely to destroy areas of working by collapse of rock, without major projection of material, except by air-blast. Where gas outbursts occur they can often project tonnes of finely powdered coal for several metres down a roadway, leaving large cavities within the coal seam, and do not necessarily cause falls of roof. Hence gas outbursts are a coal mining problem whereas rockbursts are a general problem.

Advance signs of a rockburst are usually increasing deformation of roadway supports and excessive spalling of pillar edges, or slabbing from the walls of metal mine drives. The rock itself can usually be heard to be cracking and breaking. Geoseismic phones have been used to monitor the frequency of rock breaks and to determine their loci. They have shown that peaks of seismic activity occur at the time of outbursts. Unfortunately with current techniques geoseismic monitoring cannot give sufficient advance warning of an outburst to be of use. It has been used to show that mining techniques which minimise rockbursts have produced a lower incidence of seismic activity.

At medium depths rockbursts are avoided in pillar mining by creating large pillars in first working and then adopting complete extraction on as wide a front as possible. It is preferable to use a straight extraction line with a strong waste-edge support to break the roof. Where strong hydraulic supports or timber cribs are not available, or for adjacent longwall faces (Figure 6.33), it may be necessary to use a stepped extraction line. A stepped line should always have obtuse or right-angled corners, never acute-angled openings in the mineral. An acute-angled opening raises the stress concentration locally and can act as a focus for a rockburst; similarly an acute angled pillar corner is also liable to sudden collapse. The report listed previously[19] showed that straight line pillar extraction with fast-moving continuous miners and hydraulic supports is much less prone to rockbursts than is stepped pillar extraction. In burst-prone areas all development openings should be as small as possible and pillars as large as possible.

At greater depths than about 1500-2000 m in hard rocks pillar mining methods can be very dangerous. Mines then turn to longwall methods as in Chapter 6 for mining vein-type deposits. In general pillars must not be left because if they burst they can trigger off a chain reaction of collapses. If the pillar was fairly large, and collapsed, it would transfer extra stress on to the abutments of an excavation and cause a burst there.

The deep gold workings of South Africa typify modern methods of working at depth. They use stepped longwall methods to work the 30° inclination reefs, with obtuse angles between the longwalls. Main drives are protected by substantial pillars or are put in the footwall below the reefs. The waste edge is protected by rows of substantial cribs to allow the roof to lower under control and cave off. The hard quartzites do not fracture easily and normal working precautions usually include inducer shotfiring. Shotholes are drilled in advance of the workings and are fired in between shifts, with all personnel withdrawn from the mine. Relatively large charges of explosives are used and the shock wave from the explosives will then trigger off any incipient outburst. In any case the zone of broken rock around the excavation is spread further forward by the explosion and the front abutment is both pushed forward and distributed over a wider area.

Deep Workings

The Kolar goldfields of India and the Rand (Witwatersrand) goldfields of South Africa are working at depths of around 3000 m. The methods used are longwall mining. A special committee for the Kolar goldfield made the following recommendations to avoid rockbursts, quoted by Taylor[27] who also gives details of research and mining practices.

Development: drives and crosscuts liable to rockbursts to be steel setted and lagged; holings to be made from larger excavations to smaller, and not vice versa; when approaching a fault, fissure, dyke or drive at an oblique angle, development to be turned so that the approach is at right angles.

Shafts: deep level shafts to be sunk in footwall (preferably) or hanging wall; deep level shafts to be circular or elliptical, and concrete-lined; shafts to be sited either so that stoping commences opposite the shaft and advances away from it, or well away from the ore-shoot.

Communication between shaft and reef; crosscuts to be suitably supported, especially at the junction with the reef drive; where necessary and practicable footwall drives to be provided within the destressed zone behind the stoping.

Stopes: no pillars or remnants to be left within an oreshoot; stope preparation by stripping either the back or the bottom of drives to be kept to a minimum; winzes to be kept to the minimum number for adequate ventilation; with parallel reefs, one to be stoped ahead of the other. With flatter dips the hanging wall reef to be taken first.

Stoping sequence: stope faces to conform to a longwall face as far as practicable; advance to be as rapid as possible.

Support and wall closure: permanent support (granite packwall) to be installed as rapidly as possible to resist initial closure of the walls.

Remnants and pillars: a long narrow pillar to be stoped from one of the shorter sides to work to a safer shape; stoping never to be initiated from a rise or winze through a pillar; a number of pillars to be mined in sequence; pillars to be stoped from the top downwards. (N.B. These are steeply dipping reefs.)

Dykes and faults; stoping to commence at the dyke or fault, and progress away from it.

These recommendations have been successful in reducing the incidence of rockbursts, and in any case are consistent with the best mining practice which should be followed at all depths of working.

Working Bad Ground

The procedures for coal mining and other bedded deposits were described in Chapter 7 but a brief summary is

1. progressive change from pillar mining to longwall mining
2. progressively reduce the width and arch the roof of development openings
3. reduce the number of development openings, increase their spacing and increase the strength of their supports (rock bolts→ timber sets→steel sets→steel arches)
4. introduce stronger working area supports, e.g. large hydraulic chocks.

In steeply dipping metalliferous deposits the orebody itself forms the backs (roof) and joints are frequently unfavourably orientated in the ore. The procedures are basically the same as for coal mining and are set out below. It is assumed that the usual transition will be made from open stoping down to filled square sets if necessary and economic. However in low grade ores that will not block cave, then mechanised cut and fill is fairly essential. To operate cut and fill in as weak a ground as possible a general sequence of operations should be followed.

There can be conflicts however between some of the requirements which must therefore not be followed blindly.

1. Orientate the stope backs parallel to the trend of the major joints (Figure 8.31). This gives a smooth back without wedges of rock which can drop out, and the backs are easier to bolt. This can conflict with the desirable practice of keeping the backs normal to the hanging and footwall, to give maximum support of the hanging wall.

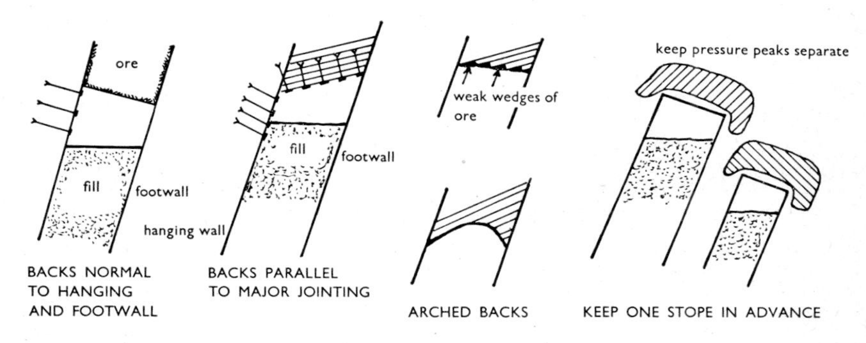

Fig. 8.31 Working bad ground

2. Arch the backs to remove the volume of ore which tends to sag causing joints to open, and blocks to drop out. The height of the arch is about 0.1 to 0.2 of the width of the stope; e.g. a 15 m wide stope would be arched from 1.5 to 3 m.
3. Mine closely adjacent stopes so that one stope is in advance of the other. This will keep the peak abutment stress due to the first stope separate from the peak stress of the second stope. The rib pillar between the stopes will then be more stable. The general precept of working the upper orebody first should be followed.
4. Introduce systematic support of the hanging wall with rock bolting and steel mesh if necessary.
5. Reduce the height of the hanging wall by changing from up-holes to flat-backing.
6. On adopting flat-backing use perimeter blasting techniques (smooth wall) with weaker explosives (such as Exactex) to reduce the effects of explosive shattering on the walls and backs.
7. Rock bolt isolated bad patches of the backs. An intrascope (periscope device for looking at rock in boreholes) can be used to check on cracks to determine the correct bolt lengths.

8. Introduce systematic bolting of the backs with sag tests on loose bolts to find the correct bolt length. In conjunction with this it may be desirable to reduce the plan area of the stope so that the backs are mined more frequently. This will prevent cracks from spreading through the ore and weakening it further.
9. Cribs (pigsties) can be introduced behind the working area to effectively narrow the plan area of the stope.
10. By this time strata control precautions are becoming expensive and it can be convenient to increase the cut-off grade to work a smaller stope. It may be possible to cave the low grade ore later after it has been broken by the first working of the stope.
11. The orebody can be mined by small transverse stopes, or E-shaped stopes, instead of a longitudinal stope.
12. If the rock is too weak to work by mechanised cut and fill in small stopes, and too low grade for square sets, serious thought would be given to block caving if the orebody were large enough to recover the development costs.

Recovery of Falls

The main knowledge in this field can only be gained from experience but there are some general points to be noted. The first and most important is that the edges of the fall must be supported immediately with strong supports between the roof and floor. This is essential both to prevent the fall spreading and to safeguard people working in the area. Thought is then given to the type of fall and the action to be taken.

If no machinery or personnel are buried, and it is a roadway that is blocked, then the cheapest alternative, particularly in coal mines using continuous miners, will probably be to drive a new road around the fall. If it is a conveyor roadway however then it may be necessary to drive through the fall. If it is a stope that has fallen in then the fall will probably be of ore from the backs, or contain a large amount of ore. The usual process is to erect temporary legs or posts between the fall and backs, and then cross-brace and rock-bolt the interior dome of the fall. When this surface is secure then the fall is loaded out and mining can continue. A similar process can be applied in bedded deposits where it is necessary to clear a relatively small fall that has secure edges. The roof is bolted and the fall is cleared. For long term security in weak rocks the line of the roadway is protected by steel sets with strong lagging and a cushion of broken rock to protect the roadway from any future falls. In gassy mines the cavity must either be efficiently ventilated or completely filled. Lightweight concrete using vermiculite as a filler is an attractive proposition for this. It has an additional advantage that vermiculite concrete is compressible and will deform as mining operations continue, instead of crushing the roadway.

If a drawpoint is to be recovered, or for some special applications, it may be economical to pump the fall full of quick-setting cement or concrete to stabilise the broken ground. The area can then be driven through with air picks, or by using small explosive charges fired singly to break large blocks.

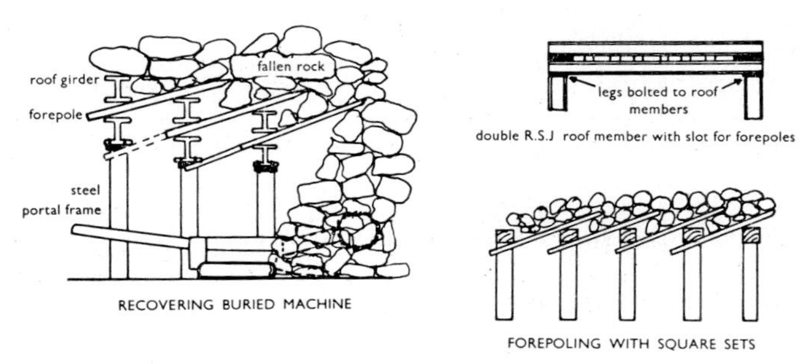

Fig. 8.32 Forepoling through a fall

Where machinery is buried in a caved waste, or under large falls, the roadway is driven through by forepoling. Special steel portals may be prefabricated and erected underground as in Figure 8.32, to recover a large machine. For a roadway wooden square sets or steel arches are used and in both cases forepoles (or spiles, or piles) of steel or wood are hammered in through the broken rock to pin it while enough is loaded out to erect another set.

References

1. Obert, L. and Duvall, W. I. *Rock Mechanics and the Design of Structures in Rock.* Wiley, New York, 1967.
2. Jaeger, J. C. and Cook, N. G. W. *Fundamentals of Rock Mechanics.* Methuen, London, 1969.
3. Thomas, L. J. *An interim assessment of strain measurements in concrete-lined shafts and insets at Wolstanton Colliery.* Int. J. Rock Mech. and Mining Science, **1**, p. 547, September 1964.
4. Hobbs, D. W. *A study of the behaviour of broken rock under triaxial compression and its application to mine roadways.* Int. J. Rock Mech. and Min. Science, **3**, p. 11, 1966.
5. Hobbs, D. W. *The behaviour of broken rock under triaxial compression.* Int. J. Rock Mech. and Min. Science, **7**, p. 125, 1970.
6. Thomas, L. J. *Pack pressure measurements.* Colliery Guardian, **218**, p. 503, October 1970.

7. LEEMAN, E. R. *Some underground observations relating to the extent of the fracture zone around excavations in some Central Rand Mines.* Papers. Assoc. Mine Mgrs S. Africa, p. 357, 1958-9.
8. SHEPHERD, R., and WILSON, A. H. *The measurement of strain in concrete shaft and roadway linings.* Trans. Inst. Mining Engineers, **119**, p. 561, 1959-60.
9. TROLLOPE, D. H. *The stability of deep circular shafts in hard rock.* Proc. 2nd Congress Int. Soc. Rock Mechanics, Belgrade 1970.
10. RADMANOVICH, M. and FRIDAY, R. G. *A preliminary rock mechanics study of the 31 (4,270 ft) level winder excavation at No. 3 shaft, North Broken Hill Limited.* Aust. I.M.M. Monograph No. 3, 'Broken Hill Mines 1968', p. 27. Aust. I.M.M., Melbourne, 1968.
11. ACKHURST, A. W. *Rock mechanics applications in the design and excavation of No. 2 crusher station at New Broken Hill Consolidated Limited.* Aust. I.M.M. Monograph No. 3, 'Broken Hill Mines 1968', p. 31. Aust. I.M.M., Melbourne, 1968.
12. HOBBS, D. W. *Scale model studies of strata movement around mine roadways: IV, Roadway shape and size.* Int. J. Rock Mech. Min. Sci., **6**, p. 365, 1969.
13. ALEXANDER, L. G. and HOSKING, A. D. *Principles of rock bolting; formation of a support medium.* Paper 1, Symposium on Rock Bolting. Illawarra Branch, Aust. I.M.M., Wollongong, 1971.
14. COATES, D. F. and COCHRANE, T. S. *Development of design specification for rock bolting from research in Canadian mines.* Research Report R224, Mines Branch, Ottawa, 1970.
15. PARKER, J. *Mining in a lateral stress field at White Pine.* Canadian Mining and Metallurgical (CIM) Bull. p. 1189, Oct. 1966. (see also U.S. Bureau of Mines Reports of Investigations, R.I. 5746 (1961) and 6372 (1964)).
16. BRYAN, SIR A., BRYAN, J. G. and FOUCHE, J. *Some problems of strata control and support in pillar workings.* The Mining Engineer, No. 41, p. 238, Feb. 1964.
17. SALAMON, D. G. and MUNRO, A. H. *A study of the strength of coal pillars.* J. S. African Inst. Min. Metallurgy, **68**, p. 55, 1967.
18. BIENIAWSKI, Z. T. *The effect of specimen size on the compressive strength of coal.* Int. J. Rock Mech. Min. Sci., **5**, p. 325, 1968.
19. BARRY, A. J., ZONA, A., GILLEY, J. L. and OITTO, R. H. *Investigations of stress distributions in burst-prone coal pillars.* U.S. Bureau of Mines Report of Investigations, R.I. 6971, 1967.
20. ORCHARD, R. J. and ALLEN, W. S. *Longwall partial extraction systems.* The Mining Engineer, No. 117, p. 523, June 1970.
21. THOMAS, L. J. *Effects of adjacent seams and methods of working in the Main Bright seam at Hucknall Colliery.* Colliery Guardian, **218**, p. 399, August 1970.
22. FAULKNER, R. and PHILLIPS, D. W. *Cleavage induced by mining.* Trans. I. Min. E., **LXXXIX***(89)*, p. 264, 1934-5.
23. DIVISIONAL STRATA CONTROL RESEARCH COMMITTEE (N.C.B. Durham & Northern Divisions). *Memorandum on the design of mine-workings to secure effective strata control.* Trans. I. Min. E., **110**, p. 252, January 1951.
24. FLETCHER, J. B. *Ground movement and subsidence from block caving at Miami Mine.* Trans. A.I.M.E., **217**, p. 413, 1960.
25. CORSON, D. R. and WAYMENT, W. R. *Load-displacement measurement in a backfilled stope of a deep vein mine.* U.S. Bureau of Mines Report of Investigations, R.I. 7038, 1967.
26. WADDELL, G. G. *In-situ measurement of rock deformation in a vein-type deep mine.* U.S. Bureau of Mines Report of Investigations, R.I. 6747, 1966.
27. TAYLOR, J. T. M. *Research on ground control and rockbursts on the Kolar Gold Field, India.* Trans. I.M.M., **72**, p. 317, 1962-3.

Chapter 9. Rock Mechanics Programs at Mines

Introduction

The scale to which a rock mechanics program is taken depends on the amount of money available. Rock mechanics has always been applied to some extent under the name of strata control, and the mining engineer learns to cope with the conditions that present themselves. However this can be an expensive approach in the long run, and a knowledge of what to expect can help in proper mine design.

Someone (probably the director of a research establishment) uttered the truism that if one waits until research is necessary before starting it, then it is too late to start. The time lag in research is considerable and very few mines are large enough to afford a research program. Most research is either paid for by governments, by nationalised industries, or by consortia of firms. A few mines, such as Mount Isa[1], manage to carry out some basic research in conjunction with a routine program. This chapter will concentrate mainly on what a large mine could consider as a routine program of investigation and forward planning.

Ideally the rock mechanics program should start with the exploration of the future mine site. The drilling records of each hole should be examined for evidence of faults, planes of weakness, cavities, soft formations, clay beds, etc. Geophysical surveys should be examined where appropriate for evidence of faults and displacements. It is suggested that before drill cores are split by the driller or geologist they should be cleaned off and photographed with colour film. Also it would be useful to carry out strength tests on the cores before analysis. Some tests are non-destructive, and in any case a broken core can be chemically analysed, but a core pulverised for analysis cannot be reconstituted for strength tests.

When an exploration shaft is sunk it should be examined for rock mechanics information as well as for ore samples. It may also be necessary to drill holes specially for rock mechanics purposes. These should be cored holes of as large a diameter as the company can afford so that joint orientation can be examined as well as mechanical strengths. Borehole core orientation is still difficult to carry out, but it should be attempted to assess whether directional mining is advisable. Fortunately for coal mining cleat directions are stable over wide areas and information can be obtained from adjacent mines.

The easiest application of rock mechanics is where a change in working practice, or extension of operations, is being planned in an existing mine. In that case access to current operations will allow a complete assessment of laboratory and field properties of the rocks including physical strength, elastic moduli, absolute stress measurements, stress change due to mining, ground movement monitoring, microfracturing, and geological discontinuities. Data can also be obtained for model tests.

Where a new mine is to be designed two approaches are used concurrently. One is standard practice based on previous experience, the other is to use design principles based on the known properties of rock and soil, and the anticipated stresses. The design principles are easiest to apply in shallow workings and in small open pits. On a larger scale they are likely to be over-ridden by the effect of discontinuities in the rock and by the interaction of mining stresses. However as more quantitative information is obtained, and more scientifically documented case studies are available, then engineering design can be improved.

Rock mechanics, as distinct from the old art of strata control, is a relatively new science and many operators are still expending considerable effort on instrumentation. Photoelastic studies have been tried but elasticity in two dimensions is not particularly helpful, and three-dimensional photoelasticity has so many practical difficulties of stress-freezing, and slicing models without disturbing the stress pattern, that it cannot be considered as other than an esoteric art. The current interest is in the field of computer-calculated finite element stress pattern design. This at present produces similar information to a photoelastic design but is more precise, and variation of parameters and designs is easier and quicker. The technique is still in a formative stage in its application to mining design and is beyond the scope of this book. The principle disadvantage of most design based on elasticity is that the mining engineer cannot afford to work within elastic limits. The mining excavations are either deliberately collapsed, or strained beyond elastic limits, to provide economic mineral extraction. Engineers at the Mining Research and Development Establishment of the National Coal Board, U.K. have recently developed nomograms for stable pillar size in relation to depth for one group of mines, and this approach is of far more use to the smaller mining companies than a long testing program of rock properties.

Rock Joint and Fracture Studies

Structural defects have a major influence on surface and underground mining. Faults and joints often limit the size and control the shape of excavations, whether in coal or metal mines. A mine geologist should be trained in engineering geology, as should a mining engineer, so that maximum information can be obtained and used. For over a hundred years the importance of coal cleat has been known to coal miners, but the logical extension of this knowledge into roof joint or open pit wall joint effects, has been much slower of recognition.

The study of joints and fractures falls into two parts, a general assessment of major structural characteristics such as faults, and a detailed study of the directions and relative magnitude of the pre-mining stress field and of rock joints to determine whether directional mining is necessary. Faults are easy to recognise, and have obvious directional weaknesses as they will move along their own shear planes. Joints present more problems but the techniques are well documented.

GRAPHICAL REPRESENTATION OF JOINTS

Before joints are described it is necessary to agree on the definitions involved. The accepted conventions are

Strike: direction of the line of intersection of the joint plane with a horizontal plane.

Dip: angle between the joint plane and a horizontal plane, measured in the vertical plane at right angles to the strike.
To describe the orientation of a plane it is necessary to specify. (i) strike angle (from North), (ii) sense of dip, (iii) dip angle, e.g. 060 SE 30

or (i) dip direction (at right angles to strike and including sense of dip) (ii) dip, e.g. 150/30 is the same plane as above.

Trend: of a line is strike of a vertical plane containing the line.

Plunge: of a line is the angle between the line and a horizontal plane, measured in the vertical plane containing the line. The sense is in the downward direction.

To assess the effect of joints they must be summarised onto a diagram and to do this they are projected onto the surface of a sphere, which is then shown in two dimensions. The presentation of the sphere cannot be perfect, one can either maintain the correct angle (orthographic) or the correct area, or the correct shape (stereographic). The two commonest projections are the equi-angular, or Wulff net; and the Lambert equal-area net, known as the Schmidt net (Figure 9.1) in connection with rock mechanics. The method of plotting planes and direction lines is the same for both projections but only the equal area projection will be described. This is the one most favoured for joint surveys because it allows the poles of joint planes to be contoured to give a frequency distribution. The Schmidt net is constructed from the formula

$$d = \frac{R \sin (\theta/2)}{0.707}$$

where

d is the distance from the centre along the equator of the projected point in question.

R is the radius of the reference sphere, and

θ is the angle of dip of the point in question.

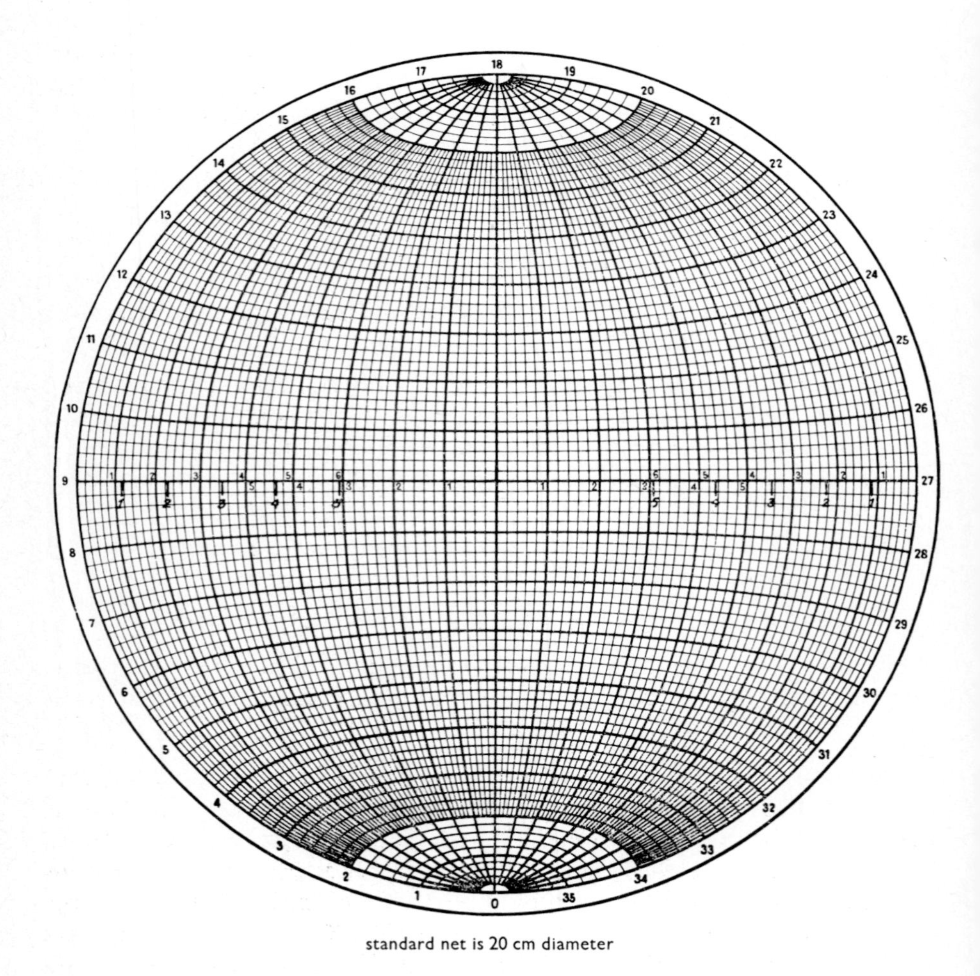

Fig. 9.1 Schmidt net for joint studies

Lower hemisphere projection is conventionally used for geological studies. This means that points and planes are plotted as in Figure 9.2A. The observer views from the zenith and points on the opposite surface of the sphere are all represented as an intersection on the equatorial plane of rays from the point to the zenith. Figure 9.2A shows a north-south striking plane dipping 60° to the east in the lower hemisphere. Its projection onto the equatorial plane of the sphere is NDS, which is shown on the Schmidt net in Figure 9.2B. Each joint plane has a pole axis normal to it. Since this is a sphere the location of the single point pole on its surface uniquely represents the joint plane. The pole **P** can also be projected as a point **P^1** onto the equatorial plane, and **P^1** will be 90° away from **D** on the equatorial scale.

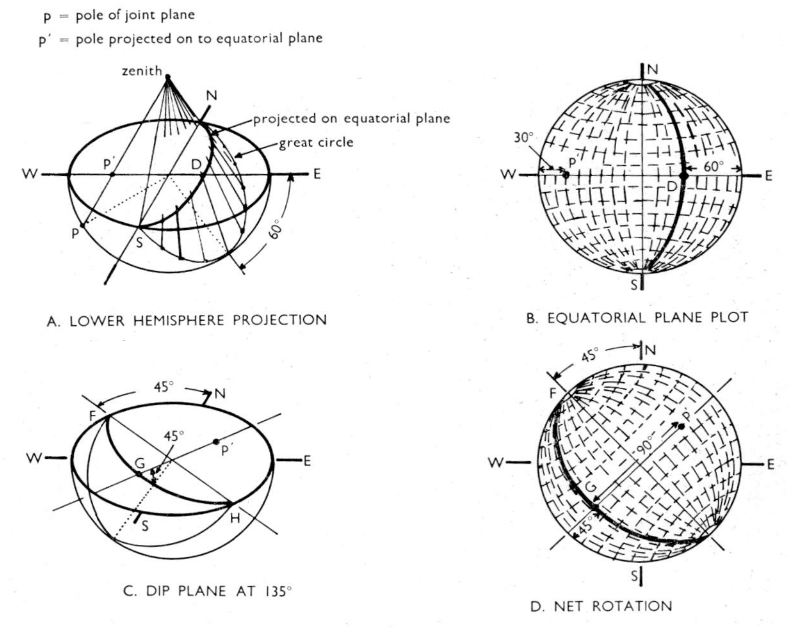

Fig. 9.2 Technique for plotting joint surveys

The general method of plotting a plane other than north-south consists of placing a transparent overlay (tracing paper) over a printed Schmidt net 20 cm in diameter, and pinning it to the centre of the circle. The overlay is marked with north-south reference points. Figure 9.2C shows a plane striking N135°E and dipping 45° to the south-west along with its projection FGH. The method of plotting it onto the net is shown in Figure 9.2D. The true dip circles are only marked onto the Schmidt net along the east-west axis, so this axis must be rotated to lie along the full dip of the joint plane which is perpendicular to the strike. In practice it is easier to rotate the tracing paper than the printed net, which is underneath. Therefore the tracing paper is marked off on the periphery of the net at a bearing of 135° (see Figure 9.1) which is in the position of F on Figure 9.2D. The mark is rotated to lie above the north point of the net, and then the projection FGH is drawn through the points North, 45° on the west side of the EW axis, and South. The projected pole P^1 then lies at 90° from G along the EW axis. The overlay is rotated back until its north-south marks coincide with the net, and then it is ready for the next joint plane to be plotted.

In most cases, to avoid confusion from many intersecting arcs, only the poles of the joint planes are plotted onto the paper. Obviously if a group of joint planes or fracture planes have similar strikes and dips their poles will form a cluster of points on the equal area diagram. This will indicate recurrent planes of weakness in the rock body and show a preferred orientation of drivages to avoid trouble from falls in roadways or stopes. Before the poles can be used purposefully it is accepted that about 100 are necessary to be statistically significant. Plotting this number can be tedious and large mines may have computer programs to perform this duty. Mount Isa Mines has a program, and another has been published by the U.S. Bureau of Mines[2].

Clustering of poles is most easily seen by contouring them (Figure 9.3). With a standard projection of 20 cm diameter a circle of 1 cm radius will have an area of 1 per cent of the whole diagram. Poles are normally contoured by the Mellis method: a circle of 1 cm radius is drawn around each pole (strictly, ellipses should be used, but the error is negligible). The boundaries around zones where two, three or more small circles overlap then correspond to the two, three, or higher percentage contours respectively and cross-hatching contoured areas will indicate prominent features. When contained poles show up clearly as in Figure 9.3 then their use is easy. However they are frequently much more scattered and then sophisticated statistical techniques are used to obtain useful information from them. Standard textbooks on engineering geology can be consulted but two recent publications on slope stability[3, 4] will be of use.

JOINT SURVEYS

Joint surveys are carried out with a dip-meter and compass, and in an open pit it may be of use to plot the features on a plane table map. Photographic techniques may also be used. The details to be recorded are strike, dip, continuity, waviness and roughness of joints, thickness and nature of any gouge material in the joint, and frequency and spacing of joints of a similar character. All these factors affect the stability of a joint, particularly in open pit mining, and will be dealt with in Chapter 10.

In an open pit the joints can be determined by examining all the walls of the pit. However if there was only one wall which ran (say) north-south, and was nearly vertical, then relatively few joints with a north-south strike would show up, and any east-west joints would assume too great a significance. This phenomenon is much more likely to cause bias in tunnels, or boreholes. A method of reducing the bias has been put forward by Terzaghi[5]. She suggests that for data collected in a linear exposure, such as a borehole or a tunnel, the diagram should be divided into *isogonic* zones (i.e. zones of equal inclination) formed between the intersections of a family of cones, drawn co-axially with the borehole or tunnel, and the projection sphere. The half angles at the apexes of the cone are chosen so that their cosines form a regular arithmetic progression. Thus any of the family of planes that intersect the borehole at a particular angle will all plot within a particular isogonic zone. Planes normal to the tunnel which have a high chance of being intersected will be in an isogonic zone with low cosine values, whilst planes parallel to the zone will have a high cosine value. The cosine values are used to weight the observed frequency to correct the bias.

The correction has the form

$$C = \cos\theta$$

and is applied as $N_c = N.C.$

where C is the correction
 θ is the inclination of the plane to the measuring line
 N is the number of poles in a counting circle
 N_c is the corrected number of poles.

The corrected number of poles is used in a normal Schmidt net.

Of course it is more precise to drill boreholes at different angles, and to survey several headings at different bearings, and then combine the results.

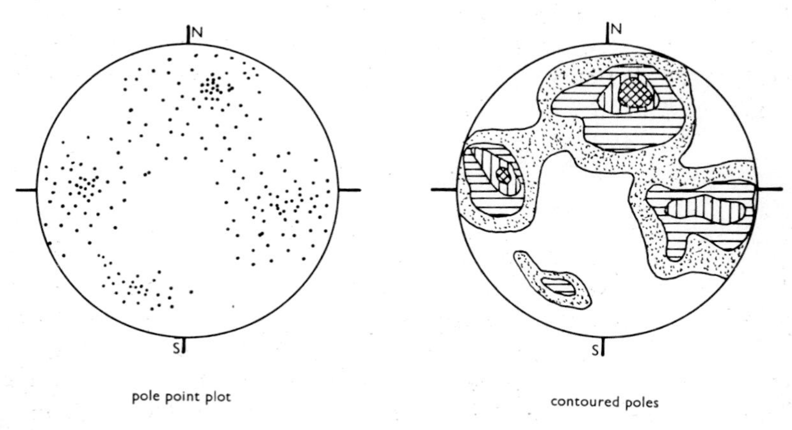

Fig. 9.3 Contoured pole plot

Rock Strength Assessment

LABORATORY TESTS

In addition to joint surveys the strength of rock must be assessed by direct crushing of drill cores, or by testing cores from lumps of rock collected from rock face exposures.

Mount Isa's system is typical of a large operation where about 300 m of drill core is treated every week on a flowline principle[1]. When a new stope is proposed the rock mechanics drilling team is set to taking NX, 10 cm, or 15 cm cores from the stope site. The cores are laid out, washed off with dilute acid, photographed, logged carefully, and then if possible one sample per 6 m of hole is cut into a parallel-ended length with a diamond saw. In some cases the cores are too broken to be sawn. (N.B. It is possible to use a 25 mm diameter corebarrel to obtain small regular specimens from larger irregular pieces of core.) Cut lengths of core are mounted in batches of up to eight on a surface grinder and their ends are ground smooth. The cores are then tested in the non-destructive dynamic moduli section and passed on to the destructive testing laboratory. There most of the cores are tested under uniaxial compression to obtain the unconfined strength, elastic modulus, and Poisson's ratio. A proportion of cores is tested under triaxial compression to obtain an approximate Mohr's envelope and internal coefficients of friction. Some direct shear tests are made in the laboratory. The Brazilian test is used for tensile strength at Mt Isa, but one can use a disk with a hole test, or an irregular lump test for badly broken rock[6,7]. All the information obtained is card coded and stored on computer tape for easy recall and use in prepared programs.

For readers who need a concise summary of the theoretical representation of all all these tests, and of the field tests to be described, the hard cover or paperback edition of the book by Jaeger[8] is useful. Alternatively the book by Obert and Duvall[9] gives a general description of testing methods and physical properties of rock. It will be assumed that readers are familiar with the elastic behaviour of materials, and understand the concept of modulus of elasticity (Young's modulus), bulk modulus and Poisson's ratio. These are the elastic moduli that can be determined by dynamic testing, or by crushing specimens. However, if the bulk modulus or Poisson's ratio are to be determined by crushing, then the rock specimen has to have deformation gauges attached to it, and the testing is more complicated. Hence a dynamic modulus testing array becomes economical with a large throughput of core.

FIELD TESTS

Simple field tests have included a very few large scale in-situ compressive strength tests. More frequently, when machinery is to be used, or supports installed on a soft floor, or under a soft roof, a hydraulic ram is used to determine the safe bearing pressure of the roof and floor. Penetration tests have been used on longwall coal faces, especially when individual hydraulic props were in frequent use. With the gradual changeover to chock-type supports the floor plates increased in size and hence bearing pressures dropped to lower values for the same depth of working and support load.

Field tests of shear strength have been made in two ways. Portable hydraulically powered equipment has been used to test fully excavated pieces of rock in a shear box. The specimen is cut out and immediately cast into a mould with its bedding planes at the desired orientation. The regular cube is placed into the shear box where the top half is forced to slide sideways over the lower half and a load-deformation curve can be obtained. It is similarly possible to take portable coring drills and triaxial test cells into the mine or open pit and obtain compressive strengths of freshly cored rock. For a very few rocks that are grossly affected by changes in moisture content this process may be justified but laboratory testing is usually more convenient.

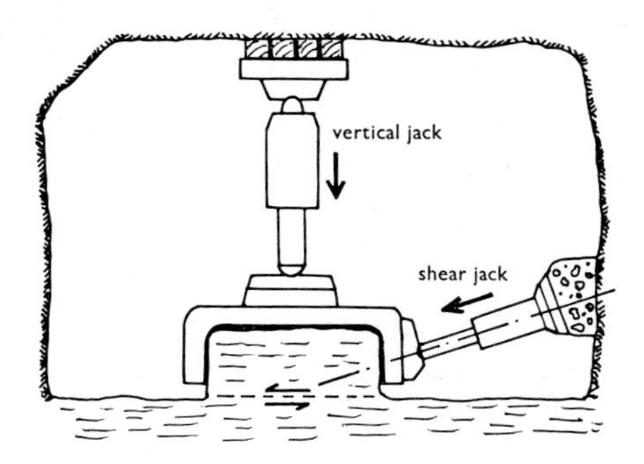

Fig. 9.4 In-situ shear test

The more usual field test of shear strength is an in-situ test as in Figure 9.4. Its use is normally restricted to major engineering construction projects such as dam abutments, where the high cost is warranted. To carry out a test the floor of a tunnel or cutting is carefully excavated by hand to expose a block of rock about 0.3 m x 0.3 m x 0.15 m high. This then has a capping cast on to it and two jacks are mounted as shown. The inclined jack is mounted at less than the angle of internal friction of the rock and with its axis passing through the centre of the shear plane. It has been suggested in conjunction with this type of test that a representative sample of the rock should have dimensions which are at least ten times greater than the average spacing of the weakness and fracture planes in the rock mass. This would minimise the effect of local variations in the geometry and physical features of the fractures. There is a practical limit however to the size of block which can be tested and samples above 0.5-0.6 m cube would be exceptional.

The principal field test is in fact physical observation and measurement of the joint spacing and continuity, both from exposed surfaces and from drill cores. In connection with drill cores Deere[10] has suggested a classification of the core known as a *rock quality designation* (RQD). This classification uses the fracture frequency in a drill core to assess the quality of the rock. The RQD is the percentage of core recovered in pieces longer than a fixed base length of core, expressed as shown. The relationship

RQD per cent	Description of Rock Quality
0–25	Very Poor
25–50	Poor
50–75	Fair
75–90	Good
90–100	Excellent

is obvious; the more broken the rock core then the weaker and more broken the strata is likely to be. Deere used a base length of 100 mm for his RQD but Mount Isa[1] used a base length of 75 mm for the Racecourse lead orebody because observed strengths underground then agreed more closely with the description of rock quality. Some attempts have been made to correlate support systems with RQD,

e.g. RQD greater than 90 per cent, rock bolt 30 per cent of roof
RQD 60 per cent, rock bolt 60 per cent of roof,

but no coherent system appears to have been developed.

In a small proportion of cases, when boreholes are drilled in zones of high stress, the engineer receives advance warning because the core may disk or the borehole can collapse. Leeman[11] reports disking and shows photographs of its occurrence in hard quartzite in deep gold mines in South Africa. The disking (Figure 9.5) is not related to the

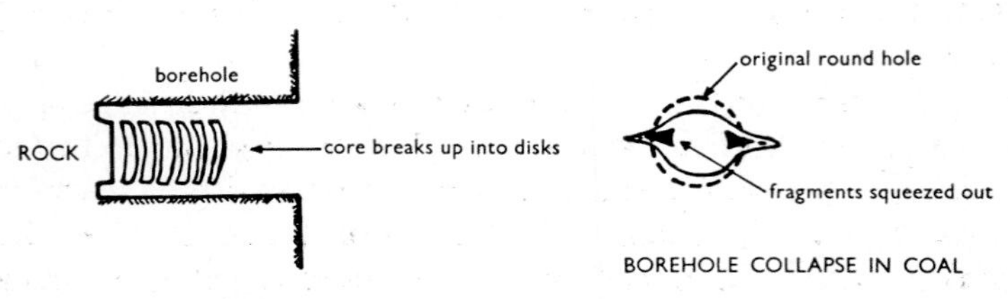

Fig. 9.5 Disking of cores and borehole collapse

bedding or natural joints of the rock and shows up best in hard homogeneous rock. When this phenomenon occurs it is advisable to de-stress the pillar or rock face by boring holes and blasting the rock to produce a protective barrier of crushed rock.

Similarly with coal pillars or ribsides: if it is noticed that boreholes are squeezing inwards vertically, with crumbling and wedge ejection at the sides, it would be advisable to de-stress the pillar or ribside by blasting it, or to keep clear of the area until the pillar fails naturally.

In connection with boreholes and visual observations it is appropriate to note that introscopes have been designed for insertion into boreholes to study crack formation and propagation, and bedding plane separation. The introscopes are variously termed stratascopes, or strata introscopes and are designed on a similar basis to a submarine periscope, except that a light source has to be provided. In metal mines the light source can be powerful and inspection and photography is easy. In mines working under safety lamp regulations the restrictions on electricity can mean that the light source is of low intensity and photography entails exposures of 60-80 secs on high speed film. The process is tedious and sometimes the information required can be more quickly and cheaply obtained by a simple tipped probe which is used to feel for cracks.

For many years engineers have wanted a simple test to determine the resistance of rock or coal to machine cutting. There are some rebound-type rock hardness testing instruments which can be used either in the field or on laboratory samples. One of these is the Shore scleroscope: a small hard metal pointed weight is dropped onto a surface from a fixed height and the rebound height is noted. The reading is grossly affected by whether the pointer drops onto a soft or hard crystal in the rock. The instrument was originally designed to test metal hardness and has not been successful with rocks. A second type of rebound tester is the Schmidt hammer (or Type N Beton-Prüfhammer) which was designed to test concrete hardness. This has also been tried for rock testing but in the experience of the author and several of his colleagues the hammer has not been successful in that the readings have too large a scatter to be useful. The rebound of the spring-loaded plunger is affected by joints and fissures in the rock unless the operator carefully chooses a large piece of solid rock and equally carefully prepares the surface. The sample is then not representative of the mass.

The latest type of field rock hardness testers resemble a pair of nutcrackers. A small sample of rock is placed between the jaws of the device and the pressure needed to force a standard indenter into the rock is measured. The instrument is calibrated in the laboratory on specimens of known strength. However these small scale instruments do not take into account the joints and cracks in rock.

Stress and Strain Measurements

Engineers tend to refer rather loosely to stress measurements, but in many cases they measure strains and convert them to stresses by using the elastic constants of the rock. They also refer to virgin stresses, but those measured should more accurately be termed *pre-mining stresses* because an excavation has already been made in the ground. Mining experience shows that the effects of even a relatively small roadway can be measured 60-90 m away so that some de-stressing will probably have occurred. Pre-mining stresses are generally measured as strains by over-coring or by re-stressing methods[9, 11].

In the overcoring method a small diameter hole, commonly 50-75 mm, is drilled into a rock face. There is then a wide choice of instruments available[11]. Basically either a diametral strain probe, with buttons mounted on gauges at 120° intervals (Figure 9.6), is inserted, or else a 120° strain gauge rosette is stuck on a carefully cleaned and dried surface at the bottom of the hole. A diamond coring drill of about 100-150 mm diameter is then used to drill concentrically around the probe or strain gauges. As the drill passes the gauges there is a relaxation of the rock to zero stress and the expansion recorded in the centre of the core is converted to give the original stress in the rock. Measurement of strains in three directions enables the principal orthogonal stresses to be calculated theoretically. The measurements are generally repeated several times in each borehole. The central small hole is cored out each time and the cores are taken to the laboratory to measure the elastic moduli so that strains can be converted to stress. If 120° rosettes or plugs are not used then three boreholes are needed with unidirectional strains measured at three different angles.

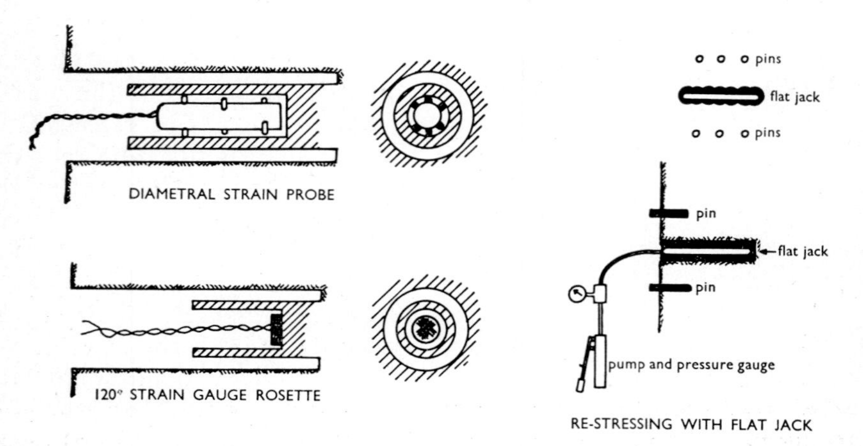

Fig. 9.6 Measurement of pre-mining stress

In re-stressing methods a rosette of measuring pins or strain gauges is fixed to the rock face and then a central slot about 0.3 m square in plan and 0.1 m thick is drilled into the rock between the pins (Figure 9.6). The relaxation in strain is measured. A flat hydraulic jack is cemented into the hole and re-pressurised until the pins have returned to their original positions. The pressure at which this occurs is assumed to be equal to the original pressure in the rock. Some investigators have used annular jacks in a central hole instead of flat jacks. Re-stressing methods are cheaper and less complicated than overcoring and have the advantage of not needing electronic equipment if used in gassy coal mines, or in wet conditions in metal mines.

To obtain virgin stresses the measurements obtained by these stress relief methods must be corrected to allow for the shape of the opening in which they are made, and for the theoretical distribution of stress away from the roadways. Possibly in this respect a series of overcoring tests at increasing depth in the same borehole are of more value than a re-stressing test at the surface of the rock. All the measurements have to rely on the assumptions made as a basis for the initial theoretical calculations of stress, and it would be wise to consider the result obtained as an indication of order of magnitude rather than as precise values. There is still not complete agreement for instance on the stress distribution at the flat end of a borehole when using the rosette method[8].

Information that is perhaps of more use in the design of limiting conditions is obtained from measuring the changes in stress and the rock deformations that occur when mineral extraction takes place: these can be called *post-mining stresses*. The main disadvantage of these measurements is that they have to be part of a long term program and may take several months, or even years to complete.

The shaft movements mentioned in Chapter 8 were obtained with vibrating wire (sonic) gauges. The frequency of vibration of a wire is a function of its length. If a wire is stretched between two points that can move relative to each other, and the wire is plucked and made to vibrate in a magnetic field, it will produce an electrical signal that is proportional to the length of the wire. Changes in length of a few microstrains can be detected with these instruments which are most useful when embedded in concrete. When installed on free surfaces they are too susceptible to changes in ambient conditions, both in themselves and because the surface skin of rock and concrete is easily affected. Vibrating wire gauges have very good long term stability and will not suffer zero drift for at least five or six years.

Stress changes can also be measured hydraulically, or electrically with resistance strain gauges, or with electro-hydraulic transducers. Generally the hydraulic instruments are cheaper and have a fail-safe action. If they leak they read zero, and if they work then they read correctly with no zero drift. Resistance strain gauges suffer from zero drift over long periods, and their adhesive cements can creep. Electro-hydraulic transducers can be unreliable when pressurised for long periods unless one pays such a high price that one cannot afford to bury them permanently in rock or concrete. Photoelastic stressmeters can also be used but they can be difficult to read, and cannot be read remotely as can hydraulic or electric instruments.

Stresses in solid rock are measured with inclusion stressmeters, often known as borehole plugs. These can be *hard* (or stiff, or rigid) with a modulus of elasticity several times that of rock. In this case the plug is usually metallic (brass, for instance) and measures changes in strain. These plugs are calibrated in a borehole in a rock sample in a press so that the strains can be read off as stress. They must also be tightly wedged or cast into place in the rock so that there is no lost strain between the rock and plug surfaces.

Soft, or compliant, stressmeters are usually hydraulic, and are easily deformed, so that pressures may be read directly. A simple and cheap borehole stressmeter has been used by the U.S. Bureau of Mines[12]. A soft copper tube is squashed flat lengthwise and soldered at the ends. Capillary tubing and an air bleed valve are soldered in at one end and the whole assembly is potted in cement or epoxy resin as in Figure 9.7. The potted plug is inserted in a borehole and then pumped up to (say) 5-7 MPa to rupture the cement and ensure good contact with the borehole walls. Subsequent pressure changes are read directly on a pressure gauge or recorded on clockwork-driven charts. Flatjacks, resembling a pancake, have been used to monitor pressure changes behind concrete walls, or in slots, but have not been very successful. Both these instruments only record stress in one direction.

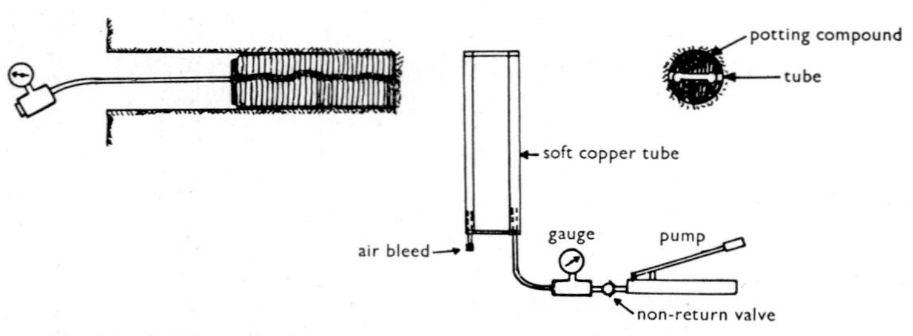

Fig. 9.7 Hydraulic borehole stressmeter

Dynamometers have been used to measure pressure in packs[13]. The pack dynamometer is a short and robust hydraulic piston which is buried in the floor under a stone pack, or sand-fill, or a stone-filled crib (cog). Pressure changes are recorded directly and, again, the instrument is unidirectional.

Pressures on supports are measured with load cells, either photo-elastic, hydraulic, or electrical. These instruments are available commercially or can be home-made, and are used over, under, or inserted in, the support as convenient. One precaution which must be taken is to ensure that the contact area of the support with the floor or roof, or bearing pad, is not altered. For example if the load on a steel RSJ leg is to be measured the cell must either be inserted between a steel-to-steel contact, or a special carrier is used. It may be necessary to shorten the RSJ and insert it in a yoke on top of the load cell, and have a matching length of RSJ below the cell as in Figure 9.8.

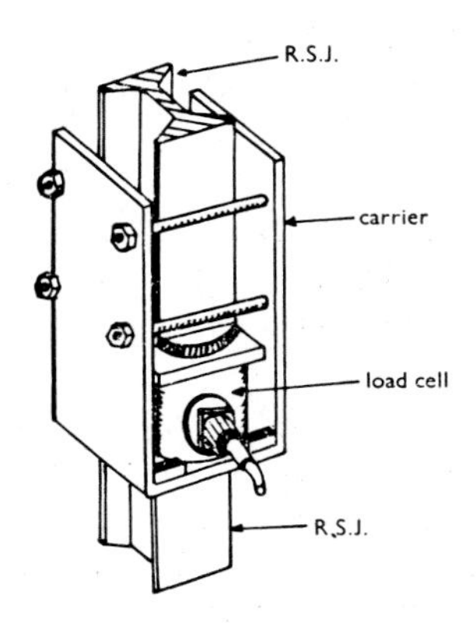

Fig. 9.8 Support load cell in carrier

Measuring stress changes only can be a pointless exercise unless the movements in the surrounding rock are measured at the same time. For instance in soft rocks that deform continuously under load it may not be possible to record high pressures,[13] whereas in hard rocks at the same depth and around the same size opening, it may be possible to measure very high pressures when no rock deformation takes place. In many cases the closures themselves can give valuable information without pressure measurements.

Closure measurements are easier and simpler to take than are stress measurements. The usual measurements taken are of vertical and horizontal closures (preferably survey-levelled to separate the movements of roof and floor, or of walls), bed separation movements, and relative lateral movement of roof over floor (ride)[14]. These measurements can be economically integrated with normal surveying duties and although sophisticated equipment has been developed, it is also possible to use a basic steel measuring tape, plumb bob, and protractor, for measuring between the ends of untensioned rock bolts inserted to various depths around the roadway (Figure 9.9). Convergence recorders are useful instruments. They consist principally of a spring-loaded telescopic strut fitted between roof and floor, or sides. A stylus fitted to one portion of the strut marks a trace on a clockwork-driven drum mounted on the other part of the strut. These recorders can be modified to measure remote closure in abandoned workings. A cheap telescopic strut of steel pipe can be mounted in the old workings and a tensioned wire led out through the pipe to a recorder drum (Figure 9.9).

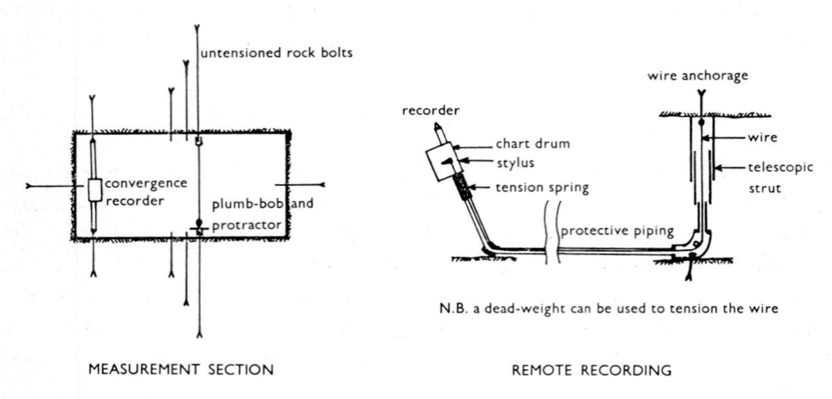

Fig. 9.9 Closure measurement section

Measurements of rock movement may be taken several tens of metres into a rock mass by the use of borehole extensometers. Boreholes are drilled from the surface or from underground workings and wires or steel tapes are anchored at various depths in the hole with expansive or explosive anchors. The expansion anchors are rock bolt shells which can be set with a string of hollow tubes. An explosive anchor (where permitted) is more convenient. A typical arrangement is sketched in Figure 9.10. The core is 25 mm steel pipe; lead weighted if it is

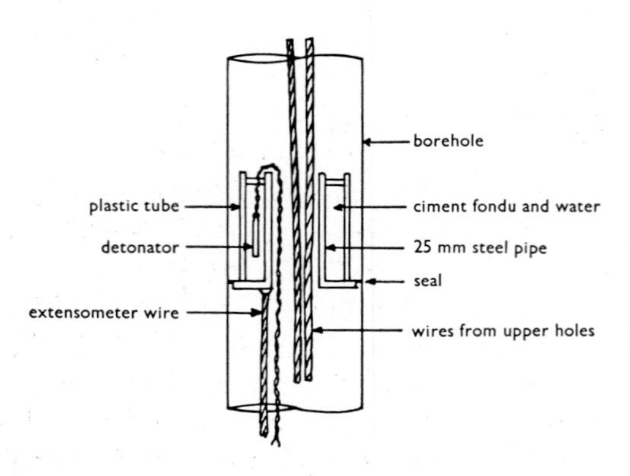

Fig. 9.10 Explosive extensometer anchor

dropped down a hole, or pushed up a tube with semi-flexible plastic tubing. Mount Isa Mines uses this type of anchor with monel metal tapes. Several tapes are brought out of one hole and can either be connected to potentiometric recorders, or to a self-tensioned mechanical extensometer which is used to read each tape separately from a reference plate at the mouth of the hole. This type of remote measurement is useful for monitoring block caves from above, or for following the progress of subsidence above mineral extraction.

Prediction of Rock Movement with Models

Photoelastic models and computer-based simulation with finite element mathematical models have already been mentioned. A visual modelling technique is possible using dimensionless analysis to scale down the factors of rock strength, physical size and the stresses involved. Prediction of movement is possible to a degree of accuracy dependent on model size, experimental technique, and the accuracy of the stress values and rock strengths obtained from measurements in the mine and scaled down for the model.

Ore flow models for shrinkage stopes, block caving and sublevel stoping, can be cheaply carried out with gravel and sand in a glass-fronted box at a mine, but model testing on a larger scale is more expensive. Charges for a single model by a commercial research laboratory may be about $10-12000 at 1972 costs, but this is cheap when one considers the commitment of a million or more dollars to a long term mining program.

The most useful type of physical model uses laminated beds of rock-like material which may be cement/sand[15] or gypsum plaster/sand based[16]. Work has been carried out at least in Australia[15], England[16], the United States[9], and in Germany by Jacobi and Everling. The models are made in various sizes and can simulate single roadways, pillar mining, stoping, or longwall mining. The smaller two dimensional models, from 0.6 m by 0.6 m to perhaps 7 m by 4 m, and from 0.15 to 0.3 m thick, suffer the usual defect of not really simulating a three dimensional orebody. They can however be stressed in increments and the effect of each increase in stress, or extraction increment, noted on simulated stope boundaries, pillars, roadways or longwall faces. These are all situations where the third dimensions can be considered infinite in relation to the local nature of the two dimensions modelled. The smallest models are quick and easy to make and to test and a half-square replicate (Latin square) experiment of, say, eight models can be completed in a month[16].

With three dimensional models, about 1.5 to 2.0 m cube, loaded in a triaxial rig, a whole underground stope along with surrounding country rock can be simulated to predict

1. strain magnitudes in pillars as measured with strain gauges attached to their sides
2. premature failure of any pillar as indicated by sudden de-stressing of a gauge (N.B. the interior of a model cannot be seen while under stress)
3. stress concentrations in boundary areas. (These three can be used to provide data for more detailed models of parts of the stope.)
4. movements of floors and backs deduced from strain gauges
5. sequence of failure of pillars and redistribution of stress
6. margins of safety.

There are some drawbacks to model testing. Operator experience is necessary to correct some anomalies that can arise through physical limits on scaling. Practical loading limits and physical size necessitate limits on the geometric scale of 500 or 600 to 1 with a resultant stope dimension of perhaps only 50 to 75 mm across. Consequently the scale strength of the rock and of mining supports may have to be adjusted outside the proper dimensionless scale for practical reasons (done both at MRE and ACIRL). There is also no true gravitational load although investigators in the United States[9] have placed models in centrifuges to simulate a gravitational stress. Thus some scepticism must be applied to scaling up mathematical results. However physical representation is often very good in terms of probable failure areas and can give useful information on weak points of design.

A study of South African literature from 1950 onwards will show that considerable effort has been put into the design of various analogue simulators for mine openings. These have included conducting paper, electrical resistances (using Laplace's theory for elastic simulation), electrolytic tanks with variable electrode arrays, and computer simulation using elastic theory. These do not seem to have met with much success, mainly because of their two dimensional character and the simplifying assumptions which have to be made. It is possible that unless the finite element method can be adapted to three dimensional stresses it too may only be applicable to small excavations and localised boundary conditions.

References

1. MATHEWS, K. E. and EDWARDS, D. B. *Rock mechanics practice at Mount Isa Mines, Ltd.* Paper 32, Ninth Commonwealth Mining and Metallurgical Congress, U.K., May 1969. Pub. by I.M.M.
2. ANON. *A computer program for the stereographic analysis of coal fractures and cleats.* U.S. Bureau of Mines, Information Circular I.C. 8454, March 1970.
3. JOHN, K. W. *Graphical stability analysis of slopes in jointed rock.* J. Soil Mech. and Found. Div. Proc. A.S.C.E., **94**, S.M.2, p. 497, 1968.
4. MCMAHON, B. K., *A statistical method for the design of rock slopes.* Proc. First Australian and New Zealand Conf. on Geomechanics. Aust. Geomechanics Soc., Vol. 1, p. 314, 1971.
5. TERZAGHI, R. D. *Sources of error in joint surveys.* Geotechnique, **15**, p. 287, 1965.
6. HOBBS, D. W. *Rock compressive strength.* Colliery Eng., p. 287, July, 1964.
7. HOBBS, D. W. *The tensile strength of rocks.* Int. J. Rock Mech. Min. Science, **1**, p. 385, 1964.
8. JAEGER, J. C. *Elasticity, Fracture and Flow: with Engineering and Geological Applications.* Methuen, London, 1969.
9. OBERT, L. and DUVALL, W. I. *Rock Mechanics and the Design of Structures in Rock.* Wiley, New York, 1967.
10. DEERE, D. U. *Technical description of rock cores for engineering purposes.* Rock Mech. and Engr. Geol., **1**, p. 16, 1964.
11. LEEMAN, E. R. *The measurement of stress in rock.* J. S. African Inst. Min. & Metallurgy, **65**, No. 2, p. 45, 1964.
12. BARRY, A. J. *et al. Investigations of stress distribution in burst-prone coal pillars.* U.S. Bureau of Mines Report of Investigations, R.I. 6971, 1967. (see also R.I. 6980, 1967, by CURTH.)
13. THOMAS, L. J. *Pack pressure measurements.* Colliery Guardian, **218**, p. 503, October 1970.
14. THOMAS, L. J. *Rock movement around roadways at Babbington Colliery.* Colliery Guardian, **212**, pp. 475 and 507, April 1966.
15. JAGGAR, F. *Designing and loading of models of underground structures.* Symposium on stress and failure around underground openings. University of Sydney, 1967 (Institute of Engineers, Australia). See also various Australian Coal Industry Research Laboratory Reports, A.C.I.R.L., Chatswood, Sydney.
16. HOBBS, D. W. *Scale model studies of strata movement around mine roadways: apparatus, technique and some preliminary results.* Int. J. Rock Mechanics and Mining Science, **3**, p. 101, 1966. (Subsequent papers are in the same Journal: **5**, p. 219, 1968; **5**, p. 237, 1968; **5**, p. 245, 1968; **6**, p. 365, 1969; **6**, p. 405, 1969.)

Chapter 10. Stability of Slopes and Tailings Dams

Introduction

Stability of slopes falls naturally into three divisions: the behaviour of soils or soft rocks where stability is critical as in road and rail cuttings or for tailings dams; the behaviour of soft rocks in bedded deposit open-cut mining; and the behaviour of hard rock in deep open-pits. In civil engineering work it is essential that long term stability is maintained, particularly where dams are involved. In road or rail cuttings a small amount of routine clean-up work may be tolerated to reduce the initial capital cost of construction. In mining excavations the slopes do not always have to possess permanent stability, and in many cases a lower factor of safety may have to be adopted to produce economic working. In a large conical shaped pit, for every degree the pit wall slope is flattened, then several million dollars may be spent in removing extra overburden. Reference to Chapter 5 will show that the important factor in open pit mining is the overburden ratio. The less waste rock that has to be removed in proportion to mineral extracted, then the lower will be the cost of mining.

There is also a practical consideration of application and scale of operations. When a bedded deposit is mined the highwall continually advances outwards. An opencast coal mine can operate with a highwall angle of 70 to 80°, and if it has been pre-blasted it will be well drained and will be stable for a mining interval of a few days or weeks. In wet areas, or with unconsolidated overburden, it doesn't matter if the slope angle (*batter*) is reduced to the natural angle of repose of 35-45° because the extra rock removed does not represent wasted money, it is only a change of mining method. Where deep overburden is stripped in separate benches then the bench tops (*berms*) can have ample widths for safety without affecting the cost of operation. When the limit of extraction is reached the last highwall is blasted to a safe angle and even this operation can be economical because it will seal off the coal seam as required by law.

In steeply pitching metalliferous ores the average slope of the side of the pit, taken through several benches, becomes important. Some of these benches may have to stay in position for several months, or even years, and optimum slope angle design is important. If the slope is too steep there may be a slide, if it is too shallow then money is being wasted. Even here there is room for compromise and consideration of scale of operations. To a small company a slide of 100 000 tonnes may be disastrous. To a large company moving a million or more tonnes a month it would hardly be noticeable, given a few hours warning to move equipment.

Soil and Soft Rock Slopes

The full theory and practice of slope design in soil and soft rock can be found in standard text books on soil mechanics (e.g. Taylor[1], or Lambe and Whitman[2]). An elementary treatment will be given here to illustrate the effects of rock joint strength and groundwater pressure. It is obvious that when one portion of a rock or soil slope slides over another that resistance to the movement will be experienced. There will be the friction between the two surfaces, and because the surfaces are not truly plane there will be other forces needed to shear off irregularities on them and to slide broken pieces of rock over each other. The rock or soil then has a shear strength which enables it to stay in position on an inclined plane.

The shear strength is normally expressed by the Coulomb equation

$$S = c + \sigma \tan \Phi$$

where S is the shear strength; c is the cohesion along the failure plane, or the shear strength under zero normal stress; σ is the stress normal to the failure plane; and Φ is the angle of friction of the material along the failure surface.

The values of S, c and Φ can be found easily by field tests with a shear box in which a sample of soil or rock is split and the split surfaces are pushed across each other. Figure 10.1A illustrates the relationship between these values, and it is a *failure criterion* because these are the values at which slip occurs. Reduction of the slope angle, or increase in σ, the normal stress, would prevent slip. In practical terms this could be achieved by lessening the angle of the slope or by imposing an artificial restraint such as a retaining wall or rock bolts.

The friction angle Φ can be visualised in two ways. It is the maximum slope at which a loose granular material (e.g. an ore stockpile) will remain at rest, i.e. its angle of repose. It is also the angle of the slope at which a loose block of rock would be just on the point of sliding. The cohesion, c, is a function of the irregularity and the initial physical bonding of the two sliding surfaces. Just before a slide the cohesion can be high and would give a peak strength as shown in Figure 10.1B. However, once the rock or soil starts to slide, the nature of the sliding surfaces alters. A layer of crushed and reorientated grains builds up and the value of Φ drops to give a residual strength for the rock. The practical significance of this is that if there is a marked difference between the value of Φ for intact rock and for broken rock, then when a failure occurs it can be sudden and violent. This is more likely to occur in hard rocks such as porphyry or granite. In soft shales or earth slopes which have little difference between peak and residual strengths, then slides will be gentle unless the slope is wet. (The effect of groundwater is noted later.)

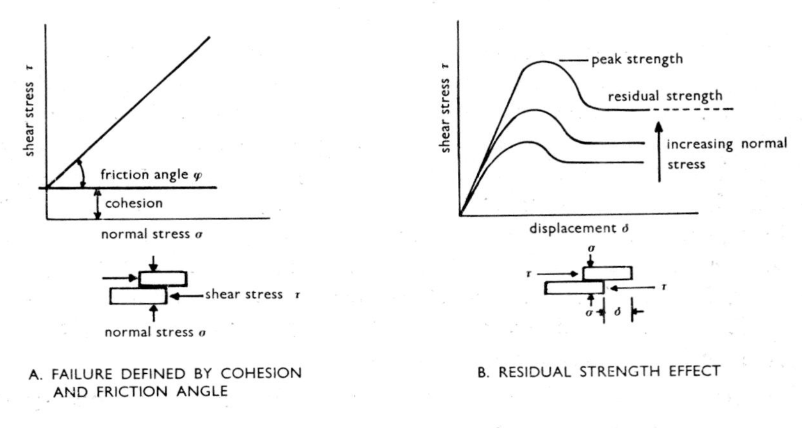

Fig. 10.1 Failure criteria for soil and rock

To extend the Coulomb equation into rock masses the effect of rock joints and roughness must be considered. Failure planes can be uneven because of the surface roughness of a joint, or because of the stepped nature of joint interfaces, as in Figure 10.2. The effect of the surface

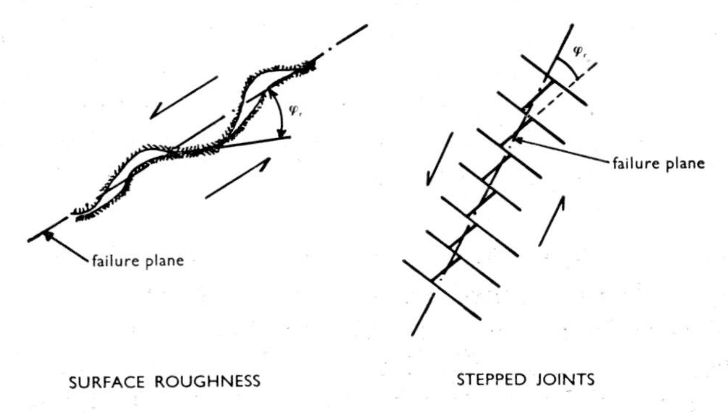

Fig. 10.2 Effect of surface roughness

roughness can be allowed for by adding an extra value Φ_r to the original friction angle, so that

$$S = c + \sigma \tan (\Phi + \Phi_r)$$

Hoek[3] gives values of Φ in his paper for a range of rocks and some of these are listed here to show their effect.

Rock Type	Value of Φ for		
	Intact Rock	Residual Strength	Joint Roughness
Granite	50–64	31–33	—
Porphyry	—	30–34	40
Sandstone	45–50	25–34	27–38
Schist	26–70	—	—
Shale	45–64	27–32	37

Resistance to sliding depends not only on friction but on the normal pressure across the sliding surfaces. Groundwater has several effects. It can lubricate clay and shale surfaces so that Φ is decreased, but more importantly it can reduce the normal stress between surfaces. It has a third effect in increasing the dead-weight loading on the rock slope. The reduction in normal stress is the most serious. The Coulomb equation is modified to

$$S = c + (\sigma - u) \tan \Phi$$

where u is the water pressure.

It must be stressed that it is the water pressure that is important, not the flow rate, although the latter will, of course, affect attempts to drain the slope to increase stability. Figure 10.3 illustrates this effect: the head of water from the water table (*phreatic line*) to any

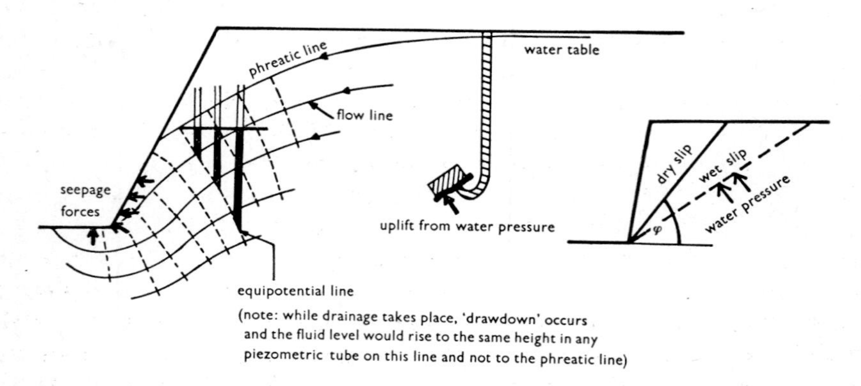

Fig. 10.3 Effect of water pressure

fissure in the rock (*piezometric head*) will tend to lift any rock above the fissure. This is not dangerous away from the slope in the position shown because of the dead-weight of the rock. However if this hypothetical position were moved against the rock face, then the rock would be burst off as soon as the water pressure exceeded the cohesion. At intermediate positions the effect of the weight of earth or rock trying to slide down any slope steeper than its angle of internal friction is helped by the pressure from the head of water. Figure 10.3 also shows part of a *flow net*. This is a pictorial representation of the effect of water on a slope. The piezometric head is determined for required places on the slope by measurement in tubes, and then *equipotential lines* are drawn connecting points of equal piezometric head. *Flow lines* are constructed at right angles to these. Where the flow emerges at the rock face then seepage forces occur. They also occur within the rock and may cause trouble at clay bands or on slip planes.

Provided that the water can drain freely then the pressure will not rise and the risk of slope failure is minimised. Artificial drainage of the slope will lower the natural water table and reduce the risk of failure. For permanent artificial slopes, and when constructing retaining walls, it is usual to install drainage channels. In opencut mines in wet areas, where the rock is not naturally well drained, then the cost of drilling drainage holes, or of driving a special gallery to intersect aquifers or to break through a clay-filled shear zone (for example), may be worthwhile.

Fortunately (or unfortunately!) for mainland Australia excessive water in open pits is not a problem. However in wet areas it is desirable to keep slopes well drained to prevent excessive seepage water which can damage haul roads and make blasting more difficult. In soft deposits it may be economic to pump from wells outside the pit area to lower the general water table. In areas where heavy frosts and winter icing are common, such as in Canada, severe slope failures can occur during winter months, although the slide may not occur until the spring thaw. As soon as a rock face freezes then the seepage channels are closed and the water table can rise behind the rock face, increasing the pressure to bursting point. Under these circumstances two lines of action are possible: pumping can be carried out from a tunnel behind the rock face, or the slope may be heavily blasted before the freeze-up. The broken rock provides good drainage and is not mined until the thaw. A final precaution for pit walls is that if tension cracks develop at the top of a slope (Figure 10.4B) then it is advisable to keep these filled in as much as possible to prevent ingress of water. A check on groundwater levels is particularly necessary in mountainous areas where the head of water pressure in cracks can be in excess of the height of the pit wall.

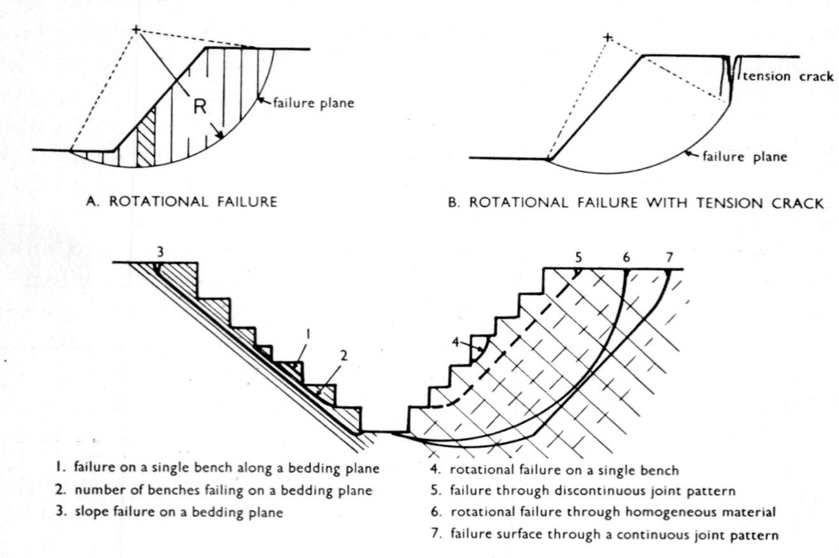

Fig. 10.4 Slope failures

PREDICTION OF FAILURE

The failure of slopes of intact and homogeneous material can be mathematically predicted. The most advanced mathematical treatment is one in which the whole zone is considered to fail in plastic flow. An easier approach is obtained if the slope is considered to fail on a cylindrical rotational path as in Figure 10.4A. The *method of slices* (Bishop, or Fellenius) is used: the failure is considered to be in simple shear along the failure plane. The segment of slope involved is divided into slices and the forces on each slice are analysed separately. Summation by integration is used to give the factor of safety against failure. Finite element analysis can also be used[4,5]. A statistical approach to the location of slips on jointed rock surfaces has been made by McMahon[6].

Most of these approaches need advanced mathematical techniques and an accurate measurement of the input parameters. For many years civil engineers have had design charts available to give them approximate safe slopes with a minimum of calculation. Hoek[3] has extended this work and has produced a design chart of the form shown in Figure 10.5 for failure in soft rocks or soils (circular failure) and in Figure 10.6 for plane failure of hard jointed rocks. Both these charts are based on an empirical approach which includes back-analysis of failed slopes to obtain the values shown. Both curves assume that the soil or rock mass is reasonably homogeneous and is without a mixture of soft and hard rock beds of varied thickness.

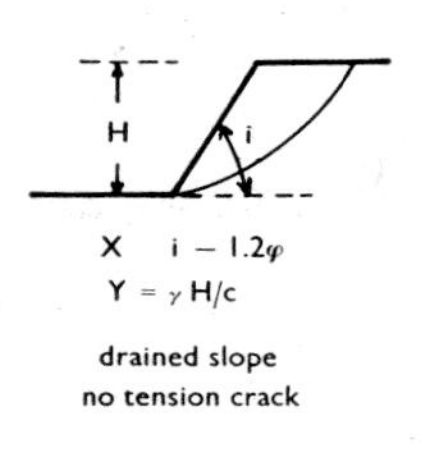

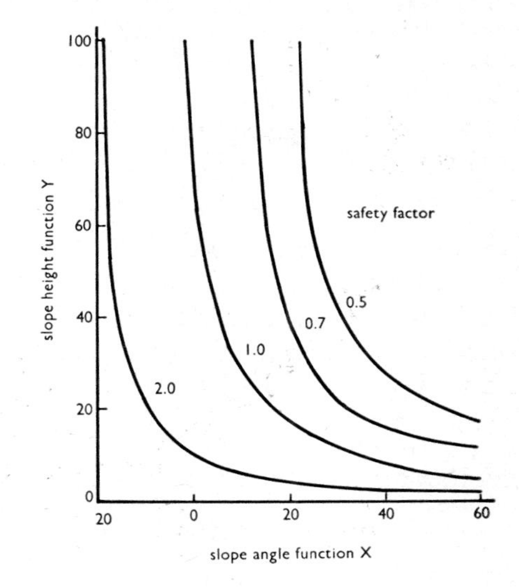

Fig. 10.5 Design chart for soft rock slopes with circular failure (Hoek[3])

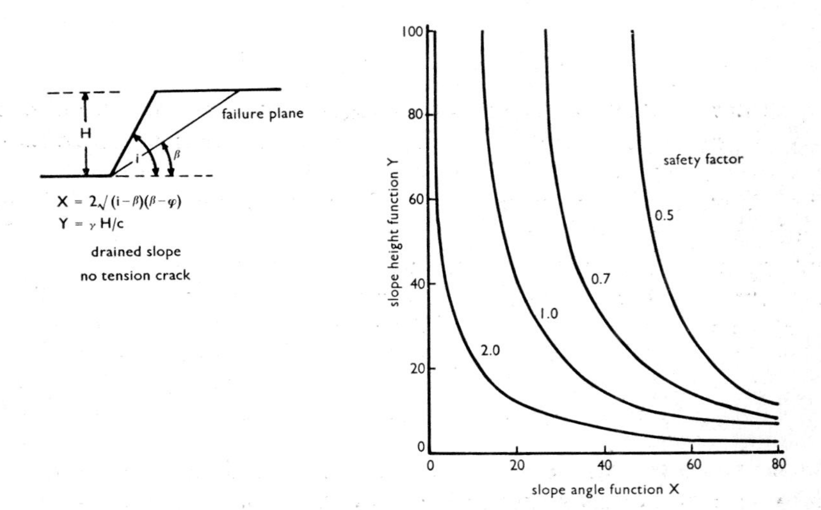

Fig. 10.6 Design chart for jointed rock slopes with plane failure (Hoek[3])

The curves shown in Figure 10.5 are for a drained slope with no tension cracks. The values of the slope height function Y and slope angle function X are modified when water and tension cracks are present. The student is referred to Hoek's paper for their full use. Use of the charts is based on dimensionless groups of parameters so that they are suitable for either the S.I. or imperial systems. For drained slopes with circular failure (Figure 10.5) the slope height function Y is given by

$$Y = \gamma H/c$$

where γ = the rock density, H is the vertical height of the slope and c is the cohesion. The slope angle function X is given by

$$X = i - 1.2\Phi$$

where i is the slope angle and Φ is the angle of internal friction. The chart can be used to find the factor of safety for a given slope angle, or the required slope angle for a given factor of safety.

For plane failure in slopes where the joints dictate the behaviour of the rock (Figures 10.4 and 10.6) the slope factor Y is given by

$$Y = \gamma H/c,$$

as before, and the slope factor X is given by

$$X = 2\sqrt{(i - \beta)(\beta - \Phi)}$$

where i is the slope inclination, β is the angle of the expected plane of failure from rock joint studies and Φ is the angle of internal friction. Again, these values are for drained slopes with no tension cracks and the values of Y and X can be reduced empirically to allow for those factors.

MONITORING SLOPES

For mining purposes pit benches are likely to be designed with low factors of safety. There is a corresponding need to maintain a close watch on the stability of these benches. If slopes are likely to be wet then it is advisable to install piezometers in boreholes behind the crest of the pit wall to measure water pressures in the rock. A rise would indicate that machinery may have to be moved off the benches to safety.

The benches themselves should be surveyed regularly to see if small movements are taking place. Gradual slump of a bench will indicate that a larger slide may take place. Survey pegs can be used with conventional levelling or angle measurement with a theodolite. If the pegs are painted white or with fluorescent paint, and arranged in straight rows, then an easy visual check can be kept on bench stability. It is also possible to use photogrammetric techniques to speed up measurements across a large open-pit mine. Movement transducers are available for installation in boreholes. Slips of strata at any depth in the borehole are registered as a change in inclination between segments and an electrical signal is transmitted to a monitoring station.

With incipient failure on a circular or partially-circular plane the toe of the slope will try to move upwards on the pit floor (Figure 10.4A). Signs of bulging on the pit floor, or a regular survey which will reveal an increase in floor lift, could indicate an imminent slide.

Geophones have also been used to monitor microseismic activity on pit slopes. The historical records indicated an expected peak activity during slides but the present state of knowledge does not enable a slide to be predicted.

Hard Rock Slopes

Most open-pit mines that are deep enough to be concerned with slope stability are working in jointed rock that is fairly hard. If values for cohesion and friction for a hard rock, say c = 250 MPa and $\Phi = 40°$, are used for an entry to the soft slope curves in Figure 10.5, then a stable vertical slope height of over 500 m would be obtained. This would be rather unusual, although high slopes can be obtained in nature if the bedding angles are suitable. Experience has shown that the soft slope relationships do not apply to hard jointed rock and a separate set of design curves is necessary.

Open pit slopes are usually about 45° as an average angle from top to bottom through all the benches. Individual benches are about 15 m high and slope at about 55° from the horizontal. Under favourable circumstances pit walls have had an average slope of about 53° from the horizontal (e.g. Mount Morgan) with benches about 30 m high and a batter of about 60°. The reason for these variable angles and relatively low bench heights is that rocks contain geological discontinuities. Figure 10.7 shows obvious examples of stable and unstable conditions in jointed rocks and Figure 10.4C shows various types of slide that can develop in open pit walls. Rock discontinuities form planes of weakness that can over-ride the probability of rotational failure, or combine with rotational failure in intermediate forms of slides.

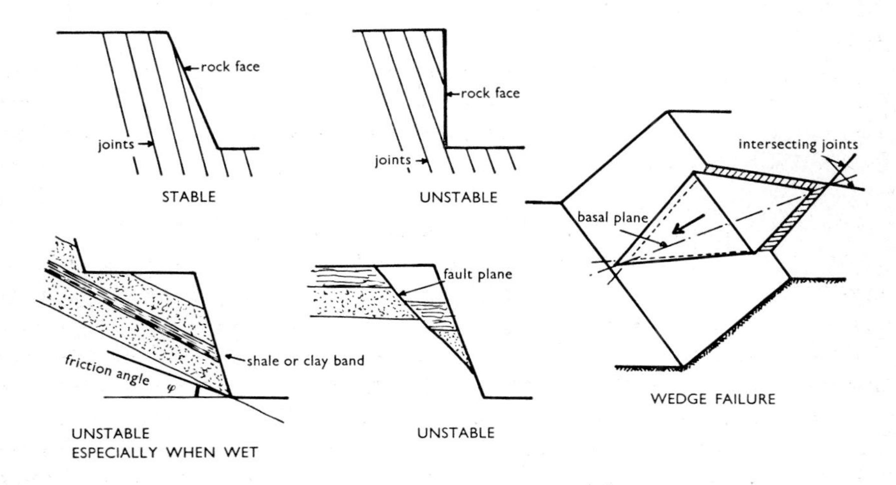

Fig. 10.7 Effect of rock discontinuities

Faults and weak strata are obvious geological discontinuities, and simple bedding plane weaknesses are easy to detect. With simple forms of jointing it may be convenient to have one steep wall and one gently sloped wall in the pit. Discontinuous joints are more difficult to detect and a joint survey as in Chapter 9 may be necessary to determine probable lines of weakness. Where joints are parallel to a pit wall a few inclined boreholes may enable their frequency and possible continuity to be determined. An example of the details of a comprehensive survey, which must include

(a) detailed knowledge of geological structures: bedding, fault planes, folds, joint structure
(b) shear strength of the materials through which the slope is cut
(c) water pressure conditions in pores and cracks
(d) surface movements
(e) magnitudes of regional stress fields,

is given in an article by Steffen and Klingman[7].

It is probable that a joint survey will reveal more than one set of joints, and intersecting joints will produce a danger of wedge failure as in Figure 10.7. Theoretical treatment of this type of failure can be complicated, although simplification can enable it to be dealt with by basic statics[8]. A conservative value for the safety factor can be obtained by considering the wedge to fail along its basal plane and using the design chart in Figure 10.6 or the basic Coulomb formula. McMahon[6] gives a more sophisticated solution.

All the foregoing material has considered a pit slope as a two-dimensional problem. However an open pit will have at least three, and probably four sides, or be a compound oval or circular shape. When the total length of pit wall is considered some general points of stability have been observed in practice, and in some cases have been theoretically deduced. Concave slopes are generally stable (Figure 10.8). They tend to have lateral compressive forces, and joints key into each other in a natural arch. In convex slopes the joints are free

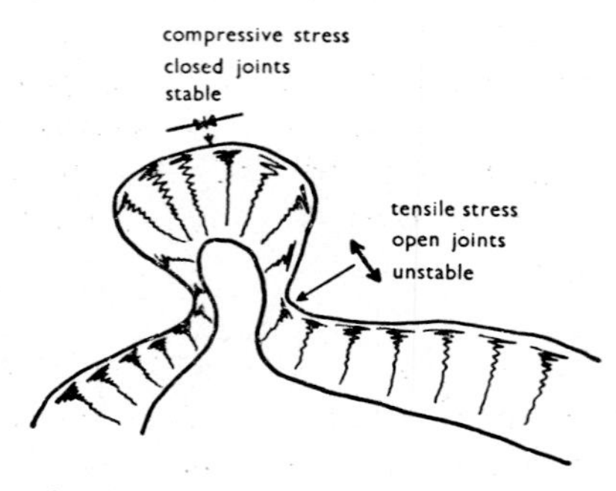

Fig. 10.8 Effect of pit shape

to move and may be under tension. Slides are much more likely to develop on convex slopes. In the absence of marked discontinuities which take precedence in slope design, a general rule-of-thumb is that a concave slope (e.g. the whole of a steep conical pit) may be about 5° steeper than average, and a convex protrusion should have a slope about 5° less than average.

PROTECTION OF PIT SLOPES

It has been mentioned that pit slopes are kept as steep as possible, consequently it is necessary to prepare them carefully as they come to their outer limit, and sometimes to repair them afterwards. As each bench reaches its final limit, if the haul road is not on top of it, its berm width may be reduced so that the batter angle can be lessened. The last row of blast holes will have reduced spacing and charges, and the final wall may be pre-split to give a smooth finish and to prevent the vibrations from the main blast damaging the wall (Figure 10.9). The outer holes of the bench may be kept above grade to prevent damage to the crest of the bench wall below.

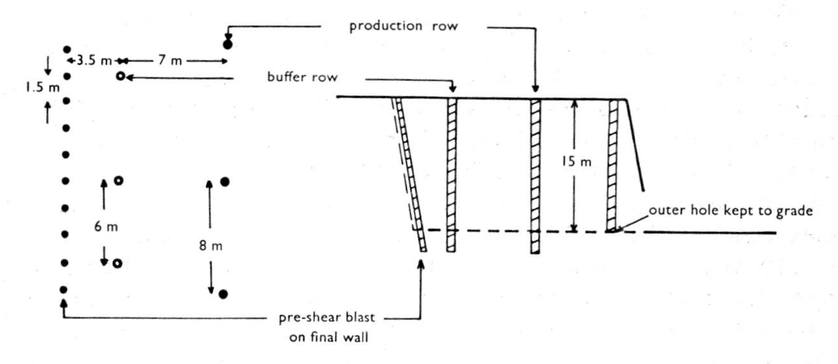

Fig. 10.9 Reduction of blast damage

Despite this care it is possible for the pit wall to weather, and to develop cracks, and it must be kept scaled down to prevent rocks from falling onto lower benches. A slusher scraper bucket, or a small dragline, may be used from above for the scaling operation. Alternatively small diameter holes may be drilled a metre or so back from the wall to blast off a thin layer from the crest so that the batter is slightly reduced and broken rock is removed (see reference 8 for example).

Studies have been made of the cost and feasibility of anchoring pit walls with wire rope bolts[9]. Two aspects can be considered: either the average slope can be steepened to reduce the cost of overburden removal, or the bolts can be used for repair work. The economics of deep anchorage are such that it is unlikely to be used as a routine procedure. However in some cases their use may be necessary, or they can provide an increased factor of safety. For example, where an open pit changes to underground working and the entry portal is in the pit, then rope bolts can be used in the wall surrounding the portal. Mesh and gunite, or a concrete mat, can be used over the ends of the rope bolts to give a secure surround. In other cases, where the stability of the main haul road is threatened, long rope bolts can be used to anchor the benches above and below the haul road to prevent loss of production from a slip. It should be noted that rope bolts are frequently used in civil engineering construction, such as for dam abutments and spillways.

Tailings Dams and Waste Tips

The waste material from both coal and metal mines falls into two size ranges: large sized rock or low grade mineral from developments, variously known as rock, dirt (coal mines) or mullock (metal mines); and finely ground material from the metallurgical plant or from the coal washery. From a metal mine mill the fines are usually known as tailings, and go to a tailings pond. From a coal washery the fines are called such, or are known as a fines discard, and go to a lagoon, or may be known as tailings and go to a tailings pond, dependent on which country the mine is in. Waste rock in coal mines can be dirt and go to a tip, or can be chitter (in Australia) and go on to a tip or bing. For convenience the terms dirt, and tip, will be used.

Coal and metal mines have similar problems; they have to find space in which to dump the waste, and must dump it safely. Although there have been frequent slides of waste tips, and failures of tailings dams, the problems were emotionally highlighted by the pit heap slide at Aberfan in South Wales, on 21st October 1966, in which 144 people lost their lives. Most of the victims were children at the local school which was over-run by the slide of mud and fine coal. Previous slides around the world had disrupted communications and blocked highways but had rarely trapped a group of victims in this way.

There was an immediate examination of mine waste disposal techniques in most countries and both Britain and the U.S.A. have published the results of investigations.[10, 11, 12] The National Coal Board investigations have been endorsed in Britain by Governmental legislation, and most States and Countries now have legislation controlling mine waste disposal. The main field of activity concerns liquefaction of fines or of argillaceous (clayey) waste, with consequent dam failure or tip slides. Other nuisances of coal tip fires, dust production, and barren eyesores, are not always as closely regulated but are indirectly or directly controlled by environmental legislation.

The references quoted report the results of a comprehensive survey of existing practices and of current research. The recommendations made are conservative because it is recognised that in mine operations the level of technical supervision may not be at the same standard as when (for instance) a large water supply dam is being built. A mine operation may work 24 hours a day for 362 days a year for the life of the mine, and cost levels are such that it is not possible to keep engineers present to constantly supervise tailings dam extensions, although daily inspections must be made.

SITE INVESTIGATIONS

Before either a dam or a waste heap is built the site should be thoroughly investigated. A case study is given by Toland[13], but the following information is based on the National Coal Board publications.[10,11] The steps to be taken are

1. Make a preliminary investigation: look at the local geology, make your own maps if necessary. Decide on possible future mining subsidence or past subsidence which can cause instability. Look for evidence of surface instability and chart the local hydrology, particularly of seasonal springs or streams.
2. Carry out tests: sink trial pits and boreholes and collect samples. Classify the surface soil by type and measure the shear strength and permeability of the surface soil, subsoil and underlying rock.
3. Calculate the foundation stability taking into consideration the tip material and its likely water table, and make sure it has an appropriate factor of safety. A tip near a built-up area needs more security than one several miles from habitation.

   Remember that
   (a) mining subsidence may open up fissures, creating paths for preferential water flow
   (b) seepage pressures will exist in tips and foundations
   (c) clay can give an unstable foundation, particularly if saturated
   (d) if there is a natural slope to the ground then it must be drained and stabilised
   (e) extending an existing tip can be dangerous if its previous history is unknown.
4. For a waste heap the final site chosen must be large enough to have a space of at least 10 m clear around the base of the tip for access and for safety. The clearance should be increased for tips on sloping ground, high tips, or tips with a low factor of safety.

FACTORS OF SAFETY

The design factor of safety depends on the consequence of failure with respect to persons and property, the reliability of the design input data, and the reliability of the design assumptions over the life of the tip. N.C.B. recommendations were that for new spoil tips on flat ground where there may be a risk to people and property the factor of safety should be 1.5, in other cases it may be 1.25. On sloping sites the factors should be 1.75 and 1.5. In some areas, for example at mines in the Andes where earthquakes can occur, then the factor of safety must be increased.

MINE WASTE MATERIALS

Mine waste must be assessed for size distribution, specific gravity, plasticity, moisure content and density. The shear strength, consolidation and permeability must also be determined. It is also necessary to assess a change of characteristics with respect to time, such as breakdown of shales, liability to spontaneous combustion of carbonaceous materials or sulphides, or cementation of some deposits. The possibility of chemical interaction of metal mine wastes should also be examined.

TIPPING SOLIDS

For the purpose of this section it is assumed that 'solids' includes all materials not deliberately comminuted (finely ground) for cleaning or for mineral extraction from ore. This material is thus likely to be relatively dry.

The first batches of rock from shaft sinking or from inclines, and from initial development, can usefully be disposed of as a protective bank around the mine site. The bank is covered with top soil and planted with grass and shrubs to act as a sight screen and to minimise noise and dust transmission. Further supplies of rock may be necessary to build a dam or retaining wall for lagoons or tailings ponds, and also to extend them from time to time.

After this a tip must be constructed. The safest way to do this is by placing the material in thin layers. The advantages of this are

(a) the risk of a flow slide is reduced, especially if the layers are compacted to fill voids to prevent a build-up of water pressure

(b) layer tipping avoids size segregation (which both increases porosity and causes impermeable layers) and the chance of hidden layers of weakness, and slumping of the crest. All of this occurs when materials are tipped over a high crest

(c) tip shaping and drainage can be easily controlled

(d) it is easy and economic on soft foundations where the rate of increase of vertical height has to be controlled

(e) for coal tips the risk of spontaneous combustion is greatly decreased

(f) chemical changes within the tip are minimised because ingress of air and water is reduced

(g) drainage water from the tip is less likely to be polluted

(h) the denser tip enables a greater tonnage to be dumped on a given area

(i) possible local government restrictions on tip heights are easier to comply with.

The material, whether for a tip or for a tailings dam, should be placed in one of three methods,

A. in layers not more than 0.3 m thick

B. in layers not more than 0.3 m thick and with each layer compacted with a minimum of four passes of a towed smooth-wheeled roller having a weight not less than 5000 kg/m width, or an equivalent roller

C. in layers not more than 5 m thick.

In method A compaction is provided by the earth moving plant. Both methods A and B are suitable for coarse discard but moisture contents should not exceed 10-14 per cent or heavy machinery will tend to sink in. If the moisture content is high then method C is used with layers of 0.1 m to 5.0 m thick as necessary. The 5 m limit is a maximum to prevent segregation when end-tipping. Where granular tip materials are liable to become saturated they should be surrounded with an embankment placed by methods A or B. In general, compaction as in B improves the shear strength of the material, reduces its permeability, increases rainfall run-off and decreases seepage so that the risk of pollution is decreased. If the material is liable to spontaneous combustion then method C should not be used.

If method C is used on any tip the maximum height of the tip should not exceed 30 m and its overall slope should not exceed 1 in 2 (26°). Special measures may have to be adopted to seal the tip against rainwater penetration.

On sloping ground only methods A and B should be used. If method C is necessary a hard compacted bank should be placed at the toe of the heap, with a trench to collect drainage water at the topside of the bank. Tip height should be initially limited to 15 m.

DISPOSAL OF COAL FINES OR MILL TAILINGS

Most of this material is very small; usually it is less than about 1 mm in size (20 mesh) and where an ore has been ground to liberate the mineral there will be a high proportion of material less than 100μm (micrometres), say 200 mesh, in size. Where tailings are returned underground for sand fill then the tailings pond will hold a higher proportion of the finest sizes, say less than 10μm, which are cycloned out of the fill. This very fine material, known as slimes, will not dry out for years, even after it has settled and surplus water has been drained off. An earth tremor[14] or excessive rainfall can liquefy the whole of the tailings deposit. At Mufulira[15] tailings that had been disposed of over 30 years earlier liquefied and ran down into a hanging wall cave area in September 1970. There was a mudrush into the mine workings and 89 men lost their lives. There was no major event which caused the mudrush, just a steady penetration of tailings through the rock, followed by sudden liquefaction when a breakthrough occurred. A disaster such as this emphasises that care must be taken in the disposal of wet slimes.

DESIGN OF TAILINGS DAMS AND LAGOONS

All disposal areas should be so constructed and drained as to prevent undue accumulation of water in them. The object is to store solids not water. (There may be some exceptions, such as when cyanide products of gold flotation are being impounded.) The moisture content of the material should be as low as practicable. With regard to drainage some of the factors to be considered are

(a) flood calculations
(b) requirements for surface drainage
(c) treatment of artesian water and springs
(d) use of ditches as against pipe drains
(e) use of culverts in particular circumstances.

Tailings dams (or dykes, or berms) are either permeable or are waterproof by design, but permeable dams may become impermeable by accident, and may fail. There is a basic difference between earth and rock fill dams for water storage, and the same type of dam for tailings storage. The slimes from the tailings can blanket off the natural or artificial drainage of a dam and cause the water pressure to rise within the dam to reduce its factor of safety. Investigators from the U.S. Bureau of Mines[16] consider that even the flow nets of tailings dams are different from water dam flow nets.

It is important that lagoons or tailings ponds should be designed as such, and that the banks or the dam design should not simply be a copy of a water storage dam. The dam construction and disposal of material should be such that maximum porosity is retained. Where possible, tailings should be disposed of below ground level, although it would be unwise to choose subsidence hollows or old open pits above current mine workings. In most cases the tailings will have to be disposed of above ground and usually a valley is chosen so that three sides of the pond are already formed. However these three sides should not be neglected, and run-off catchment drains must be installed to deflect rainwater from the pond area. If the slopes do not drain naturally then either a permeable layer or extra drains must be installed before the waste material is dumped on the slopes.

Construction of Tailings Dams.

Generally the material from coal mines will be coarser and easier to deal with than metalliferous mine tailings. It may even be possible to mix the coal mine washery fines with coarse dirt in thin layers on a single tip. In any case the fines will usually gradually dewater themselves if adequate drainage is provided. Metal mines will have more trouble; Kealy and Soderberg[12] summarise the best practices in dam construction.

It is usual to build and extend a dam by upstream methods as in Figure 10.10, which illustrates both good and bad practice. A dam is built up from a smaller starter dam or starter dyke. It is a mistake to build this of clay† which will hold back water and consequently reduce the cohesion and raise the phreatic line (water table). Considerable effort should be made to see that the starter dam is built on a strong base having a scarified surface free of all organic matter. The dam can be made of coarse rock, or of gravel and sand, as long as fine sand is banked along the upstream side to prevent piping of tailings. The starter dam is the base of a large structure and should be built as carefully as would be the foundations of an important building.

The most common upstream method, and possibly the cheapest, is to place the main tailings discharge pipe around the crest of the dam. Outlet risers are provided every 3 to 15 m and if there is a wide gradation of material sizes then the coarse material will fall first to give a wide beach that is permeable, and the slimes will run away from the dam. The free water is also driven away from the dam and can be pumped off or safely decanted. Where a large proportion of the sand has already been removed from underground fill it may be necessary to use cyclones to provide a coarse underflow for dam building, and to run the slimes overflow clear of the dam crest. The dam top should be built as wide as is convenient so that slimes are kept from under

† Chilean law, for an earthquake area, requires an impermeable starter dam to be built upstream of a permeable dam. Coarse fines are cycloned between the dams, and slimes deposited upstream of the impermeable dam.

each new berm for as long as possible. The rate of vertical increase in the tailings deposit should be limited to a few metres per year to allow maximum evaporation of water from the surface of the slimes. One very large dam or several small ponds may be necessary. One tailings pond in Arizona has a dam 1.2 km wide and 1.6 km long. In any type of construction the dam should be compacted as much as possible; its construction should proceed in layers as described under tipping solids.

An unsafe practice is to discharge the tailings at a distance from the dam so that the slimes and water run down and lie against the dam. With this method (Figure 10.10B) there will be a liquefied slimes deposit against a thin shell of dam. The phreatic line will rise in the dam wall and its shear strength and cohesion will be reduced. Dams of this nature can fail suddenly, and almost without warning.

It is possible to build a dam downstream but more and more sand is necessary as the dam height increases. It is likely to be a very safe dam but unless the mine is producing large quantities of development rock or has easy access to borrow material it is unlikely to adopt this type of construction.

Borrow material (i.e. material brought in to the site specially for the job) may have to be used for dam construction. It is not likely to be a first choice while mine dirt is available, but argillaceous dirt may be unsuitable for a dam, or all the coarse sand may be used for underground fill. As far as possible the borrow material should contain only gravel and sand, and should have a minimal clay content. An unzoned dam with a high clay content is a likely candidate for failure. The phreatic line rises, water will pipe through the dam and its face will slough and wash away. Borrow dams, rock fill dams and water-tight dams for phosphate and cyanide compounds should be built to standard civil engineering methods of construction[17]. With these dams, as with sand dams, adequate provision must be made for removal of water off the tailings, and permanent safety after the mine is abandoned.

WATER REMOVAL

Industrial water is expensive, and in Australia water may be scarce and almost irreplaceable. In any case uncontrolled water run-off may cause pollution. As free water develops on the surface of the tailings pond it must be removed for re-circulation and to increase stability. The water can be pumped directly off the surface of the pond but a breakdown of pumps could be serious. The pumps have to be barge-mounted and moved regularly. Siphons can be used to pipe the water to permanent sumps but they are difficult to keep primed and water must lie against the dam. The most convenient method is usually to

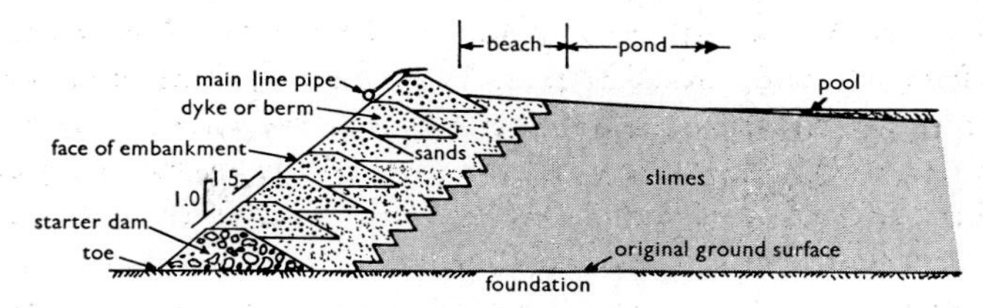

A. SAFE METHOD OF UPSTREAM DAM BUILDING

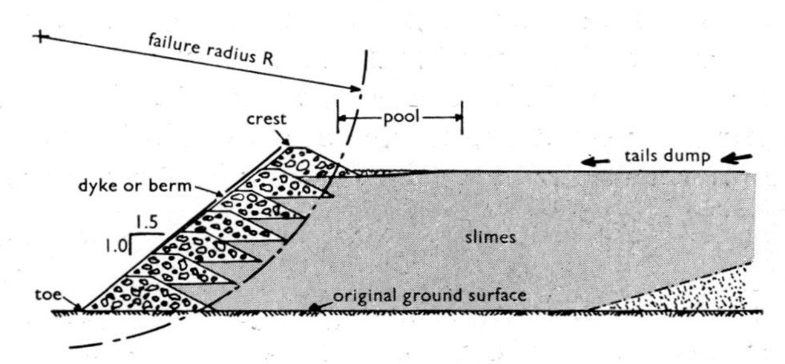

B. UNSAFE METHOD OF UPSTREAM DAM BUILDING

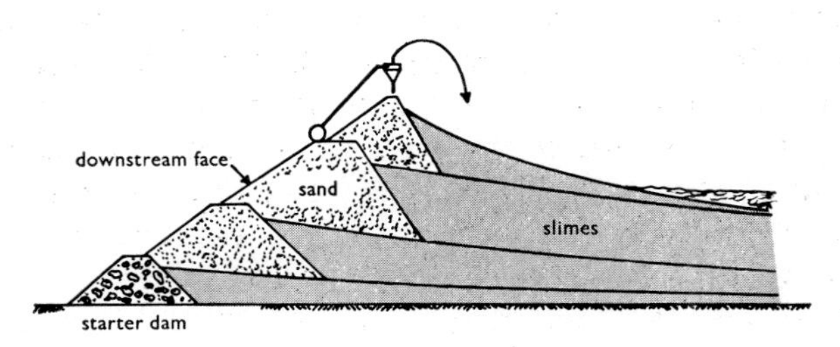

C. UPSTREAM BUILDING WITH CYCLONES

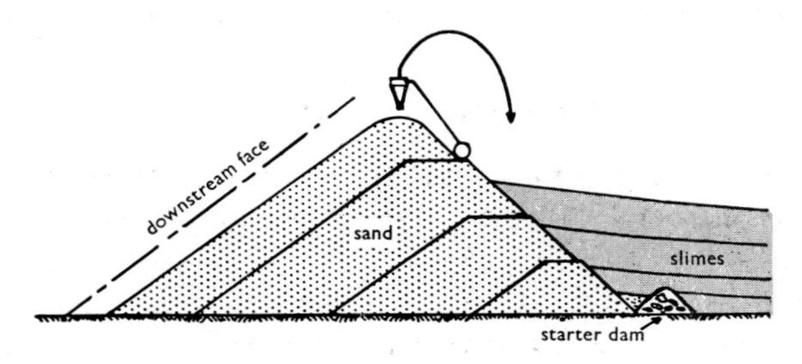

D. DOWNSTREAM BUILDING WITH CYCLONES

Fig. 10.10 Tailings dams

provide a chain of decant towers which are moved outwards and upwards as the pond increases in size. Where possible the tower culverts should be large enough for inspection and repair. They must be as strong as possible because a collapse will allow slimes to run out. A good decant system will allow permanent drainage after the pond is abandoned.

TAILINGS CONSOLIDATION

As a final note to this chapter it is appropriate to mention that when a waste dump is abandoned it must be left in a safe condition, and unless it is toxic and well guarded, it usually has to be revegetated. Research has been carried out to determine the most economic ways of stabilising tailings; a recent article[18] describes work by the U.S. Bureau of Mines. A quick, but barren, method is to cover the surface with a few centimetres of granulated slag. Straw has been harrowed in and bark has been used. Chemical stabilisation has also been tried but is complicated because of the reagents added in the metallurgical processes, and the metals remaining in the tailings. These also make revegetation difficult and costly. A full description of the methods tried and the successes in different areas would be out of place here.

References

1. TAYLOR, D. W. *Fundamentals of Soil Mechanics.* Wiley, New York, 1948.
2. LAMBE, T. W. and WHITMAN, R. V. *Soil Mechanics,* Wiley, New York, 1969.
3. HOEK, E. *Estimating the stability of excavated slopes in opencast mines.* Trans. I.M.M. **79A**, p. 109, 1970, and Discussion **80A**, p. 72, 1971.
4. WANG, F. and SUN, M. *Slope stability analysis by the finite element stress analysis and limiting equilibrium method.* U.S. Bureau of Mines Report of Investigations R.I. 7341, 1970.
5. WANG, F. and SUN, M. *A systematic analysis of pit slope structures by the stiffness matrix method.* U.S. Bureau of Mines Report of Investigations, R.I. 7343, 1970.
6. MCMAHON, B. K. *A statistical method for the design of rock slopes.* First Australia-New Zealand Conference on Geomechanics, Melbourne, 1971. Vol. 1, p. 314. Australian Geomechanics Society, Sydney, 1971.
7. STEFFEN, O. K. H. and KLINGMAN, H. L. *Slope stability at the open pit of Nchanga Consolidated Copper Mines Limited.* J. S. African Inst. Min. Metallurgy, **67**, p. 140, November 1966.
8. CALDER, P. N. and MORASH, B. J. *Pit wall control at Adams Mine.* Mining Congress Journal, p. 34, August 1971.
9. BARRON, K., COATES, D. F. and GYENGE, M. *Support for pit slopes.* The Canadian Mining and Metallurgical Bulletin, p. 113, March 1971.
10. NATIONAL COAL BOARD *N.C.B. Technical handbook on spoil heaps and lagoons.* London, 1970.
11. TOWNSEHEND-ROSE, F. H. E. and THOMSON, G. M. *Security of tips—statutory and technical requirements.* The Mining Engineer, No. 125, p. 293, Feb. 1971.
12. KEALY, C. D. and SODERBERG, R. L. *Design of dams for mill tailings.* U.S. Bureau of Mines, Information Circular I.C. 8410, 1969.
13. TOLAND, G. C. *About tailings dams—construction, sealing and stabilisation.* Mining Engineering, p. 57, December 1971.
14. DOBRY, R. and ALVAREZ, L. *Seismic failures of Chilean tailings dams.* Soil Mechanics J. (Proc. Am. Soc. Civil Engrs.), **93**, Paper No. 5582, November 1967.
15. ANON *Mufulira disaster: interim report.* Mining Magazine, **124**, p. 281, 1971.
16. KEALY, C. D. and BUSCH, R. A. *Determining seepage characteristics of mill-tailings dams by the finite element method.* U.S. Bureau of Mines Report of Investigations, R.I. 7477, 1971.
17. SHERARD, J. L. *et al. Earth and Rock Fill Dams. Wiley,* New York, 1963.
18. DEAN, K. C., HAVENS, R. and VALDEZ, E. G. *U.S.B.M. finds many routes to stabilising mineral wastes.* Mining Engineering, p. 61, December 1971.

Chapter 11. Subsidence Due to Mining

Effect of Mining Bedded Deposits

The chief source of literature and case studies on the effect of mining bedded deposits, particularly coal mining, is Europe. The high population density and the wide spread of coalfields has caused subsidence in built-up areas to be a common phenomenon. The National Coal Board in Britain has produced two official publications on subsidence[1,2], and its subsidence engineers (particularly Orchard[3]), and other private consultants have published many articles. Although these publications are based on case studies in British coalfields (157 studies from more than 50 mines) their results are consistent with individual studies made in other countries, and the conclusions appear to be internationally valid. Subsidence studies currently in progress in the New South Wales coalfields indicate a general agreement.

When a bedded mineral deposit is extracted the overlying strata subside to fill the void. Such subsidence does not occur when the extracted plan areas are small (as is often the case in metal mines), but it inevitably follows extraction over a large area. The following section deals with mining subsidence of the type which occurs in stratified rocks overlying bedded mineral deposits which are relatively flat lying; that is the strata do not dip at more than about 15 or 20°, (about 1 in 3). Subsidence which occurs in conjunction with mining metalliferous ores is given later.

Figure 11.1 illustrates a typical longwall panel in plan view and a section taken across the panel. Coal extraction has resulted in the lowering of the immediate roof of the seam followed by the lowering of the strata successively above this roof, and finally in a depression on the surface.

The subsidence of the surface is less than the thickness of the mined-out coal seam. Factors which affect the amount of ground movement include the angle of draw, width of panel extracted, depth of the seam, thickness of the seam, and the amount of support provided in the goaf.

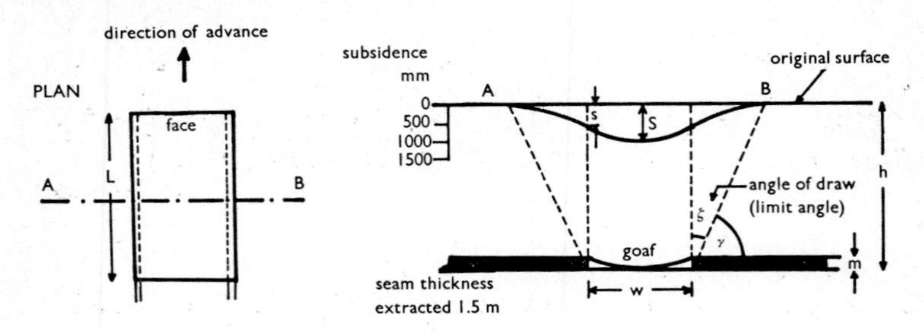

Fig. 11.1 Subsidence over a longwall face

ANGLE OF DRAW (OR LIMIT ANGLE)

The angle of draw is the angle between a vertical line from the edge of the workings and a line to the point at which the subsidence tails out to nothing. The surface area lying outside the zone of draw will not be affected by these workings.

Elastic theory predicts that the effect of draw should extend to an infinite distance from the excavation but field observations indicate that measureable surface movements do not ordinarily extend outside a zone bounded by a 35° angle of draw, and that in most cases the movement outside a zone bounded by a 25° angle of draw is so slight as to have no practical effect on surface facilities. There are indications that the angle of draw varies slightly with depth and with nature of the strata.

EFFECT OF WIDTH AND DEPTH OF WORKING

Figure 11.2 illustrates the effect on subsidence of the width of a working and its depth below the surface. If, from a point P at the surface, limit lines are drawn (at the angle of draw) to meet the seam at points A and B, these points denote the width of a panel which has to be extracted to produce maximum subsidence at point P. The area of extraction having AB as its diameter is known as the *critical area* or the *area of influence.*

Extraction of a narrower panel, i.e. of a sub-critical area, will produce less than the maximum subsidence, while extraction of a panel wider than AB, i.e. a super-critical area, will result in maximum subsidence over a wider area. Thus the amount of subsidence that results from a given working increases as the width of working increases until the critical area of extraction has been reached and, with the same width of workings, decreases with increase of depth insofar as this affects the critical area.

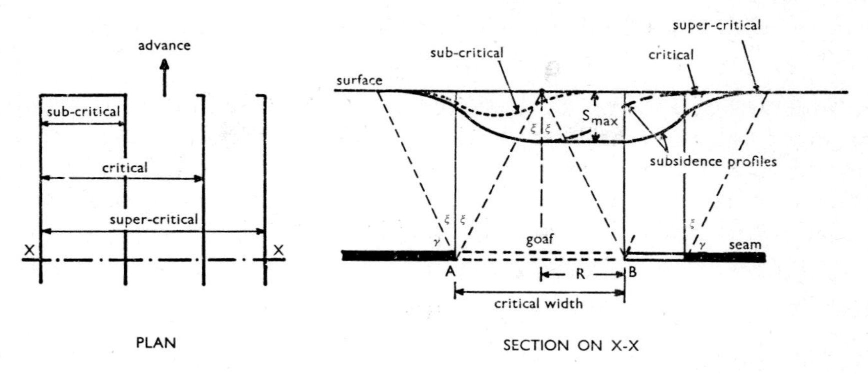

Fig. 11.2 Critical width of extraction to produce maximum subsidence

TERMINOLOGY

Subsidence measurements are commonly given as dimensionless values so that empirical curves can be used to predict closures easily. The width/depth ratio (w/h) is most important. For an area to become critical the w/h ratio is approximately equal to 1.4, e.g. at 100 m depth an area with sides of 140 m would have to be completely extracted to achieve maximum subsidence.

A list of standard symbols has been agreed in Europe, and has been adopted in Australia by at least A.I.S./B.H.P. Collieries, to denote subsidence values. The chief ones of these are:

- m = seam thickness extracted
- α = angle of dip
- h = vertical depth from surface
- w = width of panel
- L = length of panel
- a = subsidence factor = ratio of complete subsidence to seam thickness, i.e. S_{max}/m
- ζ = limit angle (angle of draw) from vertical
- γ = limit angle from horizontal
- S_{max} = maximum possible subsidence = seam thickness multiplied by a subsidence factor
- S = maximum subsidence in any profile
- s = subsidence at a particular point in a profile
- t = transition point = the point of transition between concave and convex curvature of a subsidence profile. This coincides with the half subsidence point S/2
- R = radius of critical area.

EFFECT OF SEAM THICKNESS, GOAF SUPPORT AND DEPTH

Subsidence depends on all of these factors. Provided that the depth is sufficient to break and bend the overlying rocks and that a super-critical area has been worked, then the surface will subside to a large percentage of the seam height. In the absence of any imported filling material this is about 90 per cent maximum subsidence.

The maximum subsidence can be reduced to about 45-50 per cent by solid stowing or pneumatic stowing with imported waste rock, but this is expensive. Factual evidence is scarce but hydraulic stowing with sand fill may reduce the subsidence to about 30 per cent of extracted thickness. The reason for these high minima is that vertical convergence of as much as 25 per cent of extracted seam height may have taken place before any stowing material was introduced. Thus the 10 to 20 per cent compaction of sand fill, or 35-40 per cent compaction of rock fill, has to be added to any seam closure that has already taken place. If pillars of coal are left in the mined area then these reduce subsidence in the same way as imported filling, because the extracted volume has been reduced.

Depth is concerned in two ways: the pressure (approximately 25 kPa/m of depth) must be sufficient to reconsolidate the broken rock. This happens at 1-200 m dependent on rock strength. Secondly the deeper the workings the larger the area that has to be extracted to cause the relaxed pressure arch to break through to the surface and produce subsidence.

THE SUBSIDENCE PROFILE

Figure 11.3 shows subsidence profiles for three different widths of workings. In each case the subsidence curve has a different shape. The location of the point of maximum subsidence is over the centre of the mined-out area (in flat lying strata) and the magnitude of the subsidence at this point can be predicted with about ± 10 per cent accuracy by use of suitable empirical curves. The locations of the limits of draw may also be predicted with reasonable accuracy. In order to be able to draw the subsidence profile it is necessary to know the location of at least one more point on the profile, and for this purpose it is convenient to use those at which the subsidence is equal to one-half of that at the centre of the subsidence area.

As shown in Figure 11.3 the half-subsidence point varies in its relationship to the edge of the workings in accordance with the width/depth ratio of the mined-out area (goaf).

Figure 11.3A is a typical profile over an area which has reached critical width and the half subsidence point is located over the goaf.

Figure 11.3B shows a narrower extraction area with the half-subsidence points located over the ribsides. Figure 11.3C shows a very narrow panel with the half subsidence points located over the unworked seam well outside the edges of the goaf. In this case the profile is very flat and the amount of subsidence over the rib is not much different from that over the centre of the panel.

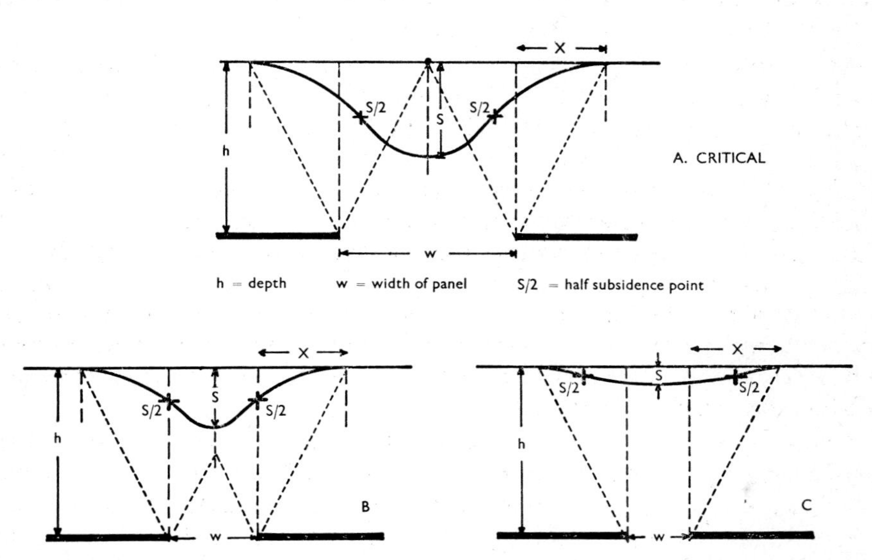

Fig. 11.3 Subsidence profiles over workings of various widths

PARTIAL SUBSIDENCE

When an area less than the critical value is extracted then the maximum subsidence is reduced progressively down to a theoretical zero above a single roadway. Case studies at British mines produced curves as shown in Figure 11.4. Use of these curves enable subsidence to be predicted to ± 10 per cent. They appear to apply almost irrespective of the overlying strata but semi-liquid deposits (e.g. peat), or thick beds of sand, have a widening effect on the subsidence trough. Indications are that these figures are reasonably valid for Australian coal mines although current research indicates that strong sandstone or conglomerate beds can reduce subsidence over shallow workings.

Figure 11.4 shows the rate of decrease of S/m (subsidence expressed as a ratio of seam height) with a narrowing of the extraction panel. In some cases a panel of given width may not advance a sufficient distance to give full subsidence and a second curve is given to correct for the limited advance. The value of S/m obtained from the first curve is multiplied by the value of s/S obtained from the second curve.

Accordingly,

$$\text{total subsidence} = \text{seam height extracted} \times \frac{S}{m} \times \frac{s}{S}$$

This is the subsidence at the centre of the rectangular panel.

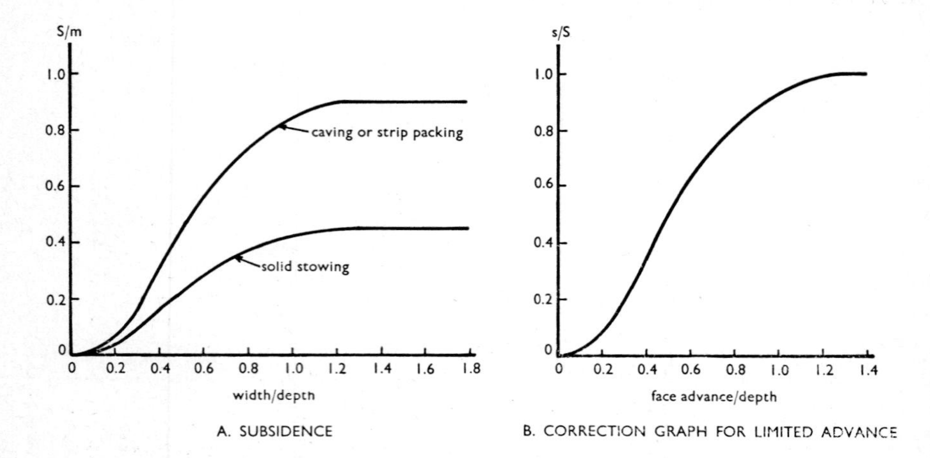

Fig. 11.4 Subsidence at various width-depth ratios[2]

More recently Orchard and Allen[3] have re-examined the evidence from some ninety cases of subsidence to obtain more accurate empirical curves. Figure 11.4A can be replotted as in Figure 11.5A which shows a linear relationship between width of panel and depth of working. Orchard and Allen show that a more careful study of the effects of shallow workings produced a non-linear relationship as in Figure 11.5B. The former line for $w/h = 0.25$ is also shown: if this is compared with the value from Figure 11.4A it can be seen that subsidence over narrow workings at shallow depths has previously been over estimated. The curve in Figure 11.4A is at maximum accuracy at a depth of 300 m.

The maximum measured subsidence in Figure 11.5B is about 90 per cent of extracted seam thickness if results affected by old workings or overlapping workings are ignored. The large w/h ratio necessary to cause 90 per cent subsidence can only be obtained in shallow workings because in deeper workings the necessary large value for w would cover several panels and there would be several pillars left in situ to protect roadways. These pillars reduce subsidence and some coalfields have been found to have no subsidence in excess of 75-80 per cent. There is also a shortage of results for subsidence at depths of less than 100 m. Dependent on rock strengths the strata may just cave-in as a plug over very shallow workings, especially if thick seams are involved.

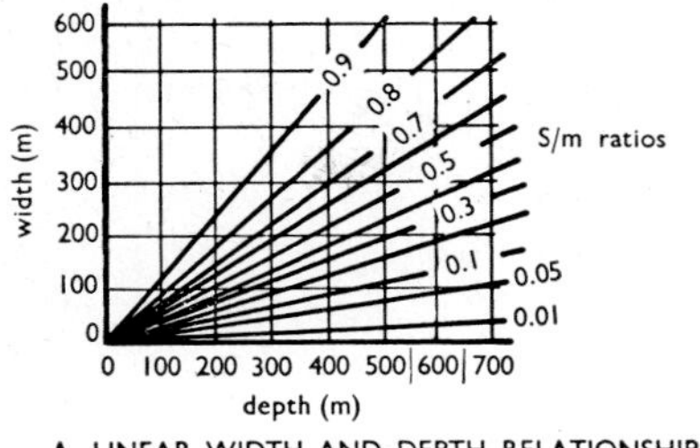

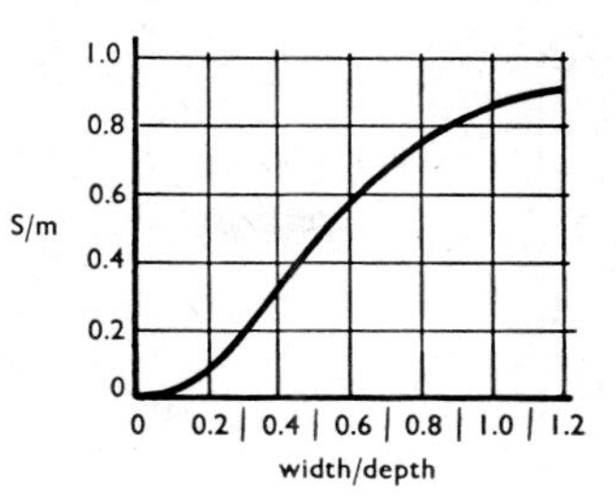

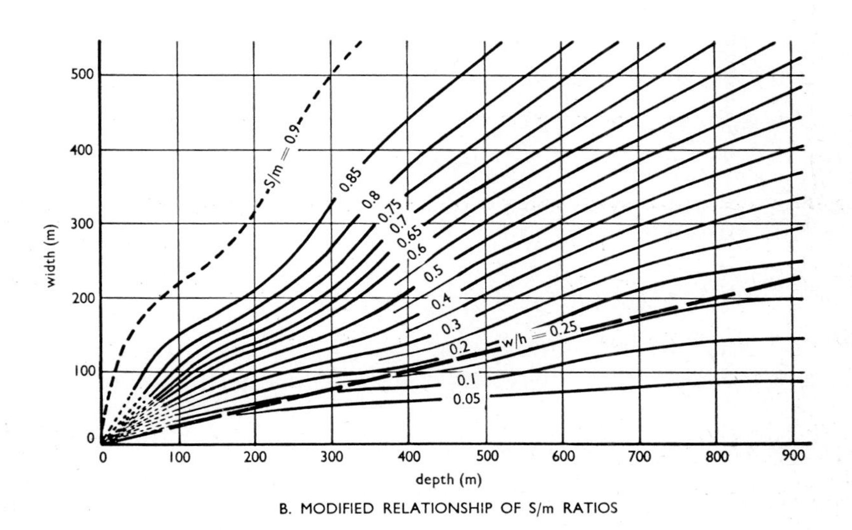

Fig. 11.5 Modified relationship between width and depth[3]

In these shallow workings it has been noticeable that subsidence is less than extrapolated empirical curves indicate. The broken rock above mine workings does not re-consolidate and a large proportion of voids results. Orchard and Allen show a pair of curves (Figure 11.6) which relate the volume of coal extracted to the volume of the resultant subsidence trough. The amount of increased void space from rock fragmentation swell, and from bed separation, is shown by the difference between subsidence volume and extraction volume. At shallow depths the volume subsided is only 21 per cent of the volume extracted. This could explain the movement over the solid coal of the half-subsidence points in Figure 11.3 Bridging, or arching, is very effective at shallow depths.

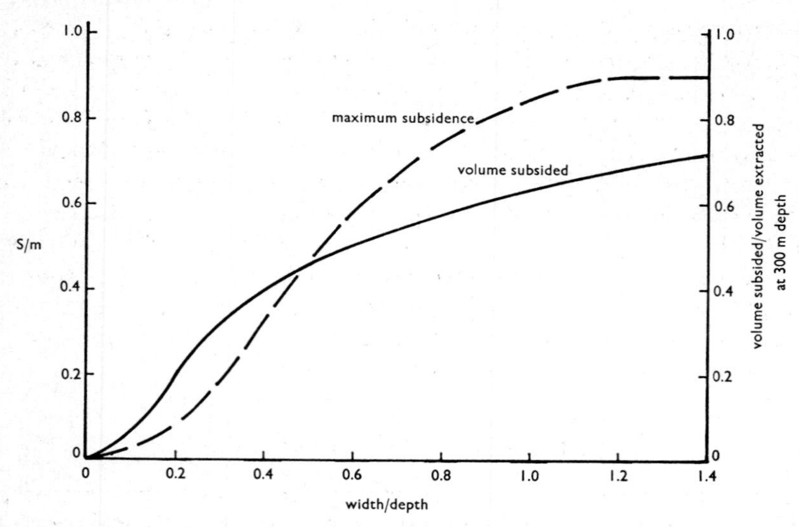

Fig. 11.6 Comparison of volume of subsidence trough with volume of coal extracted[3]

EFFECT OF INCLINED STRATA

The effect of inclination of a seam being worked is to displace the subsidence trough in the direction of the deeper part of the workings. Figure 11.7 is a cross-section showing how the maximum subsidence is displaced so that instead of being directly over the centre of the goaf area it lies on a line normal to the centre of the goaf area, and the points of half subsidence are similarly displaced toward the dip side. The angles of draw are also altered but not necessarily in direct proportion to the inclination of the seam, as are the other main points on the subsidence profile.

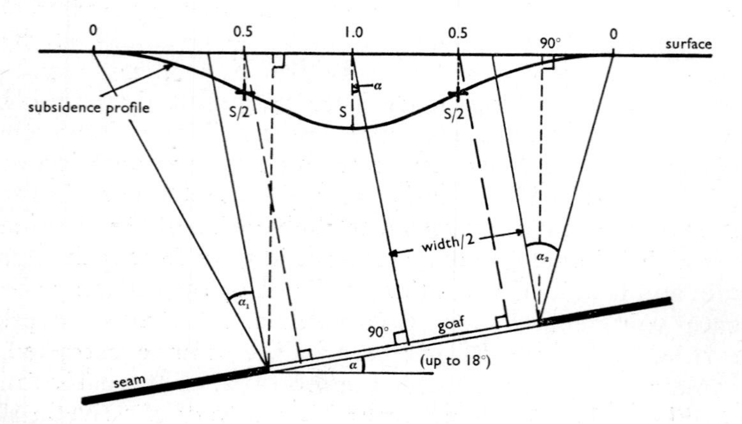

Fig. 11.7 Effect of inclination on subsidence profile

The foregoing comments apply, so far as is known, only to workings with inclinations up to 1 in 3 (18°). Presumably with steeper gradients the effect of displacement on the subsidence trough must change and eventually the displacement must reach a limit. The displacement effect is also confined to workings at moderate or considerable depth as it has been found by model experiments, and by field experience, that for shallow workings the maximum subsidence actually occurs on the rise side of the panel centre. This may be explained by the fact that the upper part of the extraction is so much shallower than the other part that a greater amount of subsidence takes place over the former. With deeper workings the difference in depth between the upper and the lower halves of an extraction becomes insignificant.

SURFACE MOVEMENTS

When a subsidence trough is formed at the surface the central part subsides vertically and the remainder moves inwards as well as downwards. The vertical components of the resultant movement are called subsidence and the horizontal movements are called displacement.

When adjacent points on the earth's surface are displaced horizontally at different rates the ground must either be extended or compressed horizontally, depending upon whether the inner point or the outer point is moving the more rapidly. Since the centre point is not moving horizontally, but the points on either side of it are moving inwards it is evident that the ground in the central portion of the subsidence trough will be subjected to lateral compression. Similarly, since the points on the outer edges of the subsidence trough are not moving horizontally, the ground between these points and the edges of the excavation must be extended horizontally.

Figure 11.8 illustrates the general limits of the zone of horizontal extension and the zone of horizontal compression of the ground surface. Figure 11.9 shows how a strain curve is drawn to indicate the unit strain (change in length expressed as unit change per unit length)

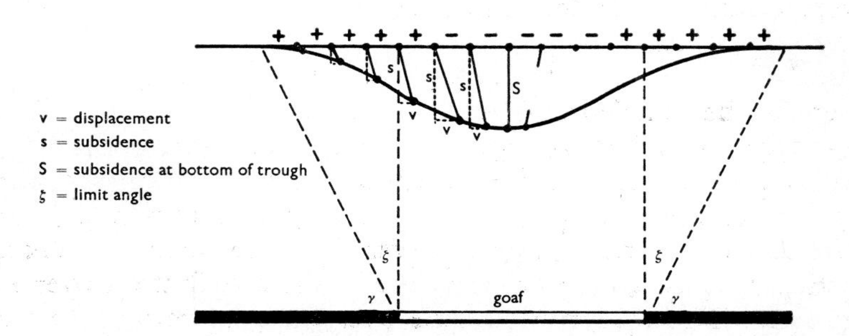

Fig. 11.8 Horizontal and vertical components of ground movement

occurring at any point on the ground surface. The surface is used as a datum plane or zero line and extension (stretching) of the ground surface is shown above the datum and compression (shortening) of the ground surface is shown below the datum.

The point at which extension is reduced to zero, and compression begins, is more or less coincident with the point at which half subsidence occurs on the subsidence profile, although this is not always the case. It appears that with subcritical widths of extraction the relationship between the two points may be less simple.

RELATIONSHIP OF STRAIN TO SUBSIDENCE AND DEPTH

The magnitude of the unit strain at any point on the ground surface is indicated by the curvature of the ground surface profile at that point. This curvature, in turn, depends upon the amount of the vertical movement and the distance over which the subsidence basin, or trough is spread.

Generally the thicker the seam the greater is the subsidence, and the shallower the seam the shorter the distance over which subsidence is spread, with both of these factors tending to produce a sharper curvature in the ground surface.

From this we see that the unit strain at any point is proportional to the maximum subsidence and is inversely proportional to the depth of the working, and we have the relationship:

$$\epsilon = \frac{S}{h}$$

where ϵ is the maximum unit strain, S is the maximum subsidence and h is the depth of working.

In different coalfields in Britain[4] the relationship has been found to vary between

$$\epsilon = \frac{0.7S}{h}$$

and

$$\epsilon = \frac{0.85S}{h}$$

An average relationship for most cases of

$$\epsilon = \frac{0.75S}{h}$$

was found to be sufficient.

The curves A and B in Figure 11.9 show the differences in strain intensities over shallow as compared with deep workings. Over the deep workings the strain is spread over a greater area and is of less intensity at its maximum. These curves show the strain at sub-critical and critical widths. Curve C shows the strain distribution over super-critical areas. The compression zone over the goaf returns to zero or near zero, over the central portions.

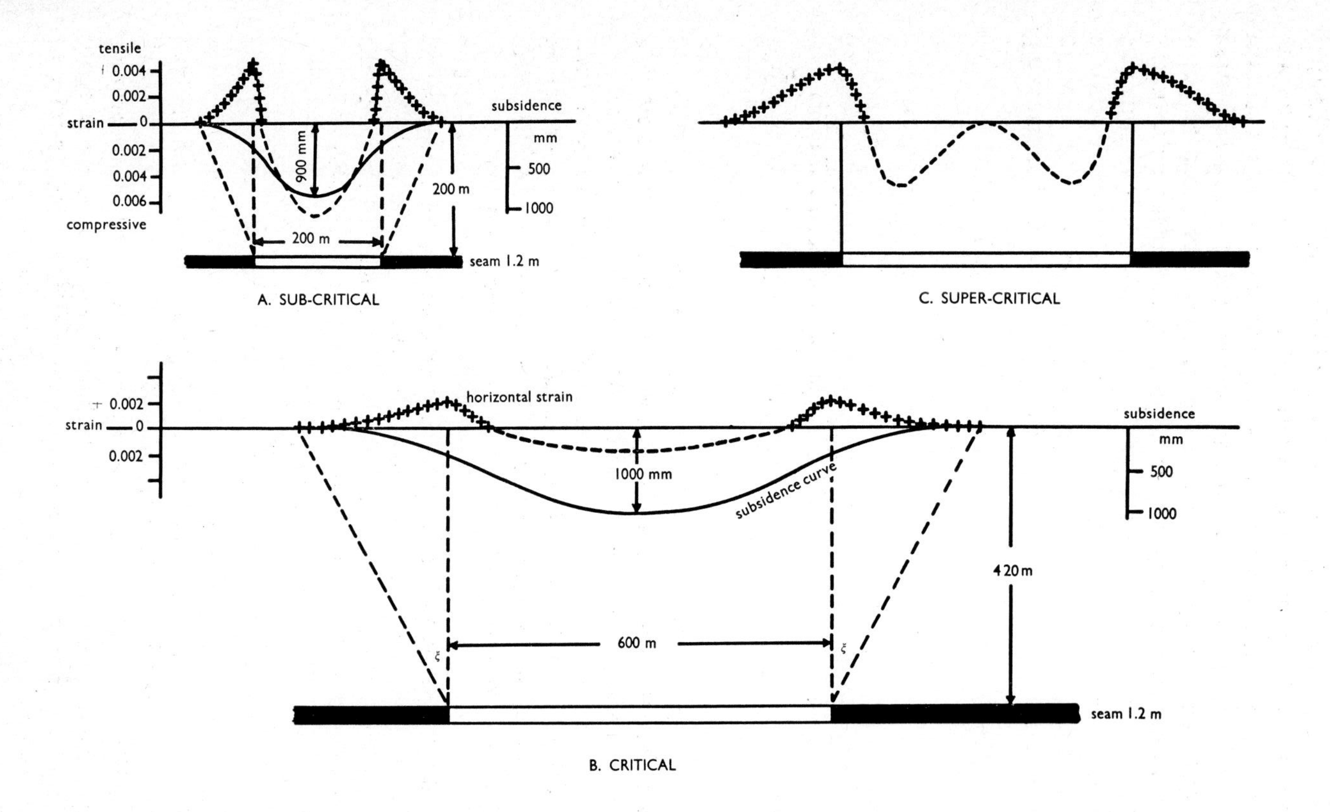

Fig. 11.9 Effect of width and depth of extraction on surface strain

COMPRESSIVE STRAIN WHEN EXCAVATION IS AT CRITICAL AREA

In narrow workings, such as that shown in curve A of Figure 11.9, there is an accentuated compression zone where the intensity of the compressive strain is approximately double that of the adjacent tensile strains. This is because the compressive strains and tensile strains must balance out (i.e. the area between the datum line and the compression curve is equal to the area under the two tensile zones).

It has been found that the maximum extension, and also the maximum normal compression, occur when the width/depth ratio is about equal to the radius of the critical area. At this width (half the critical width) subsidence has not reached its maximum development, and when the width of the extraction is increased the subsidence increases. The strains, however, have already reached their maxima and do not increase, so that the relationship $\epsilon = 0.75S/h$ is only attributable to width/depth ratios not greater than the radius of the critical area. Accordingly, many investigators, when predicting unit strain, do so by using the maximum subsidence (called critical subsidence) obtained from working up to a width/depth ratio of about 0.7 because this is the width where the angle of draw is about 35° (tan 35° = 0.7). If the angle of draw differs from 35° then the width is altered to suit. In the case of compression the maximum strain is, or can be, double the intensity of the tensile strain. (i.e. up to $1.5S/h$.) However, this applies only to widths of extraction up to the critical width for maximum strain.

For workings of greater width the compressive strains are gradually reduced until they reach the same intensity as the tensile strains at the critical width (Figure 11.9B). There is not much information available about the strain values in this range between the half critical width and the full critical width.

DAMAGE CAUSED BY GROUND MOVEMENT

The damage caused depends partly on the vertical component of displacement, and partly on the horizontal component (Figure 11.10). The vertical component, resulting in lowering of the ground, causes differential subsidence or tilt. This can be a nuisance or even a danger in structures. Gradients induced by mining subsidence do not often exceed 1 in 50, even over shallow workings, while over deep workings such gradients become very small—for example the maximum slope over the workings of a 1.5 m thick seam 900 m deep would be 1 in 400.

The most obvious effect of vertical movement is interference with the flow of canals, rivers and streams, sewers, etc. These are conditions which normally can be corrected at reasonable cost.

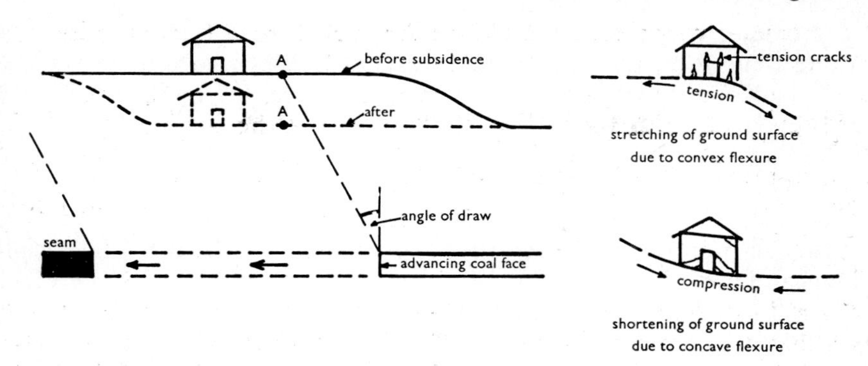

Fig. 11.10 Effects of subsidence on surface structures

Vertical movement need not be dangerous, and it can be controlled so as not to damage structures. One notable instance[5] is the controlled lowering of the inland harbour at Duisburg in Germany. To compensate for scouring of the river bed the mining companies cooperated with the harbour authorities to lower the harbour about 2 m. Buildings up to 20 m high were lowered and an 1800 m long bridge over the harbour was lowered with provision for jacking to keep it level.

Lateral movement in itself is not destructive of surface improvements. If the ground over a considerable area moves as a unit then any structures situated in the area will not be subjected to stretching or to compression. It is differential horizontal movement which results in the strain (stretching or compression) of the ground surface that accounts for most of the damage to structures. Such strain may amount to 0.8 or even 1.0 per cent in exceptional cases over shallow workings, but it is commonly 0.25 per cent or less. A 0.20 per cent change in length is more commonly referred to by subsidence specialists as 2 mm/m, which may also be written as 0.002. With that intensity of strain a building 50 m long would be stretched or shortened by 100 mm.

DAMAGE TO STRUCTURES

This is in proportion to the strain on the ground surface, but the extent of damage also depends upon the size and construction of the building, its shape, and the degree to which it is secured in the ground. Of these factors size is the most important in determining the extent of damage. Thus a building 50 m long with a strain of 2 mm/m will suffer the same damage as a building 100 m long with a strain of 1 mm/m. (This is an over-simplification, but it illustrates the relationship.)

According to the location of a particular structure with respect to the subsidence profile, the site will suffer concave or convex curvature.

In the case of very shallow workings the curvature can actually be seen on long structures, e.g. in a terrace (row) of houses where tensile strain may be sufficiently large to cause a large fracture, the curvature would cause the fracture to be wider at the top than at the bottom.

Figure 11.11 illustrates the approximate relationship between the strain of the ground surface and damage to structures. It is not necessarily exact since it does not take into account the shape of the building nor its age, nor the materials used, but most cases fit into the graph which can be used as a guide to the intensity of damage. In Figure 11.11 the length (i.e. the greater dimension of the structure) is plotted against the strain measured or calculated for the site, and the plot falls into one of five categories of damage. These five categories are listed in Table XV which also gives their accepted definitions.

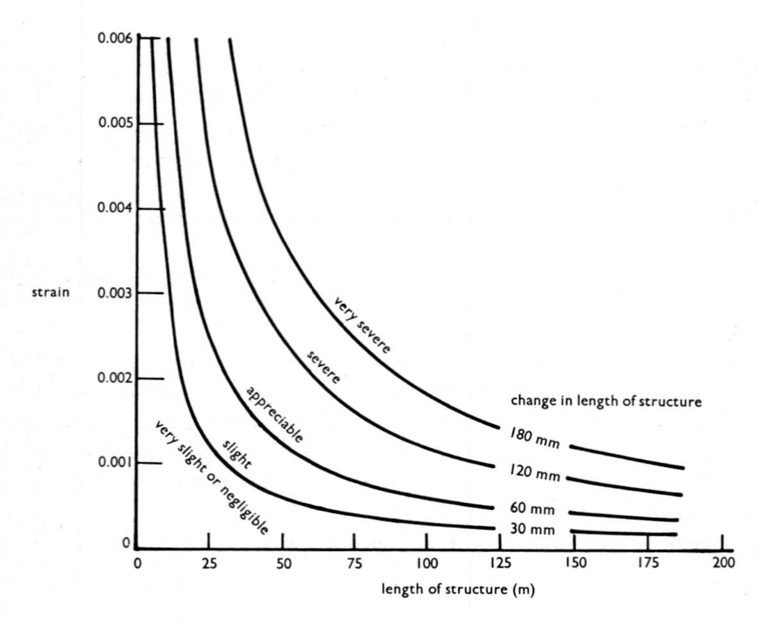

Fig. 11.11 Relationship of damage to length of structure and ground strain[2]

CHARACTERISTICS OF SUBSIDENCE DAMAGE

When a structure is lengthened by extension of the ground surface, open fractures occur, first at natural joints such as butt jointed brickwork, and then at weak points in brickwork such as the locations above and below windows and doors.

Similarly floors and roofs may open at joints or at changes of level, but if floors or walls occur in unbroken lengths then fracturing takes place at random and may take the form of one or two large cracks, or a large number of small ones.

With extension of the ground surface the joints of pipelines and cables tend to pull apart, and bridge abutments move outwards so that an unprotected bridge would sag.

Compressive strain is characterized primarily by buckling or lifting of walks and pipes, and the fracture of steps, walls, etc. In some instances the buckling action associated with the shortening process causes pipes to lift out of the ground to form short self-supporting arch spans.

TABLE XV.

NATIONAL COAL BOARD CLASSIFICATION OF SUBSIDENCE DAMAGE

Change of Length of Structure*	Class of Damage	Description of Typical Damage
Up to 30 mm	1. Very slight or negligible	Hair cracks in plaster. Perhaps isolated slight fracture in the building, not visible on outside.
30-60 mm	2. Slight	Several slight fractures showing inside the building. Doors and windows may stick slightly. Repairs to decoration probably necessary.
60-120 mm	3. Appreciable	Slight fracture showing on outside of building (or one main fracture). Doors and windows sticking; service pipes may fracture.
120-180 mm	4. Severe	Service pipes disrupted. Open fractures requiring rebonding and allowing weather into the structure. Window and door frames distorted; floors sloping noticeably; walls leaning or bulging noticeably. Some loss of bearing in beams. If compressive damage, overlapping of roof joints and lifting of brickwork with open horizontal fractures.
More than 180 mm	5. Very severe	As above, but worse, and requiring partial or complete rebuilding. Roof and floor beams lose bearing and need shoring up. Windows broken with distortion. Severe slopes on floors. If compressive damage, severe buckling and bulging of the roof and walls.

* Values rounded to millimetre equivalents by author.

EFFECTS OF DISCONTINUITIES IN THE GROUND

Faults or sudden changes in the type of rock or subsoil may cause uneven subsidence of the surface. In faulted ground the earth may slide down on one side of the fault more than on the other so that a step may be visible in the surface. This type of discontinuity is most marked when it occurs in a highway or similar smooth surface and even when repaired is still visible as a short steep ramp.

THEORETICAL CALCULATIONS OF SUBSIDENCE

Most of the previous material has been based on empirical values deduced by the N.C.B. subsidence engineers from case studies. Theoretical workers have to use curves obtained in practice to determine the validity of their approach. To date no-one has obtained a theory to cover all conditions. A summary of the various approaches is presented below.

Marr[6] lists the following as the most important factors which influence the amount of subsidence for any particular case. These are divided into two groups; constant factors and uncertain factors.

Constant factors.

1. The thickness and the dip (gradient) of the seam being worked.
2. The width of the working face.
3. The distance which the working face advances.
4. The degree of packing, or filling, applied to the face.
5. The depth of the working face below the surface.

Uncertain factors.

1. The character of the strata between the seam and the surface.
2. The presence of previously mined-out areas in the same seam.
3. The presence of goaf above or below the seam being worked.
4. The presence of large faults.

He goes on to derive an expression for the subsidence profile in which

$$s = S.\exp\left\{-\left(\frac{x^2n}{b + cx^2n}\right)\right\}$$

s and S were given earlier,
x = distance from the position of maximum subsidence
b and c are constants
n = 1 or 2.

Using Marr's formula for dipping seams a separate profile has to be calculated for the high and low sides of the extraction area.

Other theoretical approaches are those due to Bals, Knothe, and Berry and Sales. The methods of both Bals and Knothe proceed from the hypothesis that the movement of an arbitrary point on the surface is determined by the overall effects of the individual elementary parts of the extracted area (principle of superposition).

Zenc[7] gives details of the methods and compares them with measurements in the Ostrava-Karina coal basin in Czechoslovakia. Bals' method (developed in 1932 when he was a mine surveyor in Germany) assumed that the excavated area consisted of a large number of small equal masses which drew the surface into the excavated area. The influence of any such small mass on the surface point will be inversely proportional to the square of its distance from the point (Newton's law of masses). The method of calculation of subsidence is complicated and since agreement with observed movements in Australia, Britain and Czechoslovakia have not been good the student is referred to Zenc's paper for further details.

Knothe's method proceeds from the assumption that the resultant movement of a point on the surface is given by the sum of movements caused by the extraction of different areas of the seam. The area of influence is a rotational area with a vertical axis whose meridianal section is a Gaussian curve and involves the use of a double integral. As with Bal's method, the theoretical approach by Knothe only appears to work under his own conditions in Upper Silesia and cannot be widely adopted.

Australian, British, American, and Continental investigators have all found that whereas under the most favourable flat bedded deposit conditions the theoretical approaches can give a reasonable approximation the results are hopeless on even slight inclines, and can also be wrong on flat deposits. In addition they generally are only accurate on vertical displacements and can be hopelessly inaccurate on the horizontal strains. Finally they take no account of the time factor involved in subsidence.

Berry[8], and Berry and Sales[9], have attempted to predict the amount of subsidence by considering the mined area to be an elastic material. When the closure of a thin slit in an infinite, and a semi-infinite elastic, isotropic plate was considered[8] the calculated surface displacements were smaller than those obtained in practice. Further calculations by Berry and Sales[9] in which the medium was considered to be transversely anisotropic (the elastic constants for perpendicular and horizontal directions are different) produced closer agreement with observed subsidence but there were still some discrepancies. Opinions vary but the elastic approach has been discontinued in Britain.

Seldenrath[10] adopted a soil mechanics approach to mining subsidence and obtained an approximate agreement with measured subsidence. There is however no logical reason why an argument for granular materials should apply to jointed rock and the original approach has not been followed up. All these theoretical approaches have no time factor included and research has shown time to have an influence on subsidence. The extensive experience of the National Coal Board indicated that their empirical curves gave a much better approximation of both vertical subsidence and of strains. To date none of the theoretical approaches has been considered to be useful enough to supersede them.

THE TIME FACTOR IN GROUND MOVEMENTS

It is important to know how long subsidence will take to develop, and how long it will continue with relation to the extraction of the coal seam, since it is necessary to know when structures are likely to be first affected, when the likelihood of damage is most acute, and finally when the risk of damage is past so that repairs to affected structures may be undertaken. Wardell[11] summarised the relationship between surface subsidence and coal seam extraction when he stated, in effect, that as long as coal is being removed from the critical area of a given surface point, that point will continue to subside.

A typical subsidence development curve is given in Figure 11.12.

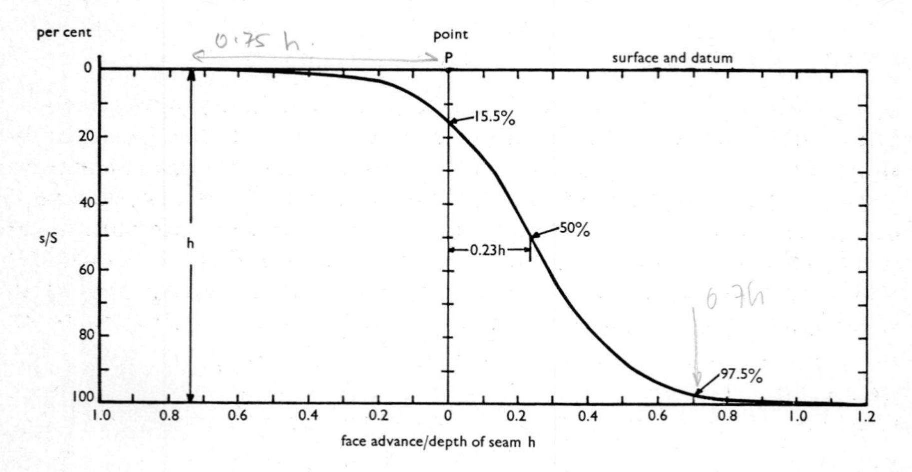

Fig. 11.12 Subsidence development curve[2]

Measurable subsidence starts when the face is within about three quarters of the depth (0.75h) from the surface point P, and reaches about 15 per cent of the maximum possible subsidence when the face is below the point on a line drawn normal from the seam to P. At about 0.8h beyond the point subsidence is virtually complete within a few per cent. The time taken for subsidence to occur depends entirely on the time it takes for a coalface to be worked through the critical area, and this in turn depends on the depth and the angle of draw, and also on the rate of advance.

For shallower seams the critical area is so much smaller that, for a given rate of face advance, the subsidence is completed much sooner than in deeper workings. It is also necessary to consider how many coal faces would be needed to extract the whole of a critical area. One face may remove only part of the area and another one or more faces may be required to complete the critical extraction. Each face on passing through the critical area causes extra subsidence, and it is probably this which in the past has led mining engineers to believe that subsidence occurs over a period of some years after mining.

It is difficult to determine exactly what length of time is required for movement to be transmitted from the workings to the surface, or how long subsidence continues after the workings have stopped. However, it has been noted that when work is stopped temporarily at a face, such as for a holiday period, the rate of subsidence within the influence area of that face slows down almost at once. It has also been noted that when an advancing face stops on reaching the limit of its working, surface subsidence slows down almost immediately and usually stops within a few days, or at the most within a few weeks time.

Subsidence over Pillar Workings

Where mining is by pillaring with pillars left permanently in place, rather than by longwall methods, the time required for surface subsidence to run its course may be quite different. Subsidence over pillar workings will usually be much slower than that over longwall workings because the lowering of the roof is a gradual process due to the slow yield of the pillars of coal which have been left in place. If the pillars are substantial and not subject to weathering then only small amounts of subsidence will occur. With smaller pillars subsidence may extend over a period of years, until, with the effects of time or weathering, or other causes, the pillars suddenly collapse resulting in considerable subsidence and damage. Such subsidence is obviously unpredictable and may vary from a few months to a hundred years or more.

Generally the movement over an area of partial extraction is much smaller than over total extraction, and partial extraction is one method used for mining where subsidence must be kept within limits to avoid excessive damage to surface improvements (i.e. houses, factories, etc.).

EFFECT OF PILLARS ON SUBSIDENCE

In room-and-pillar workings where pillars are extracted the subsidence profile over the retreating pillar line may be expected to correspond closely with subsidence profiles for longwall workings with total caving behind the face.

On the surface above first workings the amount of mineral removed is so small that movements are negligible; say a few centimetres for a seam of 3 to 4 m thickness. As more mineral is removed then movements will still be of the order of centimetres. (See later under Partial Extraction.)

However the pillars must be at least h/10 on their minimum dimensions for efficiency and stability. If floor and roof strata are very weak then the strata will flow up around the pillars and, while they remain intact and support the immediate roof, their subsidence reduction effect may be lost. In these cases, and in cases where substantial but isolated pillars are left, e.g., under a railway line, or both sides of a fault, or between a panel, then the subsidence will be modified along the lines of Figure 11.13.

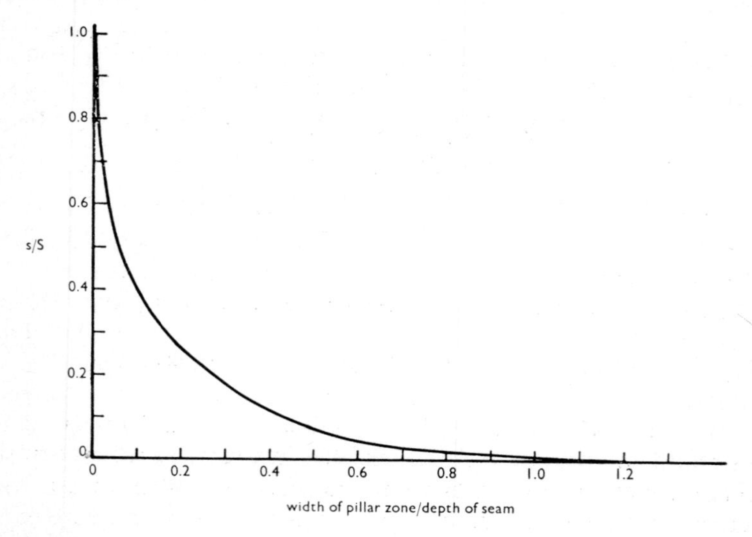

Fig. 11.13 Subsidence over solid coal pillars[2]

Methods of Control of Surface Subsidence

For hundreds of years the traditional method for protecting buildings and other surface structures has been to leave large coal pillars in the seam underneath them. Because of the angle of draw it is necessary to leave pillars which are much larger in area than the structure to be protected. The quantity of coal which is permanently locked up by these pillars ensures that only the most important and vulnerable structures are protected in this way.

PARTIAL EXTRACTION

Partial extraction of coal seams, leaving pillars for permanent support of the overlying strata, has been successfully used in many cases. Partial extraction by room-and-pillar methods has not always been effective because the relatively small pillars which are left for support tend to crush, especially when the seam is located at moderate to great depths. The roof and floor layers also tend to flow, or creep, around the small pillars.

Successful methods for partial extraction involve the leaving in place of strip pillars which may be from 30 to 100 m wide, dependent upon the depth of the seam, with the mined-out strips between the pillars being on the order of 30-50 m wide.

Prediction of the reduced amount of subsidence is obtained by superposition of the calculated subsidence profiles for the individual extraction areas. Figure 11.14 shows a case quoted by Orchard and Allen[3].

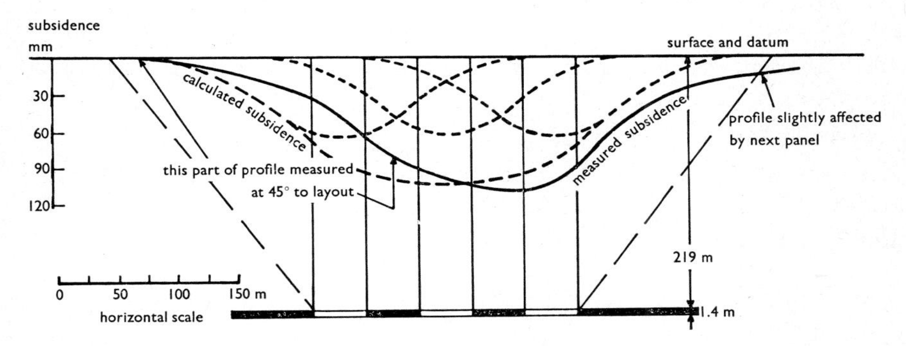

Fig. 11.14 Addition of partial extraction subsidence profiles[3]

The theoretical subsidence profile is obtained by addition of the three profiles calculated as above for the three extraction areas. The measured subsidence curve shown is different from that predicted for two reasons. The left hand side of the curve was measured over a line turned to 45° to a true cross-section. The right hand side was affected by an adjacent panel. It is a point to note that old goaf adjacent to new workings usually increases the amount of new subsidence. The principle of superposition must be applied to allow for adjacent panels; in effect the w/h ratio has been increased. Reduction in subsidence by partial extraction is a direct result of the quantity of mineral left in situ, and the stability of the pillars in which it is contained. If the pillars collapse then the w/h ratio must be recalculated, but it would be reasonable to reduce the seam thickness extracted, m, to allow for the collapsed coal.

HARMONIC EXTRACTION

Harmonic extraction is the carefully phased removal of all the coal from the critical area underneath a structure so that it is lowered smoothly in a manner in which it is more or less free from horizontal strain.

Removal of all coal from the critical area cannot be accomplished by the extraction of a single long face, because the travelling wave of subsidence, and of strain over the centre of such a panel of coal, could well be equal to the maximum possible movements for the seam in question and could cause considerable damage. Only in the case of a deep seam could one long face be beneficial because workings at great depth cause little strain at the surface.

The method evolved to date is illustrated in Figure 11.15 and involves working two or more panels so sited that the structure to be protected is just inside the edge of the leading panel; that is the first panel to be worked under the area. This is the very wide panel on the left which advanced successively to the positions shown. On June 22, 1955 this panel was just entering the critical area, on January 3, 1956 it had passed under the church, and on May 3, 1956 it had passed out of the critical area.

The following panel (the narrow panel at the right of the illustration) was allowed to lag the lead panel by about 76 m and had passed under the church by March 14, 1956, and had passed out of the critical area by August 23, 1956.

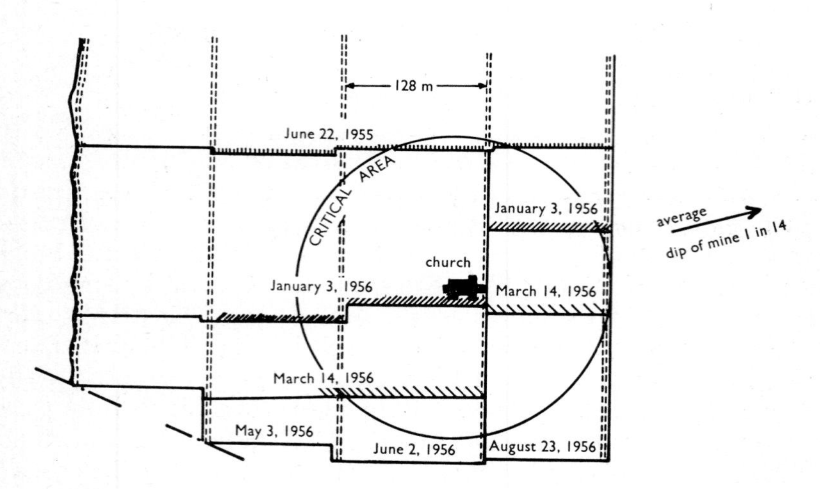

Fig. 11.15 Harmonic extraction (adapted from Reference 11)

The effect of such 'stepped' faces is that the travelling wave of subsidence is less than the maximum, since only the edge of the panel is travelling below the structure and, in fact, the amount of movement over this edge is between one third and one-half of the maximum subsidence at the centre of the panel. The subsidence from the second panel, which is following at a carefully calculated distance, is superimposed upon the ground already being affected by the first face with the result that the subsidence is spread over a greater distance. By this means the surface curvature is reduced and the transverse strains reduced in proportion.

The direction in which the greatest length of the building lies must be taken into account when calculating the distance between the two faces. If the long dimension of the building is orientated at right angles to the direction of face travel then the travelling surface wave is less significant than the transverse surface wave and measures should be taken to reduce the magnitude of the transverse wave by shortening the distance between the two faces.

HARMONIC EXTRACTION WITH MULTIPLE SEAMS

Where seams are mined simultaneously on more than one horizon the maximum horizontal strain which would be produced by the mining of only one seam may be reduced in magnitude by so spacing the advancing faces that the extensional strain produced by one extraction face is partially neutralized by the compressional strain produced by the extraction face in another seam.

Surface Precautions to Reduce Damage

The Mine Subsidence Compensation Act (1961-1967) of New South Wales set up a Compensation Fund. Coal mining companies must pay a levy into the Fund and damage from subsidence due to mining coal or shale is compensated for by payments from the Fund.

The Fund will compensate for direct damage, will contribute to protective works necessary to minimise damage, e.g.,

(1) building retaining walls or bolting together or underpinning buildings and walls

(2) altering the approaches to or the levels of land or buildings

(3) raising, lowering, diverting or making good roads, tramways, railways, pipelines, sewers, drains, etc.,

and will also pay rent for a period in which a house or building is untenantable.

However when a building is to be newly erected on land known to be liable to subsidence, then unless the owner takes precautions approved by the controllers of the Fund the Fund will not compensate for the damage.

Similar legislation exists in other countries, but (for instance in Britain with a nationalised industry) it may not always be necessary to set up a separate fund. Codes of practice may be set up by local government authorities, or by engineering institutes, for example the Institution of Civil Engineers in Britain produced a report[12].

The types of damage likely to be encountered in different situations, and methods to reduce the effect of subsidence as suggested by the N.C.B.[1,2] in conjunction with other authorities, are given below.

WATER DRAINAGE AND FLUIDS IN PIPES

This is often a case of notifying the local public authorities so that their engineers can take precautions in advance. The main problems are alteration in vertical levels which upset the natural flow in inclined pipes or on canals, rivers, etc., and changes in lengths which can buckle pipelines in the compression phase and pull them apart on contraction.

Rivers and canals may need their banks built up to increase the freeboard to the new level grade, preferably before extraction starts. Their beds may crack and leak on embankments (or in to shallow underground workings). Puddled clay can be used to seal the beds of canals, rivers or ponds, or occasionally bentonite muds (natural swelling clays) may be injected. A modern trend is to use P.V.C. welded sheets to line the affected lengths. Plastic sheeting has been particularly good in preserving farmers' stock ponds, and has been used to line small reservoirs.

Pipes of various services are protected by inserting friction seal telescopic joints of which there are several proprietary brands available. Those are usually flexible enough to deal with the travelling wave in front of an excavation, but over the side ribs of a thick seam it may be necessary to install flexible hose, particularly if there is a fault which will give a sharp step in ground level.

Temporary pumping arrangements for low pressure mains e.g. sewerage, may have to be made. The specially-jointed pipes should either be left free on the surface or buried in granular fill, e.g., gravel or loose metal, not rigidly buried, or they may be cracked.

ROADS AND BRIDGES

When shallow seams are partially extracted roads on the surface can be deformed into a severe switchback. Temporary speed restrictions will have to be imposed until repairs can be effected. Bridges are the main problem. New ones are best built on a pivot with expansion combs at each end.

SMALL BUILDINGS

These can be built on solid concrete rafts so that they can tip as the subsidence wave comes through without inducing cracks in the building. Slightly larger buildings may be built on cellular rafts. Gravel, sand or other friction-reducing material should be laid between the raft and the ground surface.

SCHOOLS, FACTORIES, ETC.

These can be built as a series of units, either with braced frames, sometimes spring loaded, or as separate blocks with weatherproof strips between. The main object is to prevent the build-up of strain in long lengths of wall by providing natural breaks to dissipate the extension or compression. This can be done for an existing building by digging a trench around it and filling it with coke, or other coarse granular material. The trench must extend below the level of the foundations to prevent extension or compression from extending into them. It may be necessary to underpin the walls of some older buildings with steel girders to stop them sagging. Stone arches will possibly need a temporary support underneath them to stop the keystone dropping out.

TILTING

When the ground level is lowered the surface will slope at some places. Tilting of factory machines and plant can be classed as damage when the inclination renders the equipment inoperable and cannot be rectified by 'built-in' methods, e.g., screw feet or packing shims. The maximum gradient (from empirical measurements) due to working a seam is 2.75 S/h.

In a 2 m seam at 200 m this amounts to

$$\frac{2.75 \times 2 \times 0.9}{200} = 0.02474$$
$$= 1 \text{ in } 40$$

On a 8 m long lathe this would give a drop of about 20 cm which would make the machine unworkable if it depended on gravity flow for lubricant, and any round object would certainly roll.

Tall buildings and chimney stacks also need careful consideration. At a gradient of 1 in 240 a 30 m high chimney would only be 125 mm out of vertical which could probably be tolerated as a temporary measure, but at 1 in 40 it would be 1.3 m out of vertical, which with wind loading in addition could make it unstable. Blocks of flats could have such things as lifts (on multi-storey blocks, say 30 m high) put out of order, but would be unlikely to fall over. It can be seen that in urban areas over coal seams at shallow depths partial extraction may be the only alternative to high compensation costs.

Extraction Under Water

Mining under the sea bed and under large bodies of water has been carried out around the world for generations. Kapp and Williams[13] give a summary of several operations. Full extraction under rivers, canals, etc. can be considered normal but a cost assessment is always made to compromise between value of coal recovered and probable damage repair costs to build embankments, raise bridges, etc.

Mining under water is an important consideration for the New South Wales coalfields, because of several large coastal lakes, water catchment areas, and dams, and the extension of coal deposits under the sea. Underwater extraction is possible because of the usual behaviour of the strata between the coal seam and the sea floor, or lake floor. The lower roof beds over the coal will cave after coal extraction and give a badly broken zone within the pressure arch (see Figure 8.27). This is unlikely to extend above about 30 m and in most seams would in fact be much less. The next rock zone, comprising the intermediate and main roof beds, will flex downwards over the extracted area without significant breakage. Tests on sandstone have shown that it will withstand a curvature in excess of the mining value without fracturing. Finally there is a thin surface zone where the lack of triaxial restraint allows fractures to develop as the surface strains from subsidence travel forward. This effect is enhanced by weathered rock. The depth of fractured rock from the surface is likely to be of the order of 15 m. If unconsolidated deposits are present on the sea bed then these must be discounted when calculations of cover (the thickness of rock overlying the workings) are made. In addition some allowance may have to be made for faults or other weaknesses.

The reasons given for successful undersea mining are:

1. cracks which occur in the surface are generally shallow and do not let water into the workings
2. any shales, mudstones or similar rocks present will swell in the presence of water and seal off percolation channels
3. the zone between fractured rock at the surface and caved rock over the mining area is relatively impermeable[14].

The proportion of coal permitted (by legislation or mines department controls) to be mined from under water is related to the depth of cover, and is basically similar internationally, although practice in Australia has not yet been clearly determined. A minimum cover of about 100-150 m is usually required. Under shallow cover extraction is limited to about 40-50 per cent and the permitted extraction is increased with depth. The most precisely defined working instructions have been laid down by the National Coal Board in Britain[15]. This instruction stipulates that there should be a minimum thickness of cover of at least 105 m which must include at least 60 m of Carboniferous strata (geological period coal measure rocks). These would of course be fairly flexible and self-sealing. In addition the cumulative tensile strain from all seams worked should not exceed 10 mm/m when calculated according to their Handbook[2] as modified by Orchard and Allen[3] (i.e. basically using Figures 11.4 to 11.9 and associated instructions). The strain of 10 mm/m has been exceeded on some early extractions without harmful effects.

If partial extraction is used then the minimum cover[15] can be reduced to 60 m with a Carboniferous strata content of 45 m. Minimum pillar size must be one-tenth the apparent depth of the seam: if the seam is more than 2 m in thickness then the width of the pillar is increased to h/10 plus the seam thickness. The apparent depth is the actual depth of the seam from the sea bed plus an allowance of half the depth of the sea water. In areas with floors which become plastic when wet the pillar size must be increased to one-sixth of the apparent depth. Solid stowing is considered to reduce the seam thickness extracted by half.

The Instruction also specifies what plans must be kept and contains the reminder that working under water is controlled by the U.K. Mines and Quarries Act (Coal Mines Regulation Act in New South Wales). It is of course essential that basic legislation is observed, but legislation usually places the responsibility for safe working onto the mining company.

Subsidence over Metalliferous Orebodies

Metalliferous orebodies are not in themselves different from bedded deposits; any difference in subsidence behaviour is due to the shape of the orebody, and often to the igneous activity or geological planes of weakness with which the orebody is associated. Dykes or faults may cause the subsidence effects to chimney to the surface instead of spreading out.

In general there has been less need to worry about subsidence in metal mining because of the more remote areas in which these bodies are worked. Apart from predicting the effect of extraction on the company's own mine buildings or shafts, or the effect of, say, block caving, on adjacent open pits, management is not usually worried. However several case studies have been made recently.[16,17,18] These indicate generally that the behaviour of strata is similar over similar sized excavations both in irregular orebodies and in bedded deposits.

In terms of coal mining subsidence it is unlikely that a critical or super-critical plan area will be worked, but the effective thickness of seam is enormous. If the mine area is bridged by competent rock then there may be no surface subsidence. However if the cap rock is weak then it will cave progressively until the stope space is filled by expansion of the broken rock. If the extracted volume is large enough then the cave will reach the surface. As in coal mining, if the stope is sand filled, then subsidence will be reduced.

For lenticular orebodies then an angle of draw of 35° from the vertical will develop as the depth of extraction increases[18]. In steeply pitching bodies, or bodies bounded by dykes[17], the strata over the orebody has a tendency to chimney rather than open out into a more normal angle of draw. This can depend however on the history of the mine operation. As mining progresses downwards in the orebody it is probable that the break-off line will run parallel to the footwall, particularly if it is stoped down through successive levels. On the hanging wall side the break-off line and angle of draw depends on the country rock at shallow depths. At first the angle of break may be normal to the surface in hard rock, but as the mine deepens then the angle will spread out to about 35°.

A graphic example of this developed at the Grängesberg mine in Sweden as shown in Figure 11.16. This case study and several others from Swedish iron ore mines are described by Vestergren[19]. For cases such as Grängesberg he suggests that a limit angle from the vertical for safety purposes should be 30° for both hanging and footwall. These are open pits which gradually deepen into block caves.

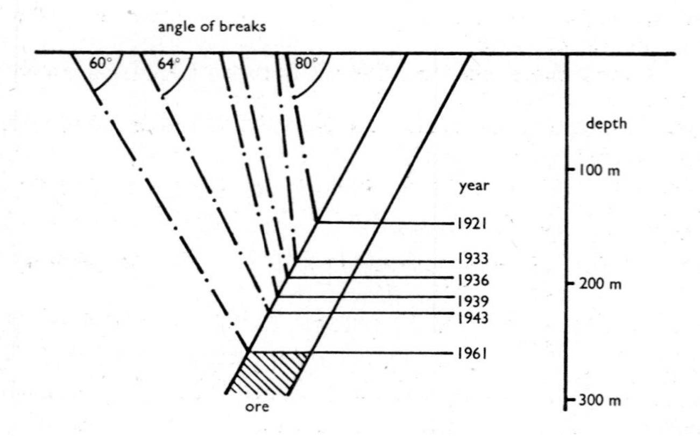

Fig. 11.16 Effect of deepening workings at Grängesberg mine

This suggests that for metalliferous mines an effective angle of draw could still be taken as 35°. It would be difficult to determine an extracted seam thickness but it is possible that subsidence could be calculated by using a device such as Figure 11.6, which would relate the volume of ore extracted to the volume subsided. To allow for sand or rock fill the compressed volume of fill added would be deducted from the volume of ore extracted. The compressed volume would be about 80 per cent for sand and 65 per cent for rock of the uncompacted volume of fill.

References

1. National Coal Board *Principles of subsidence engineering.* Production Department Information Bulletin 63/240, London, 1963.
2. National Coal Board *Subsidence Engineers' Handbook* (Revised) Production Department, London, 1966.
3. Orchard, R. J. and Allen, W. S. *Longwall partial extraction systems.* The Mining Engineer, No. 117, p. 523, June 1970.
4. Orchard, R. J. *Prediction of the magnitude of surface movements.* Colliery Engng., **34**, p. 455, 1957.
5. Bumm, H., Schweden, G. and Finke, G. *The mining subsidences in the harbours of Duisburg-Ruhrort.* Bull. Permanent Int. Ass. of Navigation Congresses, **3**, No. 21, p. 3, 1966.
   or Anon *Can't Dredge the harbour? Then lower the land.* Engineering News-Record, p. 30, January 31, 1963.
6. Marr, J. E. *The estimation of mining subsidence.* Colliery Guardian, **198**, p. 345, 1959.
7. Zenc, M. *Comparison of Bals's and Knothe's methods of calculating surface movements due to underground mining.* Int. J. Rock Mech. Min. Sci., **6**, p. 159, 1969.
8. Berry, D. S. *Ground movement considered as an elastic phenomenon.* The Mining Engineer, No. 37, p. 28, Oct. 1963.
   or *An elastic treatment of ground movement due to mining; I, isotropic ground.* J. Mech. Phys. Solids, **8**, p. 280, 1960.
9. Berry, D. S. and Sales, D. W. *An elastic treatment of ground movement due to mining: II, transversely isotropic ground.* J. Mech. Phys. Solids, **9**, p. 52, 1961.
10. Seldenrath, T. R. *Can coal measures be considered as masses of loose structure to which the laws of soil mechanics may be applied?* Proc. International Conference Rock Pressure, Liege, Belgium, 1951.
11. Wardell, K. *Some observations on the relationship between time and mining subsidence.* Trans. Inst. Min. Engns., **113**, p. 471, 1953.
12. Anon *Report on mining subsidence prepared by a committee of the Institution.* The Institution of Civil Engineers, London, 1959.
13. Kapp, W. A. and Williams, R. C. *Extraction of coal in the Sydney Basin from beneath large bodies of water.* Papers Aust. I.M.M. Conference, Newcastle, p. 77, 1972. (Aust. I.M.M., Melbourne 1972.)
14. Orchard, R. J. *The control of ground movements in undersea working.* The Mining Engineer, No. 101, p. 259, February 1969.
15. National Coal Board *Working under the sea.* Production Department Instruction PI/1968/8, London, 1968.
16. Edwards, D. B. and Just, G. D. *Ground movements due to a sub-level cave operation in the 500 orebody at Mount Isa.* Rock Mechanics Symposium, Sydney Branch Aust. I.M.M. and Institute of Engineers (Aust.) Sydney University, 1969.
17. Boyum, B. *Subsidence case histories in Michigan mines.* Fourth Rock Mechanics Symposium, Penn. State University, 1961.
18. Johnson, G. H. and Soule, J. H. *Measurements of surface subsidence, San Manuel Mine, Pinal County, Ariz.* U.S. Bureau of Mines Report of Investigations, R.I. 6204, 1963.
19. Vestergren, S. *Praktiska synpunkter pa ras vid gruvor (Practical aspects of caving-in of mines.)* Jernkont Ann., **148** (7), p. 417, 1964.
   (N.B. The Grängesberg case is analysed by Hall, B. and Hult, J., *Släntstabilitet i berg vid gruvbrytning.* Svenska Gruvforskningen serie B, **no. 72**, 1964 and quoted by Hoek, E. *Estimating the stability of opencast mines.* Trans. Section A Inst. Min. Met. **79**, p. A109, 1970.)

APPENDIXES

APPENDIX I

MINING GRADES

Note: These are examples (taken from current technical publications) of the ranges of ore grades likely to be mined. Values lower than these could be uneconomic; higher percentages would need verification. Even when an ore is found it may not be possible to recover it metallurgically. The values shown are principally for single ores: if two ores, e.g., lead and zinc, occur together, then lower grades can be mined.

ORE	GRADE (Minimum is open pit grade)	CONCENTRATE	
		Metal Recovery from Ore percent	Metal in concentrate from Mill percent
ANTIMONY	4.5 – 15%	80–85	60–70
COPPER	0.5 – 5.0%	60–95	20–34
COPPER plus Mo	0.25% Cu+0.05% Mo	85–90	Mo 50–60
COPPER plus Au and Ag.	0.4% Cu+0.06 g/t Au+ 2g/t Ag		27% Cu+22 g/t Au+70g/t Ag
CHROMIUM	35%		40–50* Cr_2O_3
GOLD	5 – 30 g/t	70–95	
IRON	28 – 65%		55–65
LEAD	2 – 15%	70–90	70–75
MANGANESE	38 – 53%	38–53	38–53* Mn O_2
NICKEL	0.5 – 9%	65–75	11–14%*
†PLATINUM (PGM)	4 – 11 g/t	80–85	
SILVER	30 – 300 g/t	70–95	
ALLUVIAL TIN	0.2 – 2.0 kg/m^3	70–95 of panning assay	50–70 (oxide ore)
LODE TIN	0.45 – 5.0%	50–70	50–65 (from sulphide ores)
TIN+COPPER	0.7%Sn+0.4%Cu	40–75	50–65
TUNGSTEN (Wolfram & scheelite)	0.5 – 5.0% WO_3	70–90	50–63 WO_3
URANIUM	1 – 5 kg/tonne		
ZINC	3 – 20%	70–90	50–55

* May be sold as ferrochrome, ferromanganese, ferronickel.
† PGM – platinum group metals : sold as combined total.

For precious metals g/t = grams per tonne
20 dwt = 1 oz troy = 31.1035 g
(ordinary measure = 1 oz avoirdupois = 28.3495 g)
1 lb/short ton = 0.5 kg/metric tonne
1 dwt/short ton = 1.714 g/t

APPENDIX II

S.I. CONVERSIONS

IMPERIAL	*S.I.*	*CONVERSION FACTOR*	
in	m	2.54	$\times 10^{-2}$
ft	m	3.048	$\times 10^{-1}$
yd	m	9.144	$\times 10^{-1}$
in^2	m^2	6.4516	$\times 10^{-4}$
ft^2	m^2	9.290304	$\times 10^{-2}$
yd^2	m^2	8.361274	$\times 10^{-1}$
in^3	m^3	1.6387064	$\times 10^{-5}$
ft^3	m^3	2.831685	$\times 10^{-2}$
yd^3	m^3	7.645549	$\times 10^{-1}$
UK gal	m^3	4.546092	$\times 10^{-3}$
US gal	m^3	3.785011	$\times 10^{-3}$
ft/min	m/s	5.08	$\times 10^{-3}$
yd/min	m/s	1.524	$\times 10^{-2}$
rev/min	rad/s	1.0471975	$\times 10^{-1}$
ft^3/min	m^3/s	4.7194748	$\times 10^{-4}$
gal/min	m^3/s	7.5768164	$\times 10^{-5}$
Mgal/day	m^3/s	5.261675	$\times 10^{-2}$
ton	t	1.0160469	
lb	kg	4.5359237	$\times 10^{-1}$
grain	kg	6.47989	$\times 10^{-5}$
lb/ft^3	kg/m^3	1.601846	$\times 10$
ft^3/ton	m^3/t	2.786963	$\times 10^{-2}$
lbf ft lbf/ft	N J/m	4.44822	
tonf	N	9.96402	$\times 10^{3}$
lbf ft ft lbf	Nm J	1.35582	
ft^2/h	m^2/s	2.58064	$\times 10^{-5}$
ft^2/min	m^2/s	1.548384	$\times 10^{-3}$
lbf h/ft^2	Pas or Ns/m^2	1.723689	$\times 10^{5}$
lbf/in^2	Pa or N/m^2	6.89476	$\times 10^{3}$
tonf/in^2	Pa or N/m^2	1.54443	$\times 10^{7}$
tonf/ft^2	Pa or N/m^2	1.07252	$\times 10^{5}$
Atm	Pa or N/m^2	1.013250	$\times 10^{5}$
in Wg	Pa or N/m^2	2.49089	$\times 10^{2}$
in Hg	Pa or N/m	3.38639	$\times 10^{3}$

IMPERIAL	*S.I.*	*CONVERSION FACTOR*	
Btu	J	1.05506	$\times 10^3$
hp	W	7.457	$\times 10^2$
Btu/h	W	2.93072	$\times 10^{-1}$
Btu/lb	J/kg	2.326009	$\times 10^3$
ft lbf/lb	J/kg	2.98907	
ft lbf/ft^3	J/m^3	4.788024	$\times 10$
Btu/lb°F	J/kg°C	4.18680	$\times 10^3$
Btu/ft^2h	W/m^2	3.154601	
Btu ft/ft^2h°F	Wm/m^2°C	1.730740	
lbf/$in^{1.5}$	N/$m^{1.5}$	1.09885	$\times 10^3$
refrigeration ton	W or J/s	3.51685	$\times 10^3$
ton mile	g m	1.63517	$\times 10^9$
ton mile	t km	1.63517	
slug	g	1.45939	$\times 10^4$
slug/ft^2	g/m^2	1.35582	$\times 10^3$
slug/ft^3	g/m^3	5.15379	$\times 10^5$
barrel (35 gal UK)	m^3 }	1.59	$\times 10^{-1}$
barrel (42 gal US)	m^3 }		
therm	J	1.05506	$\times 10^8$
in^2/s	m^2/s	6.4516	$\times 10^{-2}$

METRIC	*S.I.*	*CONVERSION FACTOR*	
kcal	J	4.187	$\times 10^3$
kcal/s	W	4.187	$\times 10^3$
kWh	J	3.6	$\times 10^6$
hectare	m^2	1	$\times 10^4$
litre	m^3	1	$\times 10^{-3}$

MISCELLANEOUS

1 chain	=	20.117 m
1 sq. chain	=	404.686 m^2
1 acre	=	4046.86 m^2
1 U.K. gal	=	4.546092 litre (l)
1 U.S. gal	=	3.785011 litre (l)
1 short ton	=	907.19 kg = 0.90719 tonne
1 barrel	=	159 litres

ASSAY — METRIC UNITS

1 centare	=	0.2990 square fathoms
1 t (tonne) (m ton)	=	0.98421 long tons (UK) (l ton)
1 t (tonne) (m ton)	=	1.1023 short tons (US) (s ton)
1 g (gram)	=	0.6430 dwt (pennyweight, Troy)
1 kg	=	32.151 oz Troy
1 g/t	=	0.653 dwt/long ton
1 g/t	=	0.583 dwt/short ton
1 cm/g	=	0.230 inch-dwt
1 kg/t	=	2.277 lb/long ton
1 kg/t	=	2.0 lb/short ton
1 cm/kg	=	0.7874 in/lb
1 square fathom	=	3.3445 centares (ca)
1 pennyweight (dwt)	=	1.5552 g
1 ounce Troy	=	31.1035 g
1 dwt/long ton	=	1.531 g/t (grams/metric ton)
1 dwt/short ton	=	1.714 g/t (grams/metric ton)
1 inch-dwt	=	4.354 cm/g
1 inch-pound	=	1.270 cm/kg
1 assay ton	=	32.6667 g

INDEX

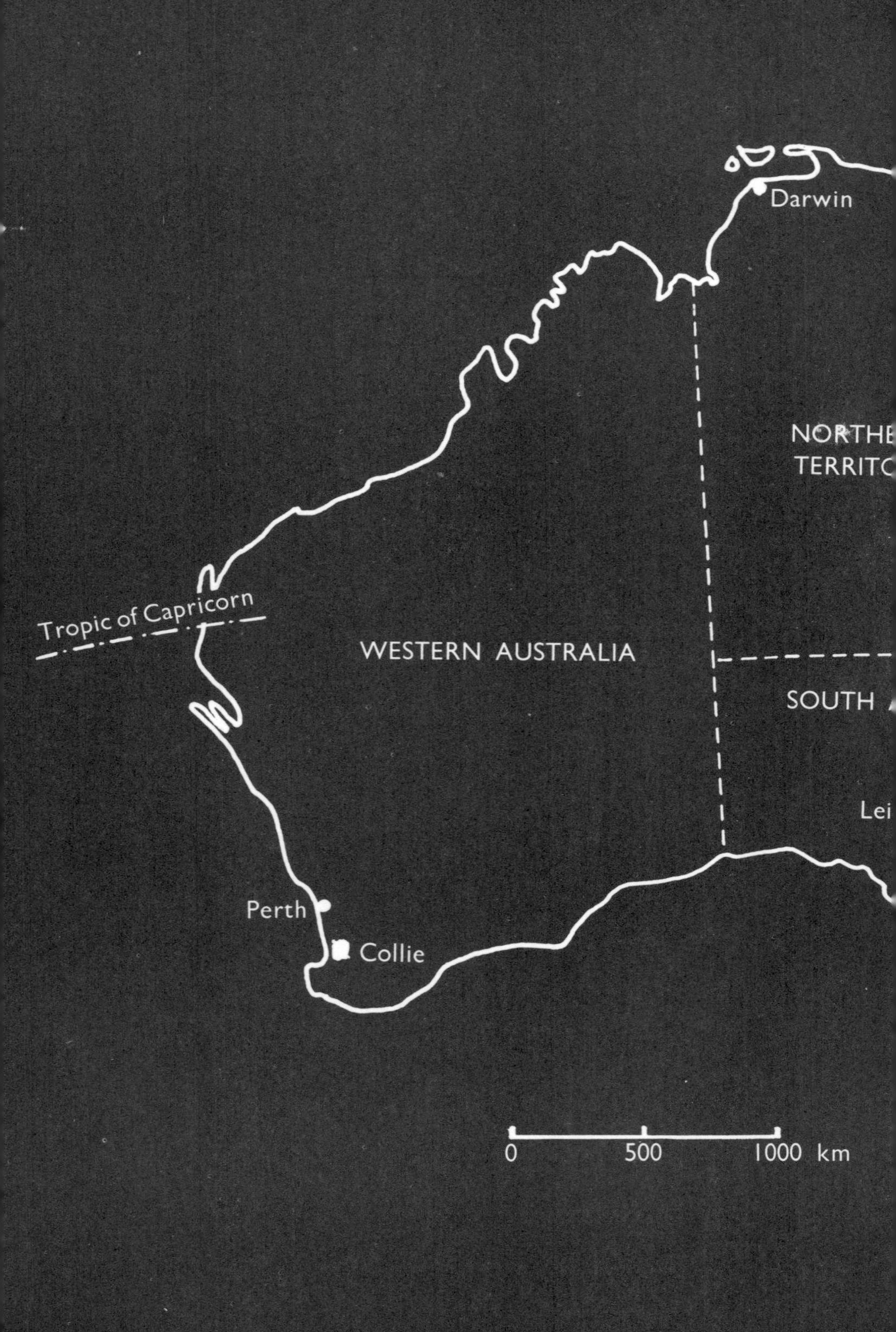
Darwin
Tropic of Capricorn
WESTERN AUSTRALIA
SOUTH
Perth
Collie
0
500
1000 km

COAL DEPOSITS OF AUSTRALIA

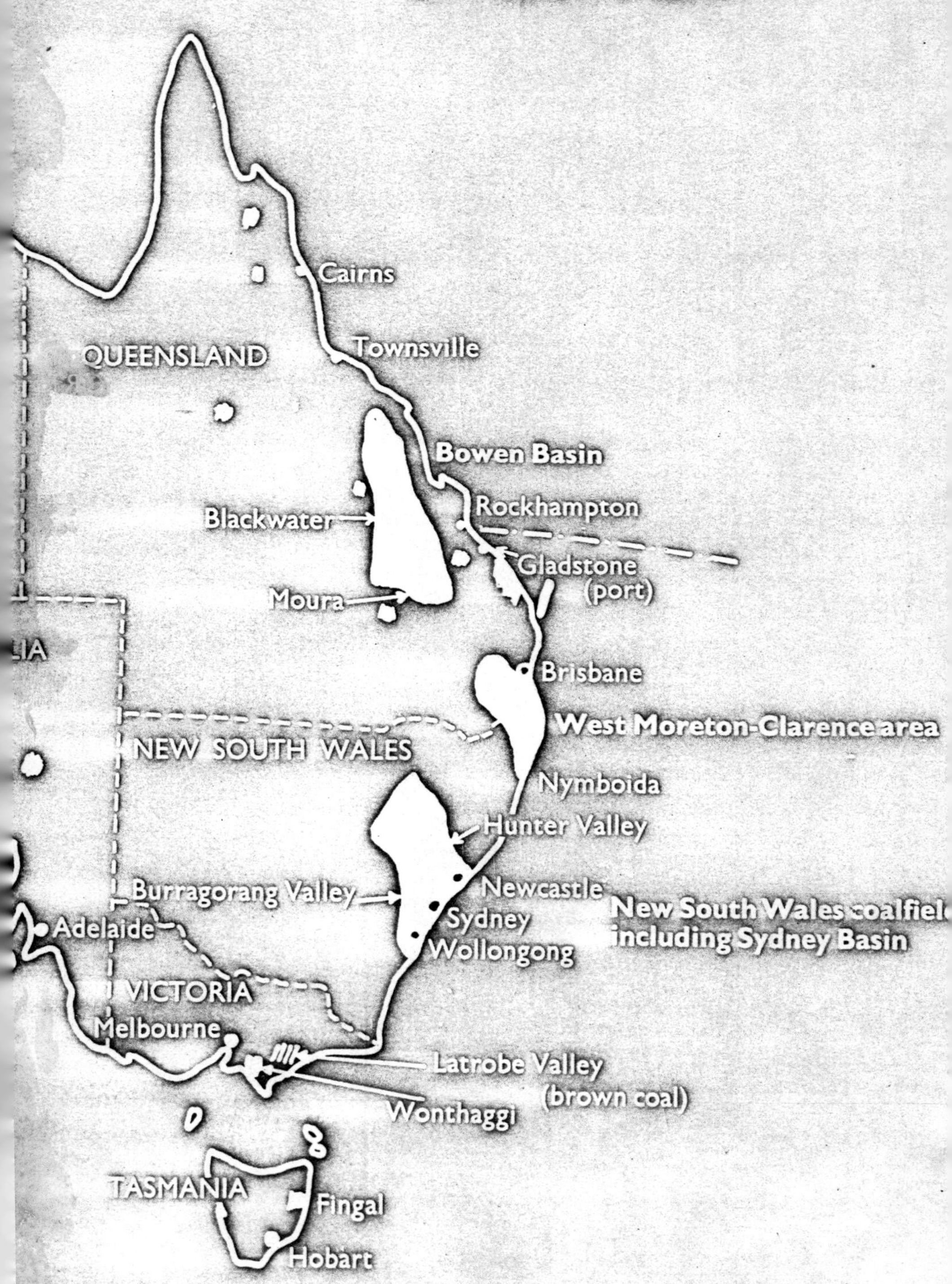